T0135109

Topics in Intelligent Engineering and Informatics

Volume 6

Editors-in-Chief

János Fodor
Imre J. Rudas

Editorial Advisory Board

Ildar Batyrshin (Mexico)
József Bokor (Hungary)
Bernard De Baets (Belgium)
Hamido Fujita (Japan)
Toshio Fukuda (Japan)
Fumio Harashima (Japan)
Kaoru Hirota (Japan)
Endre Pap (Serbia)
Bogdan M. Wilamowski (USA)

Review Board

P. Baranyi (Hungary)
U. Bodenhofer (Austria)
G. Fichtinger (Canada)
R. Fullér (Finland)
A. Galántai (Hungary)
L. Hluchý (Slovakia)
MO Jamshidi (USA)
J. Kelemen (Czech Republic)
D. Kocur (Slovakia)
P. Korondi (Hungary)
G. Kovács (Hungary)
L.T. Kóczy (Hungary)
L. Madarász (Slovakia)
CH.C. Nguyen (USA)

E. Petriu (Canada)
R.-E. Precup (Romania)
S. Preitl (Romania)
O. Prostean (Romania)
V. Puri (Italy)
GY. Sallai (Hungary)
J. Somló (Hungary)
M. Takács (Hungary)
J. Tar (Hungary)
L. Ungvari (Germany)
A.R. Várkonyi-Kóczy (Hungary)
P. Várlaki (Hungary)
L. Vokorokos (Slovakia)

For further volumes:
http://www.springer.com/series/10188

Ryszard Klempous · Jan Nikodem
Witold Jacak · Zenon Chaczko
Editors

Advanced Methods
and Applications
in Computational Intelligence

 Springer

Editors
Ryszard Klempous
Wroclaw University of Technology
Wroclaw
Poland

Witold Jacak
Upper Austria University
of Applied Sciences
Hagenberg
Austria

Jan Nikodem
Wroclaw University of Technology
Wroclaw
Poland

Zenon Chaczko
University of Technology
Sydney
Australia

ISSN 2193-9411   ISSN 2193-942X (electronic)
ISBN 978-3-319-03339-6   ISBN 978-3-319-01436-4 (eBook)
DOI 10.1007/978-3-319-01436-4
Springer Cham Heidelberg New York Dordrecht London

© Springer International Publishing Switzerland 2014
Softcover re-print of the Hardcover 1st edition 2014
This work is subject to copyright. All rights are reserved by the Publisher, whether the whole or part of
the material is concerned, specifically the rights of translation, reprinting, reuse of illustrations, recitation,
broadcasting, reproduction on microfilms or in any other physical way, and transmission or information
storage and retrieval, electronic adaptation, computer software, or by similar or dissimilar methodology
now known or hereafter developed. Exempted from this legal reservation are brief excerpts in connection
with reviews or scholarly analysis or material supplied specifically for the purpose of being entered
and executed on a computer system, for exclusive use by the purchaser of the work. Duplication of
this publication or parts thereof is permitted only under the provisions of the Copyright Law of the
Publisher's location, in its current version, and permission for use must always be obtained from Springer.
Permissions for use may be obtained through RightsLink at the Copyright Clearance Center. Violations
are liable to prosecution under the respective Copyright Law.
The use of general descriptive names, registered names, trademarks, service marks, etc. in this publication
does not imply, even in the absence of a specific statement, that such names are exempt from the relevant
protective laws and regulations and therefore free for general use.
While the advice and information in this book are believed to be true and accurate at the date of publication, neither the authors nor the editors nor the publisher can accept any legal responsibility for any
errors or omissions that may be made. The publisher makes no warranty, express or implied, with respect
to the material contained herein.

Printed on acid-free paper

Springer is part of Springer Science+Business Media (www.springer.com)

Foreword

This book includes updates and extended versions of carefully selected contributions to ACASE 2012, the 1st Australian Conference on the Applications of Systems Engineering that was held on February 6-8, 2012 at the University of Technology, Sydney, Australia. ACASE 2012 was organised and sponsored by the University of Technology, Sydney (Australia), Wroclaw University of Technology (Poland) and the University of Applied Sciences in Hagenberg (Upper Austria). ACASE 2012 conference was an opportunity for system theorists to get together to present their work in an exciting environment. This conference was modelled after a very successful series of International Conference on Computer Aided Systems Theory (Eurocast) in Spain/Austria in alternative years. Inspired by a major impact that Eurocast conferences have in the European region, the newly started series of ACASE conferences, geographically aim to provide a prestigious venue for system engineering and applied science researchers and practitioners in the Asia-Pacific region, this would include mainly Australia, Asia and both America's. The organisers of ACASE 2012 believe that the new series of conferences will offer a great opportunity for researchers to report and share recent developments in the area of intelligent systems and computation models.

In a nutshell, the concept of intelligence encompasses the capability to adapt to various situational contexts by reasoning. Interdisciplinary by nature, the intelligent systems integrate theories, models and competences from many different areas, including system analysis, modelling, design and implementation. As fundamental knowledge and technology are merging very closely, the need for introducing intelligent approaches, models and tools to various technical problems is even more apparent. These challenges are well articulated in the content of the current volume for which contributions were made by system theorists and researchers, to whom we are very grateful. We would like to thank the editors of this volume for their insightful and meritorious work in evaluating the presented talks and proceedings papers, and for the final product of this outstanding selection

Robin Braun
Zenon Chaczko
Franz Pichler

Australia
December 2012

Preface

This volume contains carefully selected and expanded papers presented at the conference. The chapters are organized into three main Parts. Part I deals with the Practical Applications of Modern Heuristic Methods. Part II deals with Network Management and Essential Problems issues associated with them. Part III deals with Intelligent System Applications. Some very interesting papers were presented in all three parts.

The first part is devoted to the heuristic approaches which are usually applied in situations where the problem cannot be solved by exact methods due to the characteristics or the dimension of the problem. The selection of chapters for this part of the book is aimed to give a state of the art overview about software solutions and exemplary applications of modern heuristic problem solving strategies.

The evolution of so many diverse heuristics results from the fact that no single method outperforms all others for all possible problems. To be a bit more precise, the so-called No-Free-Lunch theorem postulates that a universal optimization strategy is impossible and that the only way how one strategy can outperform another is to be more specialized to the structure of the tackled problem. Therefore, in practice it usually takes qualified algorithm experts to select and tune a heuristic algorithm for a concrete application. Because of these issues it is not advisable to implement only one specific method for a specific problem, as it is not feasible to determine in advance whether the chosen method is suited for the specific problem. Therefore, software frameworks like HeuristicLab, which has been used for most of the chapters presented in this part of the book, have been developed which offer a rich pool of different ready-to-use heuristic algorithms that can be tested on new problem situations.

However, even if the usage of software frameworks supports the process of selecting, analyzing, and parameterizing different heuristics, it is still not possible to perform this process in a directed and automated way. In this context, the concept of fitness landscape analysis (FLA) has become an important research topic in the last couple of years. FLA aims to describe the specific characteristics of a certain problem and its associated fitness landscape properties in a systematic way. The chapters in this part of the book cover many of the above mentioned aspects concerning the theory and application of heuristic problem solving techniques.

The chapters presented here can be grouped into three categories:

- the first group of chapters 1-6 covers application domains of heuristic search in the hypothesis space for data mining and machine learning.
- the second group of chapters 7-9 discusses the performance of different meta-heuristics on the basis of academic benchmark problems as well as on the basis of real world applications from the field of combinatorial optimization also including theoretical aspects of fitness landscape analysis.
- finally, the chapter 10 is about software tools and frameworks for algorithm design, analysis and evaluation.

Chapters 1 and 2: Grzegorz Borowik and Tadeusz Łuba, present the new algorithm for feature selection which transforms the problem of features reduction to the problem of complementation of Boolean function, well known in logic synthesis. The applications of selected methods of logic synthesis for data mining and feature selection are discussed, particularly is shown that the logic synthesis method uses Boolean function complementation is more efficient than standard data mining algorithms.

Chapters 3 and 4: Wojciech Bożejko, Mariusz Uchroński, Mieczysław Wodecki and Paweł Rajba, propose a new framework of the distributed tabu search meta-heuristic designed to be executed using a multi-GPU cluster, i.e. cluster of nodes equipped with GPU computing units. The methodology is designed to solve difficult discrete optimization problems, such as a job shop scheduling problem. Tabu search algorithm is executed in concurrent working threads, as in multiple-walk model of parallelization. Additionally, solutions stability determined by algorithms based on the tabu search method for a certain (NP-hard) one-machine arrangement problem is presented.

Chapter 5: Witold Jacak, Karin Prölll and Stephan Winkler, presents novel feature selection strategies by hybridizing supervised and unsupervised learning strategies.

Chapter 6: The team from Heuristic and Evolutionary Algorithms Laboratory, University of Applied Sciences Upper Austria, demonstrate how different heuristic and machine learning based hypothesis search techniques may be applied to the identification of virtual tumour markers and for cancer prediction incorporating evolutionary feature selection strategies based on real world data

Chapter 7: Monika Kofler, Andreas Beham, StefanWagner, and Michael Affenzeller describe a real world warehouse optimization problem and discuss the performance of different metaheuristic approaches comparing and opposing rewarehousing and healing strategies.

Chapter 8: M.A. Pérez-del-Pino, P. García-Báez, J.M. Martínez-García and C.P. Suárez-Araujo provide an introduction to practical application of modern heuristic methods. The authors explore technological infrastructure, regarding computer architecture, databases and software engineering. This issue focuses on the embedded mechanisms that allow integration between EDEVITALZH Core components and the Intelligent Systems for Diagnosis and the Intelligent Decision Support Tools, providing Computational Intelligence to the described virtual clinical environment.

Chapter 9: Erik Pitzer, Andreas Beham, Michael Affenzeller, discuss the quadratic assignment problem and incorporate techniques of fitness landscape analysis in order to categorize the different benchmark problem instances according to difficulty (hardness) of their problem.

Chapter 10 written by the team from Heuristic and Evolutionary Algorithms Laboratory, University of Applied Sciences Upper Austria, gives a comprehensive overview about the open source framework HeuristicLab, which is the platform that has been used for the first two groups of contributions presented in this part of the book.

The five chapters in the second section - Network Management Essential Problems, continue the theme of applying modern numerical methods to large intractable systems. These systems are software intensive infrastructures and communications networks. It is true to say, that the biggest problems being faced by builders and operators of software and network systems are their massive scale and their attendant complexity. There are perhaps more than 15 billion connected devices in the world today. How do we make the connection infrastructure provide the services these devices need, and how do we make it respond to changes? All of this, while ensuring that the owners and operators still keep making a profit. Our problems are not bandwidth and computational power. They are complexity and intractability. The five papers in this part address issues and applications of Wireless Sensor Networks, Next Generation Networks, Common Resource Allocation and complexities in the Signal Space. The second part of this book is organised into three into three separate thematic groups:

- Chapter 11 covers an application domain of SANET middleware infrastructure in area of autonomous control and manoeuvring of land-yacht vessels.
- Chapters 12 and 13 discusses the salient issues of wireless communication in Wireless Sensor Networks and Heterogeneous Wireless Networks.
- Chapters 14 and 15 is about intelligent models of distributed systems for next generation of networks and issues of OFDM and MIMO-OFDM network compatibility in area of measurements and simulations.

Chapter 11: Christopher Chiu and Zenon Chaczko elaborate an approach of guiding land-yachts according to a predefined manoeuvring strategy. The simulation of the path as well as the controller scheme is executed for a sailing strategy (tacking), where the scheme of the resultant path and the sailing mechanism is driven using Sensor-Actor Network (SANET) middleware infrastructure. The addition of obstacle avoidance and detection heuristics aids in the guiding process. By incorporating SANETs components in the sailing vessel, a range of sensory mechanisms can be employed to effectively manage local obstacles, while data collected from the sensory environment can be transmitted to a base station node for monitoring conditions.

Chapter 12: Jan Nikodem, Ryszard Klempous, Maciej Nikodem and Zenon Chaczko discuss theoretical work and simulation results of innovative spatial routing technique in wireless sensor network (WSN). The new framework of the routing in WSN is based on the set theory and on three elementary relations: subordination (π), tolerance (ϑ) and collision ($\varkappa$) necessary to describe behaviour of nodes in distributed system. The essence of using relations is decentralization of the process of the decisions making. Relations introduce the ability to delegate decision making process to nodes that base their actions depending on situation in the vicinity, information that is available to them and their capabilities. The proposed approach combines existing features of the spatial routing and LQI or RSSI indicators to aid in route selection within a neighborhood.

Chapter 13: Abdallah Al Sabbagh, Robin Braun, and Mehran Abolhasan discusses the evolution and challenges of wireless networks that has led to the deployment of different Radio Access Technologies (RATs) such as GSM/EDGE Radio Access Network (GERAN), UMTS Terrestrial Radio Access Network (UTRAN) and Long Term Evolution (LTE). The need and rationale of CRRM for NGWN is being discussed. Next Generation Wireless Networks (NGWNs) are predicted to interconnect various Third Generation Partnership Project (3GPP) Access Networks with Wireless Local Area Network (WLAN) and Mobile Worldwide Interoperability for Microwave Access (WiMAX). The authors of this chapter elaborate further on a major challenge how to allocate users to the most suitable RAT for them. The proposed solution could lead to efficient radio resource utilization, maximization of network operator's revenue and increasing in the users' satisfactions. Common Radio Resource Management (CRRM) is proposed to manage radio resource utilization in heterogeneous wireless networks.

Chapter 14: Pakawat Pupatwibul, Ameen Banjar, Abdallah Al Sabbagh and Robin Braun study the need of distributed systems in next generation networks and suggest that traditional network structures are inadequate to meet today's requirements. It is a centralized network which imposes on human operators to have a high experience on how to detect changes, configure new services, recover from failures and maximize Quality of Service (QoS). Therefore, network management rely on expert's operators. Hence, the adopted centralized network management is not suitable for new technologies emerging, which are complex and difficult to interact among heterogeneous networks that contain different types of services, products and applications from multiple vendors. As a result, the current network management lacks of efficiency and scalability; however, it has an acceptable performance generally. The centralized information model cannot achieve the requirements from such complex, distributed environments. This paper studies the need of distributed systems in next generation networks.

Chapter 15: Michał Kowal, Ryszard J. Zieliński, and Zenon Chaczko present the results of measurements and simulations of intrasystem compatibility of the wireless networks operated in accordance with IEEE 802.11g and IEEE 802.11n standards. Presented results of simulations were carried out using an advanced model of the MIMO-OFDM system. The simulations have been preceded by measurements in the anechoic chamber. The results of these measurements were used to define the input parameters of the simulator. The analysis of the obtained results confirmed the usefulness of presented MIMO-OFDM system simulator to performance prediction of the wireless networks in the absence and presence of interference from other networks.

There are three chapters included in the last section of the book - Intelligent System Applications. These chapters address various salient issues of complexity. These include issues that arise when multiple systems try to use the same radio space and the use of connected systems in simulation and training in the medical field.

Chapter 16: Ryszard J. Zieliński, Michał Kowal, Sławomir Kubal, and Piotr Piotrowski discuss critical issues of EMC between WIMAX 1.5GHz and WLAN 2.4GHz systems operating in the same area. Specifically, wireless communications in mines excavation is being discussed.

Chapter 17: Christopher Chiu and Zenon Chaczko discuss An Anticipatory SANET Environment for Training and Simulation of Laparoscopic Surgical Procedures, Modeling of Laparoscopic Surgery using an Agent-based Process. Application of BDI principles, heuristics and Extended Kohonen Map (EKM) techniques in context of a knowledge-based laparoscopic surgery systems are being discussed.

Chapter 18: Jan Szymański, Zenon Chaczko, and Ben Rodański provide an overview of concepts, definitions, technologies and challenges related to ubiquitous and pervasive healthcare Body Sensor Networks as Special WSNs and a brief history of Body Sensor Networks are being presented.

The conference was made possible through the efforts of the organizers: Chris Chiu, Shahrzad Aslanzadeh, Bahram Jozi, and many others. Special thanks must go to attendees who came to Australia for the conference from as far as Poland and Austria.

We would like to express our special thanks to the reviewers of this book: Werner Backfrieder, Piotr Bilski, Arkadiusz Bukowiec, Stephan Dreiseitl, Edward Hrynkiewicz, Dobler Heinz, Bartosz Jabłoński, Tomasz Janiczek, Krzysztof Kulpa, Herwig Mayr, Ewa Skubalska-Rafajłowicz, Andrzej Rusiecki, Czesław Smutnicki.

Ryszard Klempous
Jan Nikodem
Witold Jacak
Zenon Chaczko

Wrocław, Hagenberg, Sydney
January, 2013

Contents

Part I
Practical Applications of Modern Heuristic Methods

Chapter 1
Data Mining Approach
for Decision and Classification Systems
Using Logic Synthesis Algorithms

Grzegorz Borowik

Abstract. This chapter discusses analogies between decision system and logic circuit. For example, the problem of data redundancy in decision system is solved by minimizing the number of attributes and removing redundant decision rules which is analogous to the argument reduction method for logic circuits. Another issue associated with the field of data mining lies in the induction of decision rules which in result provide a basis for decision-making tasks. A similar algorithm in logic synthesis is called minimization of Boolean function. An issue of reduction of the capacity required to memorize a decision table is solved by disassembling this table to the subsystems in such a way that the original one can be recreated through hierarchical decision making. In logic synthesis it is called functional decomposition and is used for efficient technology mapping of logic circuits. Due to different interpretation and application these tasks seem totally different, however the analogies allow logic synthesis algorithms to be used in the field of data mining. Moreover, by applying specialized logic synthesis methods, these three issues, i.e. feature reduction, rule induction, and hierarchical decision making, can be successfully improved.

1.1 Introduction

In practice, logic synthesis methods are mainly used to process the digital systems which have been designed to transform binary signals. The primary task of these methods is to improve the implementation/mapping of the systems in various technology. However, it can be shown that many logic synthesis methods, in particular these used for optimization of combinational logic circuit, can be successfully used in typical tasks of storage and retrieval of information, knowledge discovery/

Grzegorz Borowik
Institute of Telecommunications, Warsaw University of Technology,
Nowowiejska 15/19, 00-665 Warsaw, Poland
e-mail: G.Borowik@tele.pw.edu.pl

R. Klempous et al. (eds.), *Advanced Methods and Applications in Computational Intelligence*, 3
Topics in Intelligent Engineering and Informatics 6,
DOI: 10.1007/978-3-319-01436-4_1, © Springer International Publishing Switzerland 2014

generalization, optimization of databases; and generally, in the field of expert systems, machine learning, and artificial intelligence.

Decision system and logic circuit are very similar. The decision system is usually described by a decision table, and combinational logic of a digital system by a truth table. Condition attributes of this information system correspond to input variables of a digital system, and attributes of the decision to the logic output variables. In both areas, there are many notions that can be mutually mapped to each other. These analogies of information systems and logic synthesis allow the use of specialized methods for logic synthesis in the field of data mining.

For example, the issue of data redundancy in information system is solved by minimizing the number of attributes (features) and removing redundant objects. Analogous task for logic synthesis is the argument reduction.

Many problems associated with decision-making lie in the so-called knowledge discovery (or generalization of knowledge), i.e. induction of decision rules. The result of induction could provide a basis for making a decision on the objects (examples, instances, cases) which do not belong to the original set of objects. It is very important in task of "learning" a system. A similar issue for logic synthesis is the minimization of Boolean function. Due to different interpretation and application these problems seem totally different, however they do not.

Reduction of the capacity required to memorize a decision table can be achieved by hierarchical decision making. This method is based on disassembling the decision table to the subsystems in such a way that the original decision table can be recreated through a series of operations corresponding to the sequential decision making. But the most important is that we can induce noticeably shorter decision rules for the resulting components to finally make the same decision as for the original decision table. A similar problem for logic synthesis is the functional decomposition of Boolean function. As a result, we obtain a tree of smaller truth tables, which can be directly mapped onto FPGA structure.

The key idea of this chapter is to focus on the subject of intersection of logic synthesis and data mining. Although almost all of the methods presented in this chapter are already established, the author would like to note that many methods of logic synthesis have not been applied so far or have rarely been applied in the field of data mining. The main reason for this is the lack of knowledge of logic synthesis methods by data mining specialists.

In particular, the author proposes to apply the compatibility relation instead of indiscernibility relation and extend the concept of r-partition [25] in the field of data mining and corresponding algorithms. Secondly, an extraordinary results can be achieved, especially in time of calculation, by applying the method of Boolean function complementation to the transformation of the discernibility function [35]. The method of complementation is a procedure of *Espresso* system [5, 40] which has been mainly developed for the minimization of Boolean functions. The details of the complementation method and its applications are discussed in the next chapter of this book. The last author's proposal is to apply the functional decomposition in hierarchical decision-making.

The structure of the chapter is as follows. First, some basic notions about information system and decision system are given. Secondly, the notion of logic synthesis called r-partition and its application in the representation of data mining system are provided. Subsequently, the complexity of the following problems: attribute reduction of information system, attribute reduction of decision system, induction of decision rules are reduced to the transformation of a conjunctive normal form into the disjunctive normal form. Then, an effective logic synthesis algorithm of the complementation of Boolean function is used in the task of searching the minimal column cover. Finally, our original theory, recently used only for digital design is applied for hierarchical decision-making.

1.2   Information Systems and Decision Systems

Many real issues, problems and situations can be described using data tables called information systems, or simply databases. For example, using these tables we can describe by selected parameters a condition of a patients at the ongoing medical research. Then, individual instances stored in rows characterize the patient by appropriate attribute values.

More formally, an *information system* is a pair $\mathscr{A} = (U, A)$, where U – is a nonempty, finite set of *objects* called the *universe* and A – is a nonempty, finite set of *attributes*, i.e. each element $a \in A$ is a function from U into V_a, where V_a is the domain of a called *value set* of a. Then, a function ρ maps the product of U and A into the set of all values and by $\rho(u_t, a_i)$, where $u_t \in U$, $a_i \in A$, we denote an attribute value for an object.

Tab. 1.1a describes values of selected parameters (attributes) for the examination of seventeen patients $\{u_1, \ldots, u_{17}\}$. We can notice that some patients (objects), e.g. $\{u_3, u_6, u_{10}\}$, have the same values for all the attributes. We say, that these objects are *pairwise indiscernible*.

Sometimes the set of attributes A is extended by one or more distinguished attributes. The purpose of these attributes is to make decisions on the basis of the information provided in the database.

Formally, a *decision system* is the information system of the form $\mathscr{A} = (U, A \cup D)$, where $A \cap D = \emptyset$. The attributes in the set A are called *condition attributes* and in the set D – *decision attributes*. Decision systems are usually described by *decision tables*. Then, a function ρ maps the product of U and $A \cup D$ into the set of all attribute values.

Example of the decision system is given in Tab. 1.1b. We can notice that this decision table classifies objects $\{u_1, \ldots, u_{17}\}$ into three different classes, i.e. dec $\in \{0, 1, 2\}$.

In both cases, i.e. when table describing information system or decision system has completely specified function ρ we call the system *completely specified*. However, in practice, input data of data mining systems/algorithms are frequently affected by missing attribute values [13, 14, 32]. In other words, the corresponding

Table 1.1. a) information system, b) decision system

	a b c d e f			a b c d e f	dec
u_1	1 * * 2 0 3		u_1	1 * * 2 0 3	0
u_2	2 0 0 0 2 2		u_2	2 0 0 0 2 2	2
u_3	0 1 0 2 2 3		u_3	0 1 0 2 2 3	1
u_4	1 0 1 0 1 1		u_4	1 0 1 0 1 1	0
u_5	2 0 1 0 2 2		u_5	2 0 1 0 2 2	2
u_6	0 1 0 2 2 3		u_6	0 1 0 2 2 3	1
u_7	1 1 * 1 2 1		u_7	1 1 * 1 2 1	1
a) u_8	2 0 1 0 2 2	b)	u_8	2 0 1 0 2 2	1
u_9	2 1 1 0 2 2		u_9	2 1 1 0 2 2	2
u_{10}	0 1 0 2 2 3		u_{10}	0 1 0 2 2 3	1
u_{11}	2 1 0 2 2 3		u_{11}	2 1 0 2 2 3	1
u_{12}	2 1 1 0 2 2		u_{12}	2 1 1 0 2 2	2
u_{13}	1 0 1 1 0 1		u_{13}	1 0 1 1 0 1	1
u_{14}	2 1 0 2 2 3		u_{14}	2 1 0 2 2 3	1
u_{15}	1 0 1 0 1 0		u_{15}	1 0 1 0 1 0	0
u_{16}	1 0 1 0 1 0		u_{16}	1 0 1 0 1 0	1
u_{17}	1 0 1 0 0 2		u_{17}	1 0 1 0 0 2	2

function ρ is incompletely specified (partial). A table with an incompletely specified function ρ is called *incompletely specified*.

In [13] Grzymała-Busse has defined four special cases of attribute values for incompletely defined decision tables, i.e. lost values have been denoted by "?", "do not care" conditions by "*", restricted "do not care" conditions by "+", and attribute-concept values by "−". Additionally, he has assumed that for each case at least one attribute value is specified.

For simplicity, in this chapter, we consider only tables with "do not care" conditions for some attribute values (Tab. 1.1a, Tab. 1.1b).

1.3 Indiscernibility and Compatibility Relation

This section presents the notion of indiscernibility relation which is fundamental in the field of data mining. A similar concept called compatibility relation works in the field of digital circuits. This concept was previously proposed by Łuba [25, 27] and mostly used for the decomposition of combinational logic [6, 33, 34] and sequential logic [24, 33]. On the basis of this relation a notion of r-partition has been constructed which has been mainly applied in the synthesis of multi-valued logic. This section extends the concept of r-partition into the field of data mining and shows that it can be successfully employed for representing the information systems and decision tables.

Let $\mathscr{A} = (U, A)$ be an information system, then with any $B \subseteq A$ there is associated an equivalence relation $\mathrm{IND}_{\mathscr{A}}(B)$:

$$\mathrm{IND}_{\mathscr{A}}(B) = \{(u_p, u_q) \in U^2 : \forall\, a_i \in B,\ \rho(u_p, a_i) = \rho(u_q, a_i)\}, \qquad (1.1)$$

$\mathrm{IND}_{\mathscr{A}}(B)$ is called *B-indiscernibility relation* [19, 30].

However, for the information system expressed in Table 1.1a (or decision system in Table 1.1b), the symbol "*" in an object may assume the value 0 or 1. This results in an object representing multiple rows. Hence, the classification by indiscernibility relation is no more valid. This is clearly explained by the fact that the object u_1 for the attributes a, b, c, i.e. $(1 * *)$, is not the same as the objects u_4 (101) and u_7 (11*). The object u_1 represents in fact a set of objects 100, 101, 110, 111. That is why we introduce a compatibility relation.

The values of the attribute a_i, i.e. $\rho_{pi} = \rho(u_p, a_i)$ and $\rho_{qi} = \rho(u_q, a_i)$ are said to be *compatible* $(\rho_{pi} \sim \rho_{qi})$ if, and only if, $\rho_{pi} = \rho_{qi}$ or $\rho_{pi} = *$ or $\rho_{qi} = *$, where "*" represents the case when attribute value is "do not care". On the other hand, if ρ_{pi} and ρ_{qi} are defined and are "different" it is said that ρ_{pi} is *not compatible* with ρ_{qi} and is denoted as $\rho_{pi} \nsim \rho_{qi}$.

The consequence of this definition is *compatibility relation* $\mathrm{COM}_{\mathscr{A}}(B)$ associated with every $B \subseteq A$:

$$\mathrm{COM}_{\mathscr{A}}(B) = \{(u_p, u_q) \in U^2 : \forall\, a_i \in B,\ \rho(u_p, a_i) \sim \rho(u_q, a_i)\}. \qquad (1.2)$$

The objects p and q, which belong to the relation $\mathrm{COM}_{\mathscr{A}}(B)$, are said to be *compatible in the set B*. Compatible objects in the set $B = A$ are simply called *compatible*.

The compatibility relation of objects is a tolerance relation (reflexive and symmetric) and hence it generates compatible classes on the set of objects U.

Compatibility relation allows us to classify objects but the classification classes do not form partitions on the set U, as it is in the case of indiscernibility relation. $\mathrm{COM}_{\mathscr{A}}(B)$ classifies objects grouping them into compatibility classes, i.e. $U/\mathrm{COM}_{\mathscr{A}}(B)$, where $B \subseteq A$.

However for the systems from Table 1.1, it is not precise to say that object u_1 is compatible with either object u_4 or object u_7. According to the $\mathrm{COM}_{\mathscr{A}}$ relation, objects of the set $U = \{u_1, \ldots, u_{17}\}$ for the attributes a, b, c can be classified as: $\{u_1, u_4, u_{13}, u_{15}, u_{16}, u_{17}\}$, $\{u_1, u_7\}$, $\{u_3, u_6, u_{10}\}$, $\{u_2\}$, $\{u_5, u_8\}$, $\{u_9, u_{12}\}$, $\{u_{11}, u_{14}\}$. This is evidently not correct because the object $1 * *$ contains the object (100) as well, and this is different from both u_4 and u_7 (not compatible). Therefore, the object u_1 has also to belong to another class. These classes ought to be as follows:

$$\{u_1\},\ \{u_1, u_4, u_{13}, u_{15}, u_{16}, u_{17}\},\ \{u_1, u_7\},\ \{u_3, u_6, u_{10}\},\ \{u_2\},\ \{u_5, u_8\},$$
$$\{u_9, u_{12}\},\ \{u_{11}, u_{14}\}.$$

That is why, any set of objects U specifying an information system or decision system can be distributed into various classes for given subset of attributes A. Such a family of classes we call *r-partition* and denote by $\Pi(B)$, where B is a selected subset of the set $A = \{a_1, \ldots, a_m\}$.

R-partition on a set U may be viewed as a collection of non-disjoint subsets of U, where the set union is equal U. All symbols and operations of *partition algebra* [6, 25] are applicable to *r*-partitions. Especially the relation *less than or equal to* holds between two *r*-partitions Π_1 and Π_2 ($\Pi_1 \leq \Pi_2$) iff for every class of Π_1, in short denoted by $\mathscr{B}_i(\Pi_1)$, there exists a $\mathscr{B}_j(\Pi_2)$, such that $\mathscr{B}_i(\Pi_1) \subseteq \mathscr{B}_j(\Pi_2)$.

The *r*-partition of a single element set $B = \{a_i\}$ is denoted as $\Pi(a_i)$ or simply Π_{a_i}. The *r*-partition generated by a set B is the product of *r*-partitions generated by the attributes $a_i \in B$:

$$\Pi(B) = \bigcap_{i:\, a_i \in B} \Pi(a_i). \tag{1.3}$$

If $B = \{a_{i_1}, \ldots, a_{i_k}\}$, the product can be expressed as: $\Pi(B) = \Pi_{a_{i_1}} \cdot \ldots \cdot \Pi_{a_{i_k}}$. For simplicity, the class $\{u_{t_1}, \ldots, u_{t_r}\}$ of *r*-partition we denote by $\overline{t_1, \ldots, t_r}$.

For the systems of Table 1.1,

$$\Pi_a = \{\overline{1,4,7,13,15,16,17};\ \overline{2,5,8,9,11,12,14};\ \overline{3,6,10}\},$$
$$\Pi_b = \{\overline{1,2,4,5,8,13,15,16,17};\ \overline{1,3,6,7,9,10,11,12,14}\},$$
$$\Pi_c = \{\overline{1,2,3,6,7,10,11,14};\ \overline{1,4,5,7,8,9,12,13,15,16,17}\}.$$

Therefore, for the set $B = \{a,b,c\}$,

$$\Pi(B) = \Pi_a \cdot \Pi_b \cdot \Pi_c = \{\overline{1};\ \overline{1,4,13,15,16,17}; \overline{1,7};\ \overline{3,6,10};\ \overline{2};\ \overline{5,8}; \overline{9,12};\ \overline{11,14}\}.$$

The above result of $\Pi(B)$ can also be calculated directly from the $\text{COM}_{\mathscr{A}}$ relation.

For the first row of the Table 1.1, i.e. $u = (1**203)$ and for the set $B = \{a,b,c\}$, the objects belonging to $\Pi(B)$ are calculated as a set difference

$$U' = (1**)\backslash\{l_1, l_2\},$$

where l_1 is the product of the objects u_1 and u_4, l_2 is the product of u_1 and u_7, i.e.

$$l_1 = (1**) \cdot (101) = 101,$$

$$l_2 = (1**) \cdot (11*) = 11*.$$

Thus, $U' = \{(100)\}$ and therefore a single element class $\{1\}$ has to be added to the $\text{COM}_{\mathscr{A}}(B)$, i.e.

$$\Pi(B) = \text{COM}_{\mathscr{A}}(B) \cup \{1\}.$$

Observation of constructing the *r*-partition is extremely valuable in the case of hierarchical decision-making. This issue, based on the method of logic synthesis, is presented in the Section 1.7.

1.4 Redundancy

Information systems and decision systems are redundant in at least two ways, i.e. the same or indiscernible objects may be represented several times or some of the attributes may be superfluous [19]. Discussed in the previous section, the indiscernibility and compatibility relations can naturally be used for performing the reduction. But we cannot treat them as competing tactics. On the contrary, they complement each other as is shown in this chapter.

1.4.1 Redundancy of Information System

The indiscernibility relation plays an important role in reducing the number of objects of an information system. On the basis of $IND_{\mathscr{A}}$ relation (1.1) we determine equivalence classes, i.e. sets of objects that are not discernible for whole set of attributes A.

Table 1.2. Object-reduced information system

$[u]_{\mathscr{A}}$	class	a b c d e f
$\{u_1\}$	$\mathscr{C}_1$	1 * * 2 0 3
$\{u_4\}$	$\mathscr{C}_2$	1 0 1 0 1 1
$\{u_{15}, u_{16}\}$	$\mathscr{C}_3$	1 0 1 0 1 0
$\{u_3, u_6, u_{10}\}$	$\mathscr{C}_4$	0 1 0 2 2 3
$\{u_7\}$	$\mathscr{C}_5$	1 1 * 1 2 1
$\{u_{11}, u_{14}\}$	$\mathscr{C}_6$	2 1 0 2 2 3
$\{u_{13}\}$	$\mathscr{C}_7$	1 0 1 1 0 1
$\{u_5, u_8\}$	$\mathscr{C}_8$	2 0 1 0 2 2
$\{u_2\}$	$\mathscr{C}_9$	2 0 0 0 2 2
$\{u_9, u_{12}\}$	$\mathscr{C}_{10}$	2 1 1 0 2 2
$\{u_{17}\}$	$\mathscr{C}_{11}$	1 0 1 0 0 2

For the Table 1.1a following sets of objects are indiscernible: $\{u_1\}$, $\{u_4\}$, $\{u_{15}, u_{16}\}$, $\{u_3, u_6, u_{10}\}$, $\{u_7\}$, $\{u_{11}, u_{14}\}$, $\{u_{13}\}$, $\{u_5, u_8\}$, $\{u_2\}$, $\{u_9, u_{12}\}$, $\{u_{17}\}$. Then, the object-reduced information system is shown in Table 1.2.

The problem of reducing the number of attributes is much more complicated. On the basis of [19, 30] let us quote some fundamentals.

A *reduct* of an information system $\mathscr{A} = (U, A)$ is a minimal set of attributes $B \subseteq A$, that $COM_{\mathscr{A}}(B) = COM_{\mathscr{A}}(A)$.

The *discernibility matrix* of $\mathscr{A}$, denoted $M(\mathscr{A})$, is a symmetric matrix $n \times n$, where n is the number of objects U, with elements (c_{pq}) defined as follows:

$$c_{pq} = \{a_i \in A : \rho(u_p, a_i) \nsim \rho(u_q, a_i)\}. \tag{1.4}$$

For the information system defined by Table 1.2, the discernibility matrix is given in Table 1.3.

Table 1.3. Discernibility matrix for the information system from Table 1.2

	$\mathscr{C}_1$	$\mathscr{C}_2$	$\mathscr{C}_3$	$\mathscr{C}_4$	$\mathscr{C}_5$	$\mathscr{C}_6$	$\mathscr{C}_7$	$\mathscr{C}_8$	$\mathscr{C}_9$	$\mathscr{C}_{10}$	$\mathscr{C}_{11}$
$\mathscr{C}_1$	–										
$\mathscr{C}_2$	d,e,f	–									
$\mathscr{C}_3$	d,e,f	f	–								
$\mathscr{C}_4$	a,e	a,b,c,d,e,f	a,b,c,d,e,f	–							
$\mathscr{C}_5$	d,e,f	b,d,e	b,d,e,f	a,d,f	–						
$\mathscr{C}_6$	a,e	a,b,c,d,e,f	a,b,c,d,e,f	a	a,d,f	–					
$\mathscr{C}_7$	d,f	d,e	d,e,f	a,b,c,d,e,f	b,e	a,b,c,d,e,f	–				
$\mathscr{C}_8$	a,d,e,f	a,e,f	a,e,f	a,b,c,d,f	a,b,d,f	b,c,d,f	a,d,e,f	–			
$\mathscr{C}_9$	a,d,e,f	a,c,e,f	a,c,e,f	a,b,d,f	a,b,d,f	b,d,f	a,c,d,e,f	c	–		
$\mathscr{C}_{10}$	a,d,e,f	a,b,e,f	a,b,e,f	a,c,d,f	a,d,f	c,d,f	a,b,d,e,f	b	b,c	–	
$\mathscr{C}_{11}$	d,f	e,f	e,f	a,b,c,d,e,f	b,d,e,f	a,b,c,d,e,f	d,f	a,e	a,c,e	a,b,e	–

A *discernibility function* $f_{\mathscr{A}}$ for an information system $\mathscr{A}$ is a Boolean function of m Boolean attributes $\widehat{a}_1, \ldots, \widehat{a}_m$ corresponding to the attributes $a_1, \ldots, a_m$, defined by the conjunction of all expressions $\vee(\widehat{c}_{pq})$, where $\vee(\widehat{c}_{pq})$ is the disjunction of all attributes $\widehat{c}_{pq} = \{\widehat{a} : a \in c_{pq}\}$, $1 \le p < q \le n$, $\widehat{c}_{pq} \ne \emptyset$.

On the basis of [5, 19] we have the following remark:

Remark. The set of all prime implicants of $f_{\mathscr{A}}$ determines the set of all minimal reducts of the information system $\mathscr{A}$.

After simplification, the discernibility function for the information system from Table 1.2 is $f_{\mathscr{A}}(a,b,c,d,e,f) = abcdf \vee abcef$ (hats were omitted for clarity).

Summarizing, finding the minimal reduct involves converting the conjunctive normal form to the disjunctive normal form. Thus, as it is known, finding a minimal reduct is NP-hard [7, 29] and one of the bottlenecks of the rough set theory [19].

However, after [7], the NP-hard problems "at their core, they are all the same problem, just in different disguises!" They have shown many search problems which can be reduced to one another, and as a consequence, they are all NP-hard. On the basis of the theorem [5] (Brayton) we extend the diagram of reductions of NP-hard problems. Finding all prime implicants of a monotone Boolean function can be reduced to the calculation of double complementation of this function, and then, it can be reduced to the problem of column cover search.

Since the problem of converting a CNF into DNF appears in this chapter repeatedly, it is discussed in Section 1.6. However, detailed information about efficiency and performance of applied logic synthesis method can be found in [4] and in the next chapter of this book.

1.4.2 Redundancy of Decision System

We can easily notice that in decision system from Table 1.1b, there are some objects/cases with the same attribute value for all condition attributes, however the value of decision attribute "dec" is different (not compatible). For example, it takes place for objects u_5 and u_8, or objects u_{15} and u_{16}.

A decision system $\mathscr{A} = (U, A \cup D)$ is *consistent* if, and only if, $\text{IND}_{\mathscr{A}}(A) \subseteq \text{IND}_{\mathscr{A}}(D)$, otherwise we call the decision system *inconsistent*. We say for a consistent decision system $\mathscr{A}$ that the set D of decision attributes depends on the set A of condition attributes.

Such inconsistent decision table can be for example the effect of different medical diagnosis for different patients, or even a medical misdiagnosis; medical examinations were made for different purposes; the condition parameters are not sufficient to make a clear diagnosis.

Generally, some subsets of objects in the decision system cannot be expressed exactly employing available attributes. However, they can be roughly defined [30].

If $\mathscr{A} = (U, A)$ is an information system, $B \subseteq A$, and $W \subseteq U$, then the sets: $\{w \in U : [w]_B \subseteq W\}$ and $\{w \in U : [w]_B \cap W \neq \emptyset\}$, where $[w]_B$ denotes the equivalence class of $\text{IND}_{\mathscr{A}}(B)$ including the object w, are called *B-lower* and *B-upper approximation* of W in $\mathscr{A}$. The lower and upper approximations is denoted by $\underline{B}W$ and $\overline{B}W$, respectively.

The set $\underline{B}W$ is the set of all elements of U which can be with certainty classified as elements of W in the knowledge represented by attributes B. Set $\overline{B}W$ is the set of elements of U which can be possibly classified as elements of W employing knowledge represented by attributes from B.

The *B-boundary region*, given by set difference $\overline{B}W - \underline{B}W$, consists of those objects that can neither be included nor excluded as members of the target set W.

The tuple $\langle \underline{B}W, \overline{B}W \rangle$ composed of the lower and upper approximation is called a *rough set*.

We also employ the following denotation:

$$\text{POS}_B(W) = \underline{B}W \tag{1.5}$$

and refer to $\text{POS}_B(W)$ as *B-positive region* of W. The positive region $\text{POS}_B(W)$ or the lower approximation of W is the collection of those objects which can be classified with full certainty as members of the set W, using classification given by $\text{IND}_{\mathscr{A}}(B)$.

By *A-positive region of D* for a decision system $\mathscr{A} = (U, A \cup D)$, denoted $\text{POS}_{\text{IND}_{\mathscr{A}}(A)}(\text{IND}_{\mathscr{A}}(D))$ or $\text{POS}_A(D)$ for simplicity, we understand the set

$$\text{POS}_A(D) = \bigcup_{w \in U/D} \underline{A}W. \tag{1.6}$$

Remark. Decision system $\mathscr{A} = (U, A \cup D)$ is consistent if, and only if, $\text{POS}_A(D) = U$.

By $D = \mathscr{F}(A)$ we denote the functional dependency of a consistent decision system $\mathscr{A} = (U, A \cup D)$, where $\mathscr{F}$ is specified by a consistent decision table.

Let consider Table 1.1b and set $W_0 = \{u_t : \rho(u_t, \text{dec}) = 0\}$. Then, we obtain following approximations of the set W_0:

- lower approximation (positive region) $\underline{A}W_0 = \{u_1, u_4\}$,
- upper approximation $\overline{A}W_0 = \{u_1, u_4, u_{15}, u_{16}\}$,
- boundary region $\overline{A}W_0 - \underline{A}W_0 = \{u_{15}, u_{16}\}$.

On the basis of the positive regions and the boundary regions of W_0 and the remaining $W_1 = \{u_t : \rho(u_t, \text{dec}) = 1\}$, $W_2 = \{u_t : \rho(u_t, \text{dec}) = 2\}$, we can make unambiguous transformation of the decision table (Tab. 1.1b) into the consistent decision table (Tab. 1.4).

Table 1.4. Consistent decision table

$[u]_{\mathscr{A}}$	class	a b c d e f	dec
$\{u_1\}$	$\mathscr{C}_1$	1 * * 2 0 3	0
$\{u_4\}$	$\mathscr{C}_2$	1 0 1 0 1 1	0
$\{u_{15}, u_{16}\}$	$\mathscr{C}_3$	1 0 1 0 1 0	{0,1}
$\{u_3, u_6, u_{10}\}$	$\mathscr{C}_4$	0 1 0 2 2 3	1
$\{u_7\}$	$\mathscr{C}_5$	1 1 * 1 2 1	1
$\{u_{11}, u_{14}\}$	$\mathscr{C}_6$	2 1 0 2 2 3	1
$\{u_{13}\}$	$\mathscr{C}_7$	1 0 1 1 0 1	1
$\{u_5, u_8\}$	$\mathscr{C}_8$	2 0 1 0 2 2	{1,2}
$\{u_2\}$	$\mathscr{C}_9$	2 0 0 0 2 2	2
$\{u_9, u_{12}\}$	$\mathscr{C}_{10}$	2 1 1 0 2 2	2
$\{u_{17}\}$	$\mathscr{C}_{11}$	1 0 1 0 0 2	2

Following notions are closely related to the task of the attribute reduction of a decision system.

We say that $a_i \in A$ is *D-dispensable* in A, if

$$\text{POS}_A(D) = \text{POS}_{A-a_i}(D), \qquad (1.7)$$

otherwise a_i is *D-indispensable* in A. If every $a_i \in A$ is *D*-indispensable, we say that A is *D-dependent*.

The set $B \subseteq A$ is called a *D-reduct* of $\mathscr{A}$ if, and only if, B is *D*-dependent subset of A and $\text{POS}_B(D) = \text{POS}_A(D)$.

The set of all *D*-indispensable attributes in A is called the *D-core* of $\mathscr{A}$, and is denoted as $\text{CORE}_D(A)$, then

$$\text{CORE}_D(A) = \bigcap \text{RED}_D(A), \qquad (1.8)$$

where $\text{RED}_D(A)$ is the family of all *D*-reducts of $\mathscr{A}$.

Let $\mathscr{A} = (U, A \cup D)$ be a consistent decision system, where set $A = \{a_1, \ldots, a_m\}$ contains m condition attributes and $D = \{d_1, \ldots, d_s\}$ contains s decision attributes. A *discernibility matrix* of $\mathscr{A}$, denoted $M_D(\mathscr{A})$, is a matrix with elements (e_{pq}):

$$e_{pq} = \begin{cases} \emptyset, & \forall j \; \rho(u_p, d_j) = \rho(u_q, d_j), \\ c_{pq}, & \text{otherwise,} \end{cases} \tag{1.9}$$

where c_{pq} is given by formula (1.4) defined for all condition attributes form the set A and $j \in \{1, \ldots, s\}$.

Adequate discernibility matrix for the decision system from Table 1.4 is given in Table 1.5.

Table 1.5. Discernibility matrix for the decision system from Table 1.4

	$\mathscr{C}_1$	$\mathscr{C}_2$	$\mathscr{C}_3$	$\mathscr{C}_4$	$\mathscr{C}_5$	$\mathscr{C}_6$	$\mathscr{C}_7$	$\mathscr{C}_8$	$\mathscr{C}_9$	$\mathscr{C}_{10}$	$\mathscr{C}_{11}$
$\mathscr{C}_1$	–										
$\mathscr{C}_2$	–	–									
$\mathscr{C}_3$	d,e,f	f	–								
$\mathscr{C}_4$	a,e	a,b,c,d,e,f	a,b,c,d,e,f	–							
$\mathscr{C}_5$	d,e,f	b,d,e	b,d,e,f	–	–						
$\mathscr{C}_6$	a,e	a,b,c,d,e,f	a,b,c,d,e,f	–	–	–					
$\mathscr{C}_7$	d,f	d,e	d,e,f	–	–	–	–				
$\mathscr{C}_8$	a,d,e,f	a,e,f	a,e,f	a,b,c,d,f	a,b,d,f	b,c,d,f	a,d,e,f	–			
$\mathscr{C}_9$	a,d,e,f	a,c,e,f	a,c,e,f	a,b,d,f	a,b,d,f	b,d,f	a,c,d,e,f	c	–		
$\mathscr{C}_{10}$	a,d,e,f	a,b,e,f	a,b,e,f	a,c,d,f	a,d,f	c,d,f	a,b,d,e,f	b	–	–	
$\mathscr{C}_{11}$	d,f	e,f	e,f	a,b,c,d,e,f	b,d,e,f	a,b,c,d,e,f	d,f	a,e	–	–	–

A *discernibility function* $f_{\mathscr{A}}$ for a decision system $\mathscr{A}$ is a Boolean function of m Boolean attributes $\widehat{a}_1, \ldots, \widehat{a}_m$ corresponding to the conditional attributes $a_1, \ldots, a_m$, defined by the conjunction of all expressions $\vee(\widehat{e}_{pq})$, where $\vee(\widehat{e}_{pq})$ is the disjunction of all attributes $\widehat{e}_{pq} = \{\widehat{a} : a \in e_{pq}\}$, $1 \le p < q \le n$, $\widehat{e}_{pq} \neq \emptyset$.

Similarly to the information system we have the following remark for decision system:

Remark. The set of all prime implicants of $f_{\mathscr{A}}$ determines the set of all minimal reducts of the decision system $\mathscr{A}$.

Similarly to the information system, finding the minimal reducts involves converting the conjunctive normal form of the discernibility function to the disjunctive normal form which is NP-hard, and is discussed in Section 1.6.

For the function considered in Table 1.4 simplified discernibility function is: $f_{\mathscr{A}}(a,b,c,d,e,f) = bcef \vee abcdf$ (hats were omitted for clarity).

1.5   Induction of Decision Rules

Noticeably, decision tables can also be represented by logic expressions. Such logic structure allows them to be used in decision-making systems, therefore the main goal in the issue of induction is the calculation of possibly shortest decision rules. However, the compression is not the most important, but the fact that on the basis of such generalized knowledge one can make decisions for such data which is not included in the original decision system.

To simplify the table, we can apply the same method as for the logic synthesis called expansion of Boolean function. This simplification in the field of data mining means "to generalize", since the method yields the decision rules we can make decision for any case, also for these cases which are not specified in the primary decision table.

Selecting one column of the indiscernibility matrix for decision system, it is the same as constructing the blocking matrix for expansion procedure. In the synthesis procedure, the matrix is formed for each row for which the function yields "1" and for the whole set of rows for which the function yields "0". Subsequently, we determine the minimum column cover of this matrix. Since blocking matrix is a binary matrix and discernibility matrix for decision system is a binary matrix as well, the process of rules induction is similar. However, in this case we obtain the minimal/generalized rules.

For example, taking the first column of the discernibility matrix only (Tab. 1.5) corresponding to the class $\mathscr{C}_1$ and simplifying suitable discernibility function, i.e. finding the set of all prime implicants (previous remark), it yields the rules which discerns objects belonging to the $\mathscr{C}_1$ from objects belonging to the other decision classes:

$$\begin{aligned} f_{\mathscr{C}_1} &= (d \vee e \vee f)(a \vee e)(d \vee e \vee f)(a \vee e)(d \vee f)(a \vee d \vee e \vee f) \\ &(a \vee d \vee e \vee f)(a \vee d \vee e \vee f)(d \vee f) = \\ &(a \vee e)(d \vee f) = ad \vee af \vee ed \vee ef. \end{aligned}$$

Nevertheless, converting the conjunctive normal form of the function to the disjunctive normal form is NP-hard and is discussed in Section 1.6.

On the other hand, our research showed that applying *Espresso* [5, 40] – the well known logic synthesis system of Boolean function minimization – to the rules induction problem we can achieve spectacular results. In comparison to the LEM2 procedure, the *Espresso* system always yields the results which include the results of data mining algorithm.

For example binary decision system (Tab. 1.6) we obtained the following rules:

- $(a_1 = 1)\&(a_5 = 1)\&(a_6 = 1)\&(a_2 = 1) \Rightarrow (d = 0)$,
- $(a_1 = 1)\&(a_2 = 0)\&(a_5 = 1)\&(a_3 = 0)\&(a_4 = 0)\&(a_6 = 0) \Rightarrow (d = 0)$,
- $(a_4 = 0)\&(a_1 = 1)\&(a_2 = 0)\&(a_7 = 0) \Rightarrow (d = 1)$,
- $(a_2 = 1)\&(a_4 = 0)\&(a_5 = 1)\&(a_6 = 0) \Rightarrow (d = 1)$,

using LEM2 procedure, and

- $(a_1 = 1)\&(a_3 = 0)\&(a_6 = 0) \Rightarrow (d = 0)$,
- $(a_2 = 1)\&(a_6 = 1) \Rightarrow (d = 0)$,
- $(a_4 = 1) \Rightarrow (d = 0)$,
- $(a_2 = 1)\&(a_6 = 0) \Rightarrow (d = 1)$,
- $(a_4 = 0)\&(a_7 = 0) \Rightarrow (d = 1)$,

applying *Espresso* system.

Table 1.6. Example decision system

a_1	a_2	a_3	a_4	a_5	a_6	a_7	d
1	0	0	0	1	0	1	0
1	0	1	1	1	1	0	0
1	1	0	1	1	1	0	0
1	1	1	0	1	1	1	0
0	1	0	0	1	0	1	1
1	0	0	0	1	1	0	1
1	0	1	0	0	0	0	1
1	0	1	0	1	1	0	1
1	1	1	0	1	0	1	1

We can not say that the solution of LEM2 is worse from the standpoint of decision-making, but we believe that the algorithms used in data mining computer systems are not the most effective.

1.6   Algorithm of Complementation

The issues from Section 1.4 and 1.5 yield a discernibility function which is a Boolean formula in a conjunctive normal form (CNF). The simplification of the discernibility function is carried out by transforming the conjunctive normal form into the disjunctive normal form (DNF). Such a transformation is usually time-consuming and therefore it is important to look for efficient algorithms which can handle this task.

An interesting approach proposed by author is based on the fast complementation algorithm [4]. The key strength of the algorithm lies in Shannon expansion procedure of monotone function f. Then,

$$f = \bar{x}_j f_{\bar{x}_j} + f_{x_j} \tag{1.10}$$

This procedure is fundamental in the field of logic synthesis, however it can successfully be applied in the field of data mining.

Proposed approach benefits from the transformation (1.11), i.e. double complementation of a Boolean function.

$$\prod_k \sum_l x_{kl} = \overline{\overline{\prod_k \sum_l x_{kl}}} = \overline{\sum_k \prod_l \overline{x_{kl}}} \tag{1.11}$$

Given that the discernibility function f_M representing the CNF is unate (monotone), it can be transformed into the F form (first complementation) and then considered as a binary matrix M (diagram at bottom of the page). In fact, the task of searching the complement of this function, i.e. $\overline{F}$, can be reduced to the concept of searching of a column cover C of the binary matrix M (second complementation).

Theorem [5]. Each row i of C, the binary matrix complement of M, corresponds to a column cover L of M, where $j \in L$ if and only if $C_{ij} = 1$.

The approach presented extremely accelerates calculations. An efficient representation of the algorithm in computational memory allows us achieving results that cannot be calculated using published methods and systems [1, 8, 9, 16, 17, 18, 28, 35, 37, 38, 41, 42, 43]. Some of the results have been published in [4] however, detailed theory and more results are given in the next chapter of this book.

Lets consider the discernibility function f_M as follows:

$$f_M = (x_2 + x_3 + x_4)(x_1 + x_2)(x_3 + x_4)(x_2 + x_3 + x_5).$$

Performing the multiplication and applying absorption law we obtain:

$$f_M = x_2 x_3 + x_2 x_4 + x_1 x_3 + x_1 x_4 x_5.$$

The same result can be obtained performing mentioned approach, i.e. double complementation of the function f_M. Then,

$$F = \overline{f}_M = \overline{x}_2 \overline{x}_3 \overline{x}_4 + \overline{x}_1 \overline{x}_2 + \overline{x}_3 \overline{x}_4 + \overline{x}_2 \overline{x}_3 \overline{x}_5,$$

and finally, applying Shannon expansion procedure, we calculate $\overline{F}$.

Illustrative diagram of the method has been shown in the diagram below and scheme of complementation using Shannon expansion in Fig. 1.1.

$$F = \begin{bmatrix} - & 0 & 0 & - \\ 0 & 0 & - & - \\ - & - & 0 & 0 \\ - & 0 & 0 & - \end{bmatrix} \quad \longrightarrow \quad M = \begin{bmatrix} 0 & 1 & 1 & 1 & 0 \\ 1 & 1 & 0 & 0 & 0 \\ 0 & 0 & 1 & 1 & 0 \\ 0 & 1 & 1 & 0 & 1 \end{bmatrix}$$

$$\downarrow$$

$$\overline{F} = \begin{bmatrix} - & 1 & 1 & - & - \\ - & 1 & - & 1 & - \\ 1 & - & 1 & - & - \\ 1 & - & - & 1 & 1 \end{bmatrix} \quad \longleftarrow \quad C = \begin{bmatrix} 0 & 1 & 1 & 0 & 0 \\ 0 & 1 & 0 & 1 & 0 \\ 1 & 0 & 1 & 0 & 0 \\ 1 & 0 & 0 & 1 & 1 \end{bmatrix}$$

$$\overline{F} = x_2 x_3 + x_2 x_4 + x_1 x_3 + x_1 x_4 x_5$$

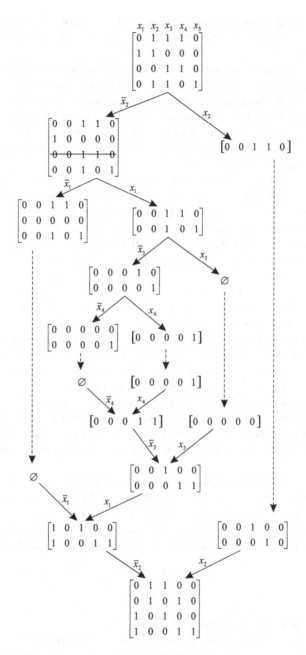

Fig. 1.1. Complementation scheme of $F = \bar{x}_2\bar{x}_3\bar{x}_4 + \bar{x}_1\bar{x}_2 + \bar{x}_3\bar{x}_4 + \bar{x}_2\bar{x}_3\bar{x}_5$

1.7 Hierarchical Decision-Making

In addition to the methods discussed previously, i.e. reduction of attributes and induction of decision rules, there is another opportunity to support data mining algorithms [10, 26, 39].

Usually we make a decision on the basis of an original decision table which is described by decision rules. But as we know data tables are usually large and difficult to store, and the decision rules are long. Therefore, we propose making decisions hierarchically. For a part of the attributes we calculate so-called intermediate decision, and then we use this intermediate decision simultaneously with the other part of attributes. Then, on the basis of both, i.e. these attributes and the intermediate decision, we make the final decision.

To this end, we apply a logic synthesis approach of functional decomposition. The main advantage of functional decomposition method in logic synthesis is that we can minimize the resources for technology mapping. It takes place in the case when a Boolean function cannot be made to fit into any single module designated for implementation. The only solution is to decompose the problem in such a way that the requirements can be met by a network of two or more devices each implementing a part of the function.

A similar problem arises in data mining systems. Application of the decomposition algorithm for decision tables yields more efficient representation of data. We can decide to break down a data table into components, depending on possible storage savings or other considerations. Then, after the decomposition of a large database into smaller components, we induce decision rules for obtained smaller databases. Thus, the rules are calculated faster and are finally shorter, but proper combination of these decision rules yields the same outcome.

To solve the problem of decision table decomposition we propose our original theory. It was elaborated mainly for digital design, however it can be successfully used in data mining. In result we obtain a complex structure of hierarchical decomposition in which the global description is broken down sequentially into smaller subtables.

Let $\mathscr{F}$ be a function representing functional dependency $D = \mathscr{F}(A)$ of the consistent decision system $\mathscr{A} = (U, A \cup D)$, where A is the set of condition attributes and D is the set of decision attributes. Let B_1, B_2 be the subsets of A such that $A = B_1 \cup B_2$ and $B_1 \cap B_2 = \emptyset$. We state the decomposition problem as the following question: when can the functional dependency $D = \mathscr{F}(A)$ be derived in two steps:

$$\mathscr{G}(B_2) = \delta,$$
$$D = \mathscr{H}(B_1, \delta), \tag{1.12}$$

where δ is some fictitious, auxiliary attribute?

Consider the decision system given in Table 1.4. As $\mathrm{IND}_{\mathscr{A}}(A) \subseteq \mathrm{IND}_{\mathscr{A}}(D)$, D functionally depends on A, i.e. all decisions are uniquely determined by conditions. Now we consider a possibility of deriving such functional dependency in hierarchical way using the tables which specify dependencies: $\delta = \mathscr{G}(a, b, c, d)$ and

$D = \mathcal{H}(e, f, \delta)$, where δ is the fictitious decision attribute playing the role of the auxiliary condition attribute. It can be easily verified that, for instance, the tuple of attribute values $1,0,1,0,1,1$ indicates the auxiliary decision $\delta = 0$. Thus, joining this result with the values of attributes e, f, we have:

$$\mathcal{H}(e, f, \delta) \Rightarrow \text{dec}$$

It is the same result as in the table specifying functional dependency $\mathcal{F}$. Verifying other valid objects of table $\mathcal{F}$ we conclude that as for all $u_i \in U$ described by attributes $a, \ldots, f$

$$\mathcal{H}(e, f, \mathcal{G}(a, b, c, d)) = \mathcal{F}(a, \ldots, f),$$

so it is possible to assemble the primary table from the subtables each describing a subset of attributes.

We say that there is a *simple hierarchical decomposition* of $\mathcal{F}(A)$ if, and only if,

$$\mathcal{F}(A) = \mathcal{H}(B_1, \mathcal{G}(B_2)) = \mathcal{H}(B_1, \delta),$$

where $\mathcal{G}$ and $\mathcal{H}$ denote functional dependencies: $\mathcal{G}(B_2) = \delta$ and $\mathcal{H}(B_1, \delta) = D$.

In other words we try to find a function $\mathcal{H}$ depending on the variables of the set B_1 as well on the outputs of a function $\mathcal{G}$ depending on the set B_2. The outputs of the function $\mathcal{H}$ are identical with the function values of $\mathcal{F}$.

The following theorem states the sufficient condition for hierarchical decomposition.

Theorem. Functions $\mathcal{G}$ and $\mathcal{H}$ represent a hierarchical decomposition of function $\mathcal{F}$, i.e. $\mathcal{F}(A) = \mathcal{H}(B_1, \mathcal{G}(B_2))$ if there exists a r-partition $\Pi_{\mathcal{G}} \geq \Pi(B_2)$ such that

$$\Pi(B_1) \cdot \Pi_{\mathcal{G}} \leq \Pi(D), \qquad (1.13)$$

where all the r-partitions are over the set of objects, and the number of values of component $\mathcal{G}$ is equal to $L(\Pi_{\mathcal{G}})$, where $L(\Pi)$ denotes the number of blocks of r-partition Π.

In the theorem, r-partition $\Pi_{\mathcal{G}}$ represents component $\mathcal{G}$, and the product of r-partitions $\Pi(B_1)$ and $\Pi_{\mathcal{G}}$ corresponds to $\mathcal{H}$. The decision tables of the resulting components can be easily obtained from these r-partitions.

Let us decompose the decision table (Tab. 1.4), where the characteristic r-partition is as follows:

$$\Pi(D) = \{\ \overline{1,2};\ \overline{3};\ \overline{4,5,6,7};\ \overline{8};\ \overline{9,10,11}\ \}.$$

For $B_1 = \{e, f\}$, $B_2 = \{a, b, c, d\}$, we obtain

$$\Pi(B_1) = \{\ \overline{1};\ \overline{2,3,11};\ \overline{4};\ \overline{5};\ \overline{6};\ \overline{7};\ \overline{8};\ \overline{9};\ \overline{10}\ \},$$

$$\Pi(B_2) = \{\ \overline{1};\ \overline{2};\ \overline{3};\ \overline{4,6};\ \overline{5};\ \overline{7};\ \overline{8,9,10};\ \overline{11}\ \},$$

$$\Pi_{\mathcal{G}} = \{\ \overline{1,2,3,8,11};\ \overline{4,5,6,7,9,10}\ \}.$$

It can be easily verified that since $\Pi(B_1) \cdot \Pi_{\mathcal{G}} \le \Pi(D)$, function $\mathcal{F}$ is decomposable as $\mathcal{F}(A) = \mathcal{H}(e, f, \mathcal{G}(a,b,c,d))$, where $\mathcal{G}$ is one-output function of four variables.

The decision tables of components $\mathcal{G}$ and $\mathcal{H}$ can be obtained using calculated above r-partitions. Encoding the blocks of $\Pi_{\mathcal{G}}$ respectively as "0" and "1", we immediately obtain the table of function $\mathcal{G}$; it is presented in Tab 1.7a. The table of function $\mathcal{H}$ can be derived by replacing the part of each object u_t with proper code of r-partition $\Pi_{\mathcal{G}}$. The truth table obtained in this way is shown in Table 1.7b.

Table 1.7. Components $\mathcal{G}$ and $\mathcal{H}$

	$a\ b\ c\ d$	δ			$\delta\ e\ f$	dec
	1 * * 2	0			0 0 3	0
	1 0 1 0	0			0 1 1	0
	0 1 0 2	1			0 1 0	{0,1}
a)	1 1 * 1	1	b)		1 2 3	1
	2 1 0 2	1			1 2 1	1
	1 0 1 1	1			1 0 1	1
	2 0 1 0	0			0 2 2	{1,2}
	2 0 0 0	1			1 2 2	2
	2 1 1 0	1			0 0 2	2

In the above example we used a two-valued fictitious attribute. As the data tables are usually stored in computer memory we are usually interested in fictitious attributes with minimum number of values.

Let Π_1 and Π_2 are r-partitions and $\Pi_1 \le \Pi_2$. Then a r-partition $\Pi_1 | \Pi_2$, whose elements are blocks of Π_2 and whose blocks are those of Π_1, is the *quotient r-partition* of Π_1 over Π_2.

Remark. The minimum number of values of fictitious attribute δ sufficient to represent function $\mathcal{F}(A)$ in the form $\mathcal{H}(B_1, \mathcal{G}(B_2))$ is equal to $\gamma(\Pi(B_1) | \Pi(D))$, where $\gamma(\Pi)$ denotes the number of elements in the largest block of r-partition Π and $\Pi(B_1) | \Pi(D)$ denotes the quotient r-partition.

The gain of the decomposition implies from the fact that two components (i.e. tables $\mathcal{G}$ and $\mathcal{H}$) generally require less memory space than non-decomposed decision table. If we express the table's relative size as $S = n \sum b_i$, where n – the number of objects, b_i – the number of bits needed to represent attribute a_i, we can compare the common size of the decomposition components with the size of the primary (non-decomposed table). In the above example the size $S_{\mathcal{F}}$ of primary table is $13 \cdot 11 = 143$ units and the size of decomposed tables is $7 \cdot 9 + 8 \cdot 9 = 135$ units, i.e. 94% of the $S_{\mathcal{F}}$.

Undoubtedly, using the functional decomposition, the typical method of logic synthesis, we can compress databases. But the most important is that we can induce noticeably shorter decision rules for the resulting components $\mathscr{G}$ and $\mathscr{H}$ to finally make a decision in a hierarchical manner. Moreover, applying the method to the real (larger) databases we get even better results.

We have made calculations for the database called *house* which was collected during election process in the United States. The table presents the answers for questions which respondents were asked. Answer to these questions were yes/no only, so table contains binary values. Using decomposition algorithm for this database we have obtained 68% reduction of memory capacity needed to store this database. Of course, compressed database yields the same final decision rules as the original database.

Acknowledgements. This work was partly supported by the Foundation for the Development of Radiocommunications and Multimedia Technologies.

References

1. Abdullah, S., Golafshan, L., Nazri, M.Z.A.: Re-heat simulated annealing algorithm for rough set attribute reduction. International Journal of the Physical Sciences 6(8), 2083–2089 (2011), doi:10.5897/IJPS11.218
2. Bazan, J., Nguyen, H.S., Nguyen, S.H., Synak, P., Wróblewski, J.: Rough set algorithms in classification problem. In: Rough Set Methods and Applications: New Developments in Knowledge Discovery in Information Systems, vol. 56, pp. 49–88. Physica-Verlag, Heidelberg (2000), doi:10.1007/978-3-7908-1840-6_3
3. Borowik, G.: Methods of logic synthesis and their application in data mining. In: 1st Australian Conference on the Applications of Systems Engineering, ACASE 2012, Sydney, Australia, pp. 56–57 (2012) (electronic document)
4. Borowik, G., Łuba, T., Zydek, D.: Features reduction using logic minimization techniques. International Journal of Electronics and Telecommunications 58(1), 71–76 (2012), doi:10.2478/v10177-012-0010-x
5. Brayton, R.K., Hachtel, G.D., McMullen, C.T., Sangiovanni-Vincentelli, A.: Logic Minimization Algorithms for VLSI Synthesis. Kluwer Academic Publishers (1984)
6. Brzozowski, J.A., Łuba, T.: Decomposition of boolean functions specified by cubes. Journal of Multi-Valued Logic & Soft Computing 9, 377–417 (2003)
7. Dasgupta, S., Papadimitriou, C.H., Vazirani, U.V.: Algorithms. McGraw-Hill (2008)
8. Dash, R., Dash, R., Mishra, D.: A hybridized rough-PCA approach of attribute reduction for high dimensional data set. European Journal of Scientific Research 44(1), 29–38 (2010)
9. Feixiang, Z., Yingjun, Z., Li, Z.: An efficient attribute reduction in decision information systems. In: International Conference on Computer Science and Software Engineering, Wuhan, Hubei, pp. 466–469 (2008), doi:10.1109/CSSE.2008.1090
10. Feng, Q., Miao, D., Cheng, Y.: Hierarchical decision rules mining. Expert Systems with Applications 37(3), 2081–2091 (March 2010), DOI: 10.1016/j.eswa.2009.06.065
11. Grzenda, M.: Prediction-oriented dimensionality reduction of industrial data sets. In: Mehrotra, K.G., Mohan, C.K., Oh, J.C., Varshney, P.K., Ali, M. (eds.) IEA/AIE 2011, Part I. LNCS, vol. 6703, pp. 232–241. Springer, Heidelberg (2011)

12. Grzymała-Busse, J.W.: LERS – a system to learning from examples based on rough sets. In: Słowiński, R. (ed.) Intelligent Decision Support – Handbook of Application and Advanced of the Rough Sets Theory. Kluwer Academic Publishers (1992)

13. Grzymała-Busse, J.W.: Incomplete data and generalization of indiscernibility relation, definability, and approximations. In: Ślęzak, D., Wang, G., Szczuka, M.S., Düntsch, I., Yao, Y. (eds.) RSFDGrC 2005. LNCS (LNAI), vol. 3641, pp. 244–253. Springer, Heidelberg (2005)

14. Grzymała-Busse, J.W., Grzymała-Busse, W.J.: Inducing better rule sets by adding missing attribute values. In: Chan, C.-C., Grzymała-Busse, J.W., Ziarko, W.P. (eds.) RSCTC 2008. LNCS (LNAI), vol. 5306, pp. 160–169. Springer, Heidelberg (2008)

15. Hedar, A.R., Wang, J., Fukushima, M.: Tabu search for attribute reduction in rough set theory. Journal of Soft Computing – A Fusion of Foundations, Methodologies and Applications 12(9), 909–918 (2008), doi:10.1007/s00500-007-0260-1

16. Jensen, R., Shen, Q.: Semantics-preserving dimensionality reduction: Rough and fuzzy-rough based approaches. IEEE Transactions on Knowledge and Data Engineering 16, 1457–1471 (2004), doi:10.1109/TKDE.2004.96

17. Jing, S., She, K.: Heterogeneous attribute reduction in noisy system based on a generalized neighborhood rough sets model. World Academy of Science, Engineering and Technology 75, 1067–1072 (2011)

18. Kalyani, P., Karnan, M.: A new implementation of attribute reduction using quick relative reduct algorithm. International Journal of Internet Computing 1(1), 99–102 (2011)

19. Komorowski, J., Pawlak, Z., Polkowski, L., Skowron, A.: Rough sets: A tutorial (1999)

20. Kryszkiewicz, M., Cichoń, K.: Towards scalable algorithms for discovering rough set reducts. In: Peters, J.F., Skowron, A., Grzymała-Busse, J.W., Kostek, B.Z., Świniarski, R.W., Szczuka, M.S. (eds.) Transactions on Rough Sets I. LNCS, vol. 3100, pp. 120–143. Springer, Heidelberg (2004)

21. Kryszkiewicz, M., Lasek, P.: FUN: Fast discovery of minimal sets of attributes functionally determining a decision attribute. In: Peters, J.F., Skowron, A., Rybiński, H. (eds.) Transactions on Rough Sets IX. LNCS, vol. 5390, pp. 76–95. Springer, Heidelberg (2008)

22. Lewandowski, J., Rawski, M., Rybiński, H.: Application of parallel decomposition for creation of reduced feed-forward neural networks. In: Kryszkiewicz, M., Peters, J.F., Rybiński, H., Skowron, A. (eds.) RSEISP 2007. LNCS (LNAI), vol. 4585, pp. 564–573. Springer, Heidelberg (2007)

23. Lin, T.Y., Yao, Y.Y., Zadeh, L.A. (eds.): Data mining, rough sets and granular computing. Physica-Verlag GmbH, Heidelberg (2002)

24. Łuba, T., Borowik, G., Kraśniewski, A.: Synthesis of finite state machines for implementation with programmable structures. Electronics and Telecommunications Quarterly 55, 183–200 (2009)

25. Łuba, T., Lasocki, R.: On unknown attribute values in functional dependencies. In: Proceedings of The Third International Workshop on Rough Sets and Soft Computing, San Jose, pp. 490–497 (1994)

26. Łuba, T., Lasocki, R., Rybnik, J.: An implementation of decomposition algorithm and its application in information systems analysis and logic synthesis. In: Ziarko, W. (ed.) Rough Sets, Fuzzy Sets and Knowledge Discovery. Workshops in Computing Series, pp. 458–465. Springer, Heidelberg (1994)

27. Łuba, T., Rybnik, J.: Rough sets and some aspects in logic synthesis. In: Słowiński, R. (ed.) Intelligent Decision Support – Handbook of Application and Advances of the Rough Sets Theory. Kluwer Academic Publishers (1992)

28. Nguyen, D., Nguyen, X.: A new method to attribute reduction of decision systems with covering rough sets. Georgian Electronic Scientific Journal: Computer Science and Telecommunications 1(24), 24–31 (2010)
29. Papadimitriou, C.H.: Computational complexity. Academic Internet Publ. (2007)
30. Pawlak, Z.: Rough Sets. Theoretical Aspects of Reasoning about Data. Kluwer Academic Publishers (1991)
31. Pei, X., Wang, Y.: An approximate approach to attribute reduction. International Journal of Information Technology 12(4), 128–135 (2006)
32. Peters, J.F., Skowron, A., Chan, C.-C., Grzymała-Busse, J.W., Ziarko, W.P. (eds.): Transactions on Rough Sets XIII. LNCS, vol. 6499. Springer, Heidelberg (2011)
33. Rawski, M., Borowik, G., Łuba, T., Tomaszewicz, P., Falkowski, B.J.: Logic synthesis strategy for FPGAs with embedded memory blocks. Electrical Review 86(11a), 94–101 (2010)
34. Selvaraj, H., Sapiecha, P., Rawski, M., Łuba, T.: Functional decomposition – the value and implication for both neural networks and digital designing. International Journal of Computational Intelligence and Applications 6(1), 123–138 (2006), doi:10.1142/S1469026806001782
35. Skowron, A., Rauszer, C.: The discernibility matrices and functions in information systems. In: Słowiński, R. (ed.) Intelligent Decision Support – Handbook of Application and Advances of the Rough Sets Theory. Kluwer Academic Publishers (1992)
36. Słowiński, R., Sharif, E.: Rough sets analysis of experience in surgical practice. In: Rough Sets: State of The Art and Perspectives, Poznań-Kiekrz (1992)
37. Wang, C., Ou, F.: An attribute reduction algorithm based on conditional entropy and frequency of attributes. In: Proceedings of the 2008 International Conference on Intelligent Computation Technology and Automation, ICICTA 2008, vol. 1, pp. 752–756. IEEE Computer Society, Washington, DC (2008), doi:10.1109/ICICTA.2008.95
38. Yao, Y., Zhao, Y.: Attribute reduction in decision-theoretic rough set models. Information Sciences 178(17), 3356–3373 (2008), doi:10.1016/j.ins.2008.05.010
39. Zupan, B., Bohanec, M., Bratko, I., Cestnik, B.: A dataset decomposition approach to data mining and machine discovery. In: KDD, pp. 299–302 (1997)
40. Espresso – multi-valued PLA minimization,
    http://embedded.eecs.berkeley.edu/pubs/downloads/espresso
41. ROSE2 – Rough Sets Data Explorer,
    http://idss.cs.put.poznan.pl/site/rose.html
42. ROSETTA – A Rough Set Toolkit for Analysis of Data,
    http://www.lcb.uu.se/tools/rosetta/
43. RSES – Rough Set Exploration System, http://logic.mimuw.edu.pl/~rses/

Chapter 2
Fast Algorithm of Attribute Reduction Based on the Complementation of Boolean Function

Grzegorz Borowik and Tadeusz Łuba

Abstract. In this chapter we propose a new method of solving the attribute reduction problem. Our method is different to the classical approach using the so-called discernibility function and its CNF into DNF transformation. We have proved that the problem is equivalent to very efficient unate complementation algorithm. That is why we propose new algorithm based on recursive execution of the procedure, which at every step of recursion selects the splitting variable and then calculates the cofactors with respect to the selected variables (Shannon expansion procedure). The recursion continues until at each leaf of the recursion tree the easily computable rules for complement process can be applied. The recursion process creates a binary tree so that the final result is obtained merging the results in the subtrees. The final matrix represents all the minimal reducts of a decision table or all the minimal dependence sets of input variables, respectively. According to the results of computer tests, better results can be achieved by application of our method in combination with the classical method.

2.1 Introduction

Often in the area of machine learning, artificial intelligence, as well as logic synthesis, we deal with some functional dependencies (for example, in the form of decision tables), in which not all attributes are necessary, i.e. some of them could be removed without loss of any information. The problem of removing redundant input attributes is known as the argument reduction problem. Its applications in the area of artificial intelligence have been studied by several researchers [22, 26].

Grzegorz Borowik · Tadeusz Łuba
Institute of Telecommunications, Warsaw University of Technology,
Nowowiejska 15/19, 00-665 Warsaw, Poland
e-mail: {G.Borowik,T.Luba}@tele.pw.edu.pl

R. Klempous et al. (eds.), *Advanced Methods and Applications in Computational Intelligence*,   25
Topics in Intelligent Engineering and Informatics 6,
DOI: 10.1007/978-3-319-01436-4_2, © Springer International Publishing Switzerland 2014

Methods of attribute reduction have proved very useful in many applications. As an example, a data base containing information about 122 patients with duodenal ulcers, treated by highly selective vagotomy, was analyzed by Słowinski and Sharif [27]. This database contained information needed to classify patients according to the long term results of the operation. From 11 attributes describing the patient's health state, after reduction it turned out that only 5 of them were enough to ensure a satisfactory quality of classification.

Such results indicate that application of argument reduction could also be useful in logic synthesis [20], where circuits performance can be presented as truth tables which are in fact decision tables with two valued attributes, where condition attributes are in fact input variables, and decision ones are to represent output variables of the circuit. In the practical application of Boolean algebra the key problem is to represent Boolean functions by formulas which are as simple as possible. One approach to this simplification is to minimize the number of variables appearing in truth table explicitly. Then, the reduced set of input variables is used in other optimization algorithms, e.g. logic minimization and logic decomposition. Combined with other design techniques, argument reduction allows for great size reduction of implemented circuits [19, 20].

A number of methods for discovering reducts have already been proposed in the literature [1, 2, 7, 8, 10, 11, 12, 13, 14, 15, 17, 21, 23, 26, 28, 29]. The most popular Algorithms for Discovering Rough Set Reducts methods are based on discernibility matrices [26]. Besides mutual information and discernibility matrix based attribute reduction methods, they have developed some efficient reduction algorithms based on CI tools of genetic algorithm, ant colony optimization, simulated annealing, and others [11]. These techniques have been successfully applied to data reduction, text classification and texture analysis [17].

Interestingly, another potentially very promising area of application of arguments reduction algorithm is logic systems designing. This is because the novel hardware building blocks impose limitations on the size of circuits that can be implemented with them. The concept of argument reduction was introduced and effectively applied in the balanced decomposition method [16, 24]. Based on redundant variable analysis of each output of a multi-output Boolean function, parallel decomposition separates F into two or more functions, each of which has as its inputs and outputs a subset of the original inputs and outputs. It was proved that parallel decomposition based on argument reduction process plays a very important role in FPGA based synthesis of digital circuits.

Recently, we have proposed the Arguments/Attributes Minimizer [4] which computes reducts of a set of attributes of knowledge representation in information systems or reduces the number of input variables in logic systems. The algorithm is based on unate complementation concept [5]. This chapter extends the results obtained in [4]. Here, we propose new algorithm based on recursive execution of the procedure, which at every step of recursion selects the splitting variable and then calculates the cofactors with respect to the selected variables (Shannon expansion procedure). The recursion continues until, at each leaf of the recursion tree, the easily computable rules for complement process can be applied. The recursion process creates a binary tree so that the final result is obtained merging the results obtained in the subtrees (merging procedure). The final matrix represents all the minimal reducts of a decision table or all the minimal dependence sets of input variables, respectively.

The chapter begins with an overview of basic notions of information systems, functional dependencies, decision tables and reducts. In section 2.2 and 2.3 we discuss relations between multi-valued logic and decision table systems with respect to classification of data with missing values. Particularly, it is shown that elimination of attributes can be easily obtained using standard procedures used in logic synthesis. New contribution is presented in section 2.4, where we describe how to apply complementation algorithm and provide new variant of the attribute reduction process.

2.2   Preliminary Notions

The information system contains data about objects characterized by certain attributes, where two classes of attributes are often distinguished: condition and decision attributes (in logic synthesis they are usually called input and output variables). Such an information system is called a decision system and it is usually specified by a decision table (in logic synthesis it is called a truth table). The decision table describes conditions that must be satisfied in order to carry out the decisions specified for them. More formally, an *information system* is a pair $\mathscr{A} = (U, A)$, where U is a nonempty set of objects (in logic synthesis: minterms) called the universe, and A is a nonempty set of attributes (variables). If we distinguish in an information system two disjoint classes of attributes, condition (A) and decision (D) attributes (input and output variables), where $A \cap D = \emptyset$, then the system is called a *decision system* $\mathscr{A} = (U, A \cup D)$. Any information or decision table defines a function ρ that maps the direct product of U and A (U and $A \cup D$, respectively) into the set of all values.

The attribute values $\rho_{pi} = \rho(u_p, a_i)$ and $\rho_{qi} = \rho(u_q, a_i)$ are called *compatible* ($\rho_{pi} \sim \rho_{qi}$) if, and only if, $\rho_{pi} = \rho_{qi}$ or $\rho_{pi} = *$ or $\rho_{qi} = *$, where "$*$" represents the case when attribute value is "do not care". On the other hand, if ρ_{pi} and ρ_{qi} are defined and are "different" it is said that ρ_{pi} is *not compatible* with ρ_{qi} and is denoted as $\rho_{pi} \nsim \rho_{qi}$. The consequence of this definition is a COM relation defined as follows:

Let $B \subseteq A$ and $u_p, u_q \in U$. The objects $p, q \in \text{COM}(B)$ if and only if $\rho(u_p, a_i) \sim \rho(u_q, a_i)$ for every $a_i \in B$.

The objects u_p and u_q, which belong to the relation $\text{COM}(B)$, are said to be *compatible in the set B*. Compatible objects in the set $B = A$ are simply called *compatible*.

The compatibility relation of objects is a tolerance relation (reflexive and symmetric) and hence it generates compatible classes on the set of objects U.

Compatibility relation allows us to classify objects but the classification classes do not form partitions on the set U, as it is in the case of indiscernibility relation

(IND) [9]. COM(B) classifies objects grouping them into compatibility classes, i.e. $U/\text{COM}(B)$, where $B \subseteq A$.

For the sake of simplicity, collection of subsets $U/\text{COM}(B)$ we call *r-partition* on U and denote as COM(B).

R-partition can be used as a tool to classify objects of a data table description. It can be shown that the *r*-partition concept is a generalization of the ideas of partitioning a set into consistent classes and partitioning a set into tolerance classes. Therefore, partitioning a set U into consistent classes of certain relation $\mathscr{R}$ and partitioning a set into tolerance classes of a certain relation $\mathscr{T}$ can be treated as special cases of *r*-partitioning.

R-partition on a set U may be viewed as a collection of non-disjoint subsets of U, where the set union is equal U; and all symbols and operations of partition algebra are applicable to *r*-partitions. Therefore, convention used for denoting *r*-partitions and their typical operators are the same as in the case of partitions. We assume the reader's familiarity with these *r*-partition concepts which are simple extensions of partition algebra [18].

Especially the relation *less than or equal to* holds between two *r*-partitions Π_1 and Π_2 ($\Pi_1 \leq \Pi_2$) iff for every block of Π_1, in short denoted by $\mathscr{B}_i(\Pi_1)$, there exists a $\mathscr{B}_j(\Pi_2)$, such that $\mathscr{B}_i(\Pi_1) \subseteq \mathscr{B}_j(\Pi_2)$. If Π_1 and Π_2 are partitions, this definition reduces to the conventional ordering relation between two partitions.

The *r*-partition generated by a set B is the product of *r*-partitions generated by the attributes $a_i \in B$:

$$\Pi(B) = \bigcap_i \Pi(a_i).$$

If $B = \{a_{i_1}, \ldots, a_{i_k}\}$, the product can be expressed as: $\Pi(B) = \Pi(a_{i_1}) \cdot \ldots \cdot \Pi(a_{i_k})$.

Let Π_1 and Π_2 are *r*-partitions and $\Pi_1 \leq \Pi_2$. Then a *r*-partition $\Pi_1|\Pi_2$, whose elements are blocks of Π_2 and whose blocks are those of Π_1, is the *quotient r-partition* of Π_1 over Π_2.

Example. For the decision system shown in Table 2.1

$$\Pi(a_1) = \{\overline{1,3,5,6,7};\ \overline{2,4,5};\ \overline{5,8}\},$$

$$\Pi(a_6) = \{\overline{1,4,7,8};\ \overline{2,3,5,6,7}\},$$

$$\Pi(d) = \{\overline{1,5};\ \overline{2,3,4};\ \overline{6};\ \overline{7,8}\}.$$

Therefore, for the set $B = \{a_1, a_6\}$,

$$\Pi(B) = \Pi(a_1) \cdot \Pi(a_6) = \{\overline{1,7};\ \overline{3,5,6,7};\ \overline{4};\ \overline{2,5};\ \overline{8};\ \overline{5}\},$$

and the quotient *r*-parttion $\Pi(B)|\Pi(d)$ is

$$\Pi(B)|\Pi(d) = \{\overline{(1),(7)};\ \overline{(3),(5),(6),(7)};\ \overline{(4)};\ \overline{(2),(5)};\ \overline{(8)};\ \overline{(5)}\}.$$

Table 2.1. Example of a decision system

	a_1	a_2	a_3	a_4	a_5	a_6	d
1	0	0	1	1	0	0	1
2	1	*	2	0	1	1	2
3	0	1	1	0	0	1	2
4	1	2	2	*	2	0	2
5	*	2	2	2	0	1	1
6	0	0	1	1	0	1	3
7	0	1	0	3	2	*	4
8	2	2	2	3	2	0	4

2.3   Elimination of Input Variables

In this section the process of searching and elimination of redundant arguments is described using concepts of logic systems, however appropriate simplifications caused by functional dependency features as well as the generalization to the case of r-partition have been efficiently applied.

An argument $x \in X$ is called *dispensable* in a logic specification of function F iff $\Pi(X - \{x\}) \leq \Pi(F)$, otherwise, i.e. $\Pi(X - \{x\}) \nleq \Pi(F)$, an argument is called *indispensable* (i.e. an essential variable).

The meaning of an indispensable variable is similar to that of a core attribute, i.e. these are the most important variables. In other words, no indispensable variable can be removed without destroying the consistency of the function specification. Thus, the set of all indispensable arguments is called a *core* of X and is denoted as CORE(X).

In order to find the core set of arguments we have to eliminate an input variable and then verify whether the corresponding partition inequality holds. A key theorem is stated below to make this procedure more efficient.

First of all we reformulate this problem to apply more useful tools which are efficiently used in switching theory [18, 19, 20].

A set $B = \{b_1, \dots, b_k\} \subseteq X$ is called a *minimal dependences set*, i.e. *reduct*, of a Boolean function F iff $\Pi(B) \leq \Pi(F)$, and there is no proper subset B' of B, such that $\Pi(B') \leq \Pi(F)$.

It is evident that an indispensable input variable of function F is an argument of every minimal dependence set of F.

Now we introduce two basic notions, namely discernibility set and discernibility function, which help us to construct an efficient algorithm for attribute reduction process.

Let $\mathscr{A} = (U, A \cup D)$ be a decision system. Let u_p, u_q ($u_p \neq u_q$) are objects of U, such that $d \in D$ is a decision attribute and $\rho(u_p, d) \nsim \rho(u_q, d)$. By $\{c_{pq}\}$, we denote a set of attributes called a *discernibility set* which is defined as follows:

$$c_{pq} = \{a_i \in A : \rho(u_p, a_i) \nsim \rho(u_q, a_i)\}. \tag{2.1}$$

A *discernibility function* $f_{\mathscr{A}}$ for a decision system $\mathscr{A}$ is a Boolean function of m Boolean attributes $\widehat{a}_1, \ldots, \widehat{a}_m$ corresponding to the attributes $a_1, \ldots, a_m$, defined by the conjunction of all expressions $\vee(\widehat{c}_{pq})$, where $\vee(\widehat{c}_{pq})$ is the disjunction of all attributes $\widehat{c}_{pq} = \{\widehat{a} : a \in c_{pq}\}, 1 \le p < q \le n$.

A strong relation between the notion of a reduct of function $F(RED(X))$ and prime implicant of the monotonic Boolean function f_F:

$$\{x_{i_1}, \ldots, x_{i_k}\} \in RED(X) \text{ iff } x_{i_1} \wedge \cdots \wedge x_{i_k} \text{ is a prime implicant of } f_F.$$

was investigated among others by Skowron and Kryszkiewicz [14, 26]:

Example. Using the previously calculated quotient r-parttion $\Pi(B)|\Pi(d)$:

$$\Pi(B)|\Pi(d) = \{\overline{(1),(7)}; \ \overline{(3),(5),(6),(7)}; \ \overline{(4)}; \ \overline{(2),(5)}; \ \overline{(8)}; \ \overline{(5)}\}$$

we can conclude that the only pairs of objects which should be separated by condition attributes are shown in the Table 2.2. For each pair u_p, u_q the set of all distinguishing attributes is calculated. It is easy to observe that the discerniblity function expressed in CNF is as

$$f_M = (a_2 + a_4)(a_4 + a_5).$$

Table 2.2. Pairs of objects and corresponding separations

p,q	attributes
1,7	$\{a_2, a_3, a_4, a_5\}$
3,5	$\{a_2, a_3, a_4\}$
3,6	$\{a_2, a_4\}$
3,7	$\{a_3, a_4, a_5\}$
5,6	$\{a_2, a_3, a_4\}$
5,7	$\{a_2, a_3, a_4, a_5\}$
6,7	$\{a_2, a_3, a_4, a_5\}$
2,5	$\{a_4, a_5\}$

Transforming CNF into DNF

$$f_M = a_4 + a_2 a_5$$

we conlude that all reducts for DT from Table 2.1 are

- $\{a_1, a_4, a_6\}$,
- $\{a_1, a_2, a_5, a_6\}$.

Noticeably, minimization of the discernibility function is carried out by transforming the conjunctive normal form (in which it is originally constructed) into the disjunctive normal form and finding a minimum implicant. Such a transformation is usually time-consuming. Therefore, it is important to look for efficient algorithms which can handle this task.

2.4   Computing Minimal Sets of Attributes Using COMPLEMENT Algorithm

A proposed approach is based on the fact that transformation of a conjunctive normal form into the disjunctive normal form can be reduced to the task of complementation of monotone/unate Boolean function which is intimately related to the concept of a column cover of the binary matrix. The unate complementation was proposed by Brayton [5], and is used in logic minimization algorithms as unate recursive paradigm. The unate recursive paradigm exploits properties of unate functions, while performing recursive decomposition. Therefore, to obtain discernibility function in the minimal DNF we apply the fast complementation algorithm adopted from *Espresso* system [5].

To this end, we describe the collection $\{c_{pq}\}$ from (2.1), of all c_{pq} sets in the form of the binary matrix M for which an element M_{ij} is defined as follows:

$$M_{ij} = \begin{cases} 1, & \text{if } a_j \in c_{pq_i}, \\ 0, & \text{otherwise}, \end{cases} \tag{2.2}$$

$i = 1,\ldots,t = \text{CARD}(\{c_{pq}\})$, $j = 1,\ldots,m = \text{CARD}(A)$. Then,

Theorem [5]. Each row i of $\overline{M}$, the binary matrix complement of M, corresponds to a column cover L of M, where $j \in L$ if and only if $\overline{M}_{ij} = 1$.

Column cover means that every row of M contains a "1" in some column which appears in L. The rows of $\overline{M}$ include the set of all minimal column covers of M. If $\overline{M}$ was minimal with respect to containment, then $\overline{M}$ would precisely represent the set of all minimal column covers of M.

More precisely, a *column cover* of binary matrix M is defined as a set L of columns such that for every i

$$\sum_{j \in L} M_{ij} \geq 1. \tag{2.3}$$

Our goal is to select an optimal set L of arguments corresponding to columns of M. Covers L of M are in one-to-one correspondence with the reduced subsets of arguments, i.e. reducts.

The task of searching the minimal column cover is based on the Shannon expansion of Boolean function f:

$$f = \overline{x}_j f_{\overline{x}_j} + x_j f_{x_j}, \tag{2.4}$$

where $f_{x_j}, f_{\overline{x}_j}$ are *cofactors* of f with respect to splitting variable x_j, i.e. the results of substituting "1" and "0" for x_j in f. It is the key to a fast recursive complementation process.

On the basis of [5] a logic function f is *monotone increasing* (*monotone decreasing*) in a variable x_j if changing x_j from 0 to 1 causes all the outputs of f that change, to increase also from 0 to 1 (from 1 to 0). A function that is either monotone increasing or monotone decreasing in x_j is said to be *monotone* or *unate* in x_j.

A function is *unate* (*monotone*) if it is unate in all its variables. For example, the function $f = x_1\bar{x}_2 + \bar{x}_2 x_3$ is unate since it is increasing in x_1 and x_3, and decreasing in x_2.

An important aspect of the algorithm is that properties of unate functions are exploited to simplify or terminate the recursion. Unate functions have special properties which make them especially useful.

Applying the property of unateness the equation (2.4) can be expressed as simplified formulas, i.e. when function is monotone decreasing, then $f_{x_j} \subseteq f_{\bar{x}_j}$, and therefore

$$f = \bar{x}_j f_{\bar{x}_j} + f_{x_j} \tag{2.5}$$

and when function is monotone increasing, then $f_{x_j} \supseteq f_{\bar{x}_j}$, and therefore

$$f = f_{\bar{x}_j} + x_j f_{x_j} \tag{2.6}$$

Stating that discernibility function f_M is in CNF, proposed approach benefits from the transformation (2.7), i.e. double complementation of a Boolean function.

$$\prod_k \sum_l x_{kl} = \overline{\overline{\prod_k \sum_l x_{kl}}} = \overline{\sum_k \prod_l \overline{x_{kl}}} \tag{2.7}$$

Given that the discernibility function f_M representing the CNF is unate, then by applying De Morgan's law it can be transformed into $F = \bar{f}_M$ (first complementation) and then considered as a binary matrix M. Then, each row of M corresponds to the conjunction of negative literals $\bar{x}_j$ of F, i.e. then $M_{ij} = 1$. In fact, the task of searching the complement of function F, i.e. $\bar{F}$, can be reduced to the concept of searching of a column covers (represented by matrix C) of the binary matrix M (second complementation).

Given an initial matrix M – the initial cover of a unate Boolean function F, for simplicity we call the *cover of function F* – the algorithm recursively splits the matrix into smaller pieces until, at each leaf of the recursion tree, the easily computable termination rules (described in Subsection 2.4.1) can be applied. If a basic termination rule applies, then the appropriate cover is returned immediately. If no basic termination rule applies, the initial cover of F is cofactored by both $\bar{x}_j$ and x_j, and the algorithm is recursively called. Each recursive call returns a complemented cover, i.e. $\bar{F}_{\bar{x}_j}$ and $\bar{F}_{x_j}$.

Then, the results are reassembled into a final solution. For the complementation algorithm, the result is the complement of the initial cover M.

To assemble the final result, the complemented covers are merged using following formulas, i.e. for monotone decreasing function

$$\bar{f} = \bar{f}_{\bar{x}_j} + x_j \bar{f}_{x_j} \tag{2.8}$$

and for monotone increasing function

$$\bar{f} = \bar{x}_j \bar{f}_{\bar{x}_j} + \bar{f}_{x_j} \tag{2.9}$$

Noticeably, function F is monotone decreasing for all literals, therefore only equations (2.5) and (2.8) are considered in the presented approach.

2.4.1   Unate Complementation

In the unate complementation algorithm calculations are organized in a top-down synthesis process to obtain the required final complement of function F. At each node of the binary recursion tree the splitting variable x_j is chosen and both cofactors F_{x_j}, $F_{\bar{x}_j}$ are calculated.

In order to identify the splitting variables we choose them among the shortest terms in F, i.e. the object with the maximum number of "do not care" values. Having identified the splitting variables, we have to decide the order in which these variables are processed. This order is made by choosing first the variables that appear most often in the other terms of F. Such procedure eliminates the largest number of rows of matrix M in one of the branches of the recursion.

The cofactor with respect to $\bar{x}_j$ is obtained by setting up the j-th column to "0", and the cofactor with respect to x_j, is obtained by excluding all the rows for which the j-th element is equal to 1.

Stating that matrix M is the cover function F, the recursion continues until at each leaf of the recursion tree one of the basic termination rules is encountered:

- The cover of F is empty. A cover is empty if F contains no terms. In this case, its complement is a tautology. Hence, a cover containing the universal cube [5] is returned. Then, the resulting cover contains one row of all 0's.
- F includes the universal cube, i.e. M contains a row of all 0's. Here, F is a tautology, so its complement is empty. The empty cover is returned (no terms).
- F contains a single term. Here, the complement of F can be computed directly using De Morgan's law. After complementation, $\bar{F}$ contains term(s) of one variable only. Then, corresponding cover is returned.

In each step of the recursive complementation algorithm, termination conditions are checked; if they are not satisfied, recursion is performed. The complementation algorithm returns an actual cover (the complement of the initial cover).

Example. The influence of the unate recursive algorithm on the final result of the attribute reduction process is explained with function f_M represented by a following CNF:

$$f_M = x_4 x_6 (x_1 + x_2)(x_3 + x_5 + x_7)(x_2 + x_3)(x_2 + x_7).$$

Hence, omitting indispensable variables:

$$f'_M = (x_1 + x_2)(x_3 + x_5 + x_7)(x_2 + x_3)(x_2 + x_7),$$

and complementing

$$F = \overline{f'}_M = \overline{x}_1\overline{x}_2 + \overline{x}_3\overline{x}_5\overline{x}_7 + \overline{x}_2\overline{x}_3 + \overline{x}_2\overline{x}_7.$$

Then, the initial cover

$$
M = \begin{array}{c} x_1\ x_2\ x_3\ x_4\ x_5\ x_6\ x_7 \\ \begin{bmatrix} 1 & 1 & 0 & 0 & 0 & 0 & 0 \\ 0 & 0 & 1 & 0 & 1 & 0 & 1 \\ 0 & 1 & 1 & 0 & 0 & 0 & 0 \\ 0 & 1 & 0 & 0 & 0 & 0 & 1 \end{bmatrix} \end{array}
$$

In order to identify the splitting variables we choose them among the shortest terms in F. Here we select the first term, yielding variables x_1 and x_2. Since the variable that appears most often in the other terms of F is x_2, we decide to choose this one.

Now we compute the cofactors of F with respect to the variable x_2. This is illustrated in Fig. 2.1 by two arrows with the common starting point going to different directions. The current tree has two leaves: the larger represents cofactor $F_{\overline{x}_2}$ and the smaller one represents F_{x_2}.

The smaller cofactor is the subject to special case resulting in three objects of the complement, because F_{x_2} cofactor has only one row. The larger cofactor is again decomposed yielding cofactor for $\overline{x}_1$ (left hand side matrix), and for x_1 (right hand side matrix). For the left hand side component we again can apply special case, now resulting in empty cover – there is a row of all 0's. Thus, the next step deals with the right hand side function matrix, which can be expanded onto two matrices, with easily calculated complements.

To illustrate the merging operation of the unate recursive process, consider the actions taken at the nodes x_7, x_1, x_2. Applying formula (2.8) at the nodes x_7, x_1, x_2 we calculate following complements, denoted C_7, C_1, C_2:

$$C_7 = x_7 \cdot [0\ 0\ 1\ 0\ 0\ 0\ 0] + \emptyset = [0\ 0\ 1\ 0\ 0\ 0\ 1]$$

$$C_1 = x_1 \cdot [0\ 0\ 1\ 0\ 0\ 0\ 1] + \emptyset = [1\ 0\ 1\ 0\ 0\ 0\ 1]$$

$$C_2 = x_2 \cdot \begin{bmatrix} 0 & 0 & 1 & 0 & 0 & 0 & 0 \\ 0 & 0 & 0 & 0 & 1 & 0 & 0 \\ 0 & 0 & 0 & 0 & 0 & 0 & 1 \end{bmatrix} + [1\ 0\ 1\ 0\ 0\ 0\ 1] = \begin{bmatrix} 0 & 1 & 1 & 0 & 0 & 0 & 0 \\ 0 & 1 & 0 & 0 & 1 & 0 & 0 \\ 0 & 1 & 0 & 0 & 0 & 0 & 1 \\ 1 & 0 & 1 & 0 & 0 & 0 & 1 \end{bmatrix}$$

Resulted complement C_2 together with indispensable variables x_4 and x_6 represents the following reducts:

- $\{x_2, x_3, x_4, x_6\}$,
- $\{x_2, x_4, x_5, x_6\}$,
- $\{x_2, x_4, x_6, x_7\}$,
- $\{x_1, x_3, x_4, x_6, x_7\}$.

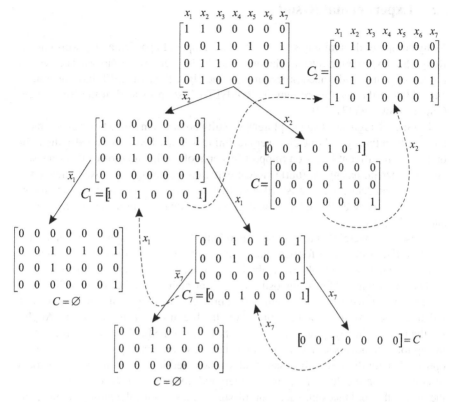

Fig. 2.1. Complementation scheme of $F = \bar{x}_1\bar{x}_2 + \bar{x}_3\bar{x}_5\bar{x}_7 + \bar{x}_2\bar{x}_3 + \bar{x}_2\bar{x}_7$

This agrees with the result that can be obtained applying fundamental transformations of Boolean algebra:

$$f_M = (x_1 + x_2)(x_3 + x_5 + x_7)(x_2 + x_3)(x_2 + x_7),$$

hence, performing the multiplication and applying absorption law

$$(x_1 + x_2)(x_3 + x_5 + x_7)(x_2 + x_3)(x_2 + x_7) =$$

$$(x_2 + x_1)(x_2 + x_3)(x_2 + x_7)(x_3 + x_5 + x_7) =$$

$$(x_2 + x_1x_3x_7)(x_3 + x_5 + x_7) =$$

$$x_2x_3 + x_2x_5 + x_2x_7 + x_1x_3x_7$$

we obtain the same set of reducts.

2.5 Experimental Results

Many computer data mining systems were developed. In particular, the well-known Rough Set Exploration System elaborated at the University of Warsaw. The system implements many advanced procedures. Some algorithms of RSES have also been embedded in the even more famous ROSETTA system located in the Biomedical Center in Sweden [31, 32].

ROSE2 (Rough Sets Data Explorer) is a software implementing basic elements of the rough set theory and rule discovery techniques. It has been developed at the Laboratory of Intelligent Decision Support Systems of the Institute of Computing Science in Poznań, basing on fourteen-year experience in rough set based knowledge discovery and decision analysis. All computations are based on rough set fundamentals introduced by Z. Pawlak [22] with some modifications proposed by Słowiński and Ziarko [30].

These tools were used to compare them with the presented synthesis method (Table 2.3). Experiments performed show that despite many efforts directed to the designing of an effective tools for attribute reduction, existing tools are not efficient.

The confirmation of this supposition are experiments with the system RSES. Our research has shown that this system cannot process data tables with large number of indeterminacy. An important example from the literature is *trains* database. Applying the new method we generate 689 reducts, however RSES 333 only (not selecting the option "Do not discern with missing values"). While running the system with the option "Do not discern with missing values" selected it ends up after several hours of computation yielding "Not enough memory" message. Note, that not selecting the option the tool takes into account missing values when calculating the discernibility matrix, which yields a result that is different from the set of all minimal sets of attributes.

Another example confirming the absolute superiority of the proposed method is *kaz* function. It is the binary function of 21 arguments, used when testing advanced logic synthesis tools. RSES calculates all the 5574 reducts within 70 minutes. In

Table 2.3. Results of analysis the proposed method in comparison to RSES data mining system

database	attributes	instances	RSES	compl. method
house	17	232	1s	187ms
breast-cancer-wisconsin	10	699	2s	823ms
kaz	22	31	70min	234ms
trains	33	10	out of memory (5h 38min)	6ms
agaricus-lepiota-mushroom	23	8124	29min	4min 47s
urology	36	500	out of memory (12h)	42s 741ms
audiology	71	200	out of memory (1h 17min)	14s 508ms
dermatology	35	366	out of memory (3h 27min)	3min 32s
lung-cancer	57	32	out of memory (5h 20min)	111h 57min

comparison, the new procedure developed and implemented by the authors calculates the set of all reducts in 234ms.

Presented method was additionally proved on the typical databases of medicine, i.e. audiology database, dermatology database, urology database, breast cancer database and lung cancer database. Table 2.3 shows the computation time for all the minimum sets of attributes.

The experiments performed confirm that logic synthesis algorithms developed for the design of digital systems are much more effective than currently used algorithms in data mining systems.

Undoubtedly, in logic synthesis systems and hardware realizations we are almost always looking for these sets of arguments (reducts) which are both: minimal and least. However, the decision systems depend on all the minimal sets of attributes. For example, when considering a reduct of the least cardinality, it can include an attribute that its implementation is actually expensive. In particular, when we consider calculations for a medical diagnosis, it may be a parameter which express complicated or expensive examination, or a test which may have a negative impact on the health of the patient and it is not possible to carry out. Therefore, the reducts of higher cardinality could be sometimes easier to be applied/used in practice.

The importance of calculating the minimum reducts is explained on simple decision table representing the results of medical examination and diagnosis for the seven patients (Tab. 2.4).

Table 2.4. Example of a decision system

	$exam_1$	$exam_2$	$exam_3$	$exam_4$	$exam_5$	$exam_6$	$exam_7$	$exam_8$	diagnosis
p_1	1	1	1	1	1	0	0	0	d_3
p_2	0	1	1	1	1	1	0	1	d_3
p_3	1	1	2	1	0	0	0	0	d_3
p_4	0	1	1	0	0	0	0	0	d_1
p_5	0	0	0	1	0	0	1	0	d_2
p_6	0	1	1	1	0	1	0	0	d_3
p_7	0	1	2	1	0	0	0	1	d_3

The use of such data involves the induction of decision rules. Based on the rules induced the initial decisions of patient's health status are made. For Table 2.4 decision rules induced with Rough Set Exploration System [32] are as follows:

- $(exam_2 = 1)\&(exam_4 = 1) \Rightarrow (diagnosis = d_3)$,
- $(exam_4 = 1)\&(exam_7 = 0) \Rightarrow (diagnosis = d_3)$,
- $(exam_3 = 1)\&(exam_4 = 1) \Rightarrow (diagnosis = d_3)$,
- $(exam_1 = 1) \Rightarrow (diagnosis = d_3)$,
- $(exam_5 = 1) \Rightarrow (diagnosis = d_3)$,
- $(exam_6 = 1) \Rightarrow (diagnosis = d_3)$,
- $(exam_8 = 1) \Rightarrow (diagnosis = d_3)$,

- $(\text{exam}_3 = 2) \Rightarrow (\text{diagnosis} = d_3)$,
- $(\text{exam}_4 = 0) \Rightarrow (\text{diagnosis} = d_1)$,
- $(\text{exam}_1 = 0)\&(\text{exam}_3 = 1)\&(\text{exam}_6 = 0) \Rightarrow (\text{diagnosis} = d_1)$,
- $(\text{exam}_1 = 0)\&(\text{exam}_2 = 1)\&(\text{exam}_6 = 0)\&(\text{exam}_8 = 0) \Rightarrow (\text{diagnosis} = d_1)$,
- $(\text{exam}_1 = 0)\&(\text{exam}_6 = 0)\&(\text{exam}_7 = 0)\&(\text{exam}_8 = 0) \Rightarrow (\text{diagnosis} = d_1)$,
- $(\text{exam}_3 = 1)\&(\text{exam}_5 = 0)\&(\text{exam}_6 = 0) \Rightarrow (\text{diagnosis} = d_1)$,
- $(\text{exam}_2 = 0) \Rightarrow (\text{diagnosis} = d_2)$,
- $(\text{exam}_3 = 0) \Rightarrow (\text{diagnosis} = d_2)$,
- $(\text{exam}_7 = 1) \Rightarrow (\text{diagnosis} = d_2)$,
- $(\text{exam}_1 = 0)\&(\text{exam}_4 = 1)\&(\text{exam}_6 = 0)\&(\text{exam}_8 = 0) \Rightarrow (\text{diagnosis} = d_2)$.

On the basis of the rule $(\text{exam}_2 = 1)\&(\text{exam}_4 = 1) \Rightarrow (\text{diagnosis} = d_3)$ it can be concluded that the patient whose all eight examinations are "1" should be diagnosed with d_3. However, the rule $(\text{exam}_7 = 1) \Rightarrow (\text{diagnosis} = d_2)$ suggests the diagnosis d_2. This situation means that in order to make a correct diagnosis additional examination should be performed.

In fact, we may find that some of the tests are expensive or difficult to be carried out. However, reduction of attributes can enable the elimination of troublesome examinations. For example, after reducing the attributes in this example decision system we get following minimum sets of attributes:

- $\{\text{exam}_1, \text{exam}_4, \text{exam}_6, \text{exam}_8\}$,
- $\{\text{exam}_1, \text{exam}_6, \text{exam}_7, \text{exam}_8\}$,
- $\{\text{exam}_3, \text{exam}_4\}$,
- $\{\text{exam}_3, \text{exam}_5, \text{exam}_6\}$,
- $\{\text{exam}_4, \text{exam}_7\}$,
- $\{\text{exam}_2, \text{exam}_4\}$,
- $\{\text{exam}_1, \text{exam}_3, \text{exam}_6\}$,
- $\{\text{exam}_1, \text{exam}_2, \text{exam}_6, \text{exam}_8\}$.

For example, if exam_4 is difficult to be performed, we can choose a set (reduct) that does not include this examination, i.e. $\{\text{exam}_3, \text{exam}_5, \text{exam}_6\}$. It is worth noting that the choice of the smallest set of attributes is not always a good option and it is important to calculate all the minimal sets of attributes. Selection of the set with greater number of attributes can cause that elements included are easier to be put into practice.

2.6 Conclusion

The argument reduction problem is of a great importance in logic synthesis. It is the basis of some functional transformations, such as parallel decomposition [24]. Combined with some other design techniques it allows us to reduce the size of implemented circuits.

In this chapter we have described an important problem of attribute reduction. This concept, originating from artificial intelligence (namely the theory of rough sets), helps to deal with functional dependencies having redundant input attributes. We have presented a new exact algorithm for attribute reduction which is based on the unate complementation task. Experimental results which have been obtained using this approach proved that it is a valuable method of processing the databases.

Acknowledgements. This work was partly supported by the Foundation for the Development of Radiocommunications and Multimedia Technologies.

References

1. Abdullah, S., Golafshan, L., Nazri, M.Z.A.: Re-heat simulated annealing algorithm for rough set attribute reduction. International Journal of the Physical Sciences 6(8), 2083–2089 (2011), doi:10.5897/IJPS11.218
2. Bazan, J., Nguyen, H.S., Nguyen, S.H., Synak, P., Wróblewski, J.: Rough set algorithms in classification problem. In: Rough Set Methods and Applications: New Developments in Knowledge Discovery in Information Systems, vol. 56, pp. 49–88. Physica-Verlag, Heidelberg (2000), doi:10.1007/978-3-7908-1840-6_3
3. Borowik, G., Łuba, T.: Attribute reduction based on the complementation of boolean functions. In: 1st Australian Conference on the Applications of Systems Engineering, ACASE 2012, Sydney, Australia, pp. 58–59 (2012) (electronic document)
4. Borowik, G., Łuba, T., Zydek, D.: Features reduction using logic minimization techniques. International Journal of Electronics and Telecommunications 58(1), 71–76 (2012), doi:10.2478/v10177-012-0010-x
5. Brayton, R.K., Hachtel, G.D., McMullen, C.T., Sangiovanni-Vincentelli, A.: Logic Minimization Algorithms for VLSI Synthesis. Kluwer Academic Publishers (1984)
6. Brzozowski, J.A., Łuba, T.: Decomposition of boolean functions specified by cubes. Journal of Multi-Valued Logic & Soft Computing 9, 377–417 (2003)
7. Dash, R., Dash, R., Mishra, D.: A hybridized rough-PCA approach of attribute reduction for high dimensional data set. European Journal of Scientific Research 44(1), 29–38 (2010)
8. Feixiang, Z., Yingjun, Z., Li, Z.: An efficient attribute reduction in decision information systems. In: International Conference on Computer Science and Software Engineering, Wuhan, Hubei, pp. 466–469 (2008), doi:10.1109/CSSE.2008.1090
9. Grzenda, M.: Prediction-oriented dimensionality reduction of industrial data sets. In: Mehrotra, K.G., Mohan, C.K., Oh, J.C., Varshney, P.K., Ali, M. (eds.) IEA/AIE 2011, Part I. LNCS, vol. 6703, pp. 232–241. Springer, Heidelberg (2011)
10. Hedar, A.R., Wang, J., Fukushima, M.: Tabu search for attribute reduction in rough set theory. Journal of Soft Computing – A Fusion of Foundations, Methodologies and Applications 12(9), 909–918 (2008), doi:10.1007/s00500-007-0260-1
11. Jensen, R., Shen, Q.: Semantics-preserving dimensionality reduction: Rough and fuzzy-rough based approaches. IEEE Transactions on Knowledge and Data Engineering 16, 1457–1471 (2004), doi:10.1109/TKDE.2004.96
12. Jing, S., She, K.: Heterogeneous attribute reduction in noisy system based on a generalized neighborhood rough sets model. World Academy of Science, Engineering and Technology 75, 1067–1072 (2011)

13. Kalyani, P., Karnan, M.: A new implementation of attribute reduction using quick relative reduct algorithm. International Journal of Internet Computing 1(1), 99–102 (2011)
14. Kryszkiewicz, M., Cichoń, K.: Towards scalable algorithms for discovering rough set reducts. In: Peters, J.F., Skowron, A., Grzymała-Busse, J.W., Kostek, B.Z., Swiniarski, R.W., Szczuka, M.S. (eds.) Transactions on Rough Sets I. LNCS, vol. 3100, pp. 120–143. Springer, Heidelberg (2004)
15. Kryszkiewicz, M., Lasek, P.: FUN: Fast discovery of minimal sets of attributes functionally determining a decision attribute. In: Peters, J.F., Skowron, A., Rybiński, H. (eds.) Transactions on Rough Sets IX. LNCS, vol. 5390, pp. 76–95. Springer, Heidelberg (2008)
16. Lewandowski, J., Rawski, M., Rybiński, H.: Application of parallel decomposition for creation of reduced feed-forward neural networks. In: Kryszkiewicz, M., Peters, J.F., Rybiński, H., Skowron, A. (eds.) RSEISP 2007. LNCS (LNAI), vol. 4585, pp. 564–573. Springer, Heidelberg (2007)
17. Lin, T.Y., Yao, Y.Y., Zadeh, L.A. (eds.): Data mining, rough sets and granular computing. Physica-Verlag GmbH, Heidelberg (2002)
18. Łuba, T., Lasocki, R.: On unknown attribute values in functional dependencies. In: Proceedings of The Third International Workshop on Rough Sets and Soft Computing, San Jose, pp. 490–497 (1994)
19. Łuba, T., Lasocki, R., Rybnik, J.: An implementation of decomposition algorithm and its application in information systems analysis and logic synthesis. In: Ziarko, W. (ed.) Rough Sets, Fuzzy Sets and Knowledge Discovery. Workshops in Computing Series, pp. 458–465. Springer (1994)
20. Łuba, T., Rybnik, J.: Rough sets and some aspects in logic synthesis. In: Słowiński, R. (ed.) Intelligent Decision Support – Handbook of Application and Advances of the Rough Sets Theory. Kluwer Academic Publishers (1992)
21. Nguyen, D., Nguyen, X.: A new method to attribute reduction of decision systems with covering rough sets. Georgian Electronic Scientific Journal: Computer Science and Telecommunications 1(24), 24–31 (2010)
22. Pawlak, Z.: Rough Sets. Theoretical Aspects of Reasoning about Data. Kluwer Academic Publishers (1991)
23. Pei, X., Wang, Y.: An approximate approach to attribute reduction. International Journal of Information Technology 12(4), 128–135 (2006)
24. Rawski, M., Borowik, G., Łuba, T., Tomaszewicz, P., Falkowski, B.J.: Logic synthesis strategy for FPGAs with embedded memory blocks. Electrical Review 86(11a), 94–101 (2010)
25. Selvaraj, H., Sapiecha, P., Rawski, M., Łuba, T.: Functional decomposition – the value and implication for both neural networks and digital designing. International Journal of Computational Intelligence and Applications 6(1), 123–138 (2006), doi:10.1142/S1469026806001782
26. Skowron, A., Rauszer, C.: The discernibility matrices and functions in information systems. In: Słowiński, R. (ed.) Intelligent Decision Support – Handbook of Application and Advances of the Rough Sets Theory. Kluwer Academic Publishers (1992)
27. Słowiński, R., Sharif, E.: Rough sets analysis of experience in surgical practice. In: Rough Sets: State of The Art and Perspectives, Poznań-Kiekrz (1992)
28. Wang, C., Ou, F.: An attribute reduction algorithm based on conditional entropy and frequency of attributes. In: Proceedings of the 2008 International Conference on Intelligent Computation Technology and Automation, ICICTA 2008, vol. 1, pp. 752–756. IEEE Computer Society, Washington, DC (2008), doi:10.1109/ICICTA.2008.95

29. Yao, Y., Zhao, Y.: Attribute reduction in decision-theoretic rough set models. Information Sciences 178(17), 3356–3373 (2008), doi:10.1016/j.ins.2008.05.010
30. ROSE2 – Rough Sets Data Explorer,
    `http://idss.cs.put.poznan.pl/site/rose.html`
31. ROSETTA – A Rough Set Toolkit for Analysis of Data,
    `http://www.lcb.uu.se/tools/rosetta/`
32. RSES – Rough Set Exploration System,
    `http://logic.mimuw.edu.pl/~rses/`

Chapter 3
Multi-GPU Tabu Search Metaheuristic for the Flexible Job Shop Scheduling Problem

Wojciech Bożejko, Mariusz Uchroński, and Mieczysław Wodecki

Abstract. We propose a new framework of the distributed tabu search metaheuristic designed to be executed using a multi-GPU cluster, i.e. cluster of nodes equipped with GPU computing units. The methodology is designed to solve difficult discrete optimization problems, such as a job shop scheduling problem, which we introduce to solve as a case study for the framework designed.

3.1 Introduction

The job shop problem can be briefly presented as follows (see [4]). There is a set of jobs and a set of machines. Each job consists of a number of operations which have to be processed in a given order, each one on a specified machine during a fixed time. The processing of an operation cannot be interrupted. Each machine can process at most one operation at a time. We want to find a schedule (the assignment of operations to time intervals on machines) that minimizes the *makespan*. The job shop scheduling problem, although relatively easily stated, is strongly NP-hard and it is considered as one of the hardest problems in the area of combinatorial optimization.

Because of NP-hardness of the problem heuristics and metaheuristics are recommended as 'the most reasonable' solution methods. The majority of these

Wojciech Bożejko
Institute of Computer Engineering, Control and Robotics,
Wrocław University of Technology Janiszewskiego 11-17, 50-372 Wrocław, Poland
e-mail: `wojciech.bozejko@pwr.wroc.pl`

Mariusz Uchroński
Wrocław Centre of Networking and Supercomputing, Wyb. Wyspańskiego 27,
50-370 Wrocław, Poland
e-mail: `mariusz.uchronski@pwr.wroc.pl`

Mieczysław Wodecki
Institute of Computer Science, University of Wrocław
Joliot-Curie 15, 50-383 Wrocław, Poland
e-mail: `mwd@ii.uni.wroc.pl`

R. Klempous et al. (eds.), *Advanced Methods and Applications in Computational Intelligence*,   43
Topics in Intelligent Engineering and Informatics 6,
DOI: 10.1007/978-3-319-01436-4_3, ⓒ Springer International Publishing Switzerland 2014

methods refer to the makespan minimization. We mention here a few recent studies: Jain, Rangaswamy, and Meeran [20]; Pezzella and Merelli [28]; Grabowski and Wodecki [16]; Nowicki and Smutnicki [26]; Bożejko and Uchroński [6]. Heuristics algorithms based on dispatching rules are also proposed in papers of Holthaus and Rajendran [18], Bushee and Svestka [10] for the problem under consideration. For the other regular criteria such as the total tardiness there are proposed metaheuristics based on various local search techniques: simulated annealing [32], tabu search [2] and genetic search [25].

Here we propose the new framework of the distributed tabu search metaheuristic designed to solve difficult discrete optimization problems, such as the job shop problem, using a multi-GPU cluster with distributed memory. We also determine theoretical number of processors, for which the speedup measure has a maximum value. We experimentally determine what parallel execution time T_p can be obtained in real-world installations of multi-GPU clusters (i.e. nVidia Tesla S2050 with a 12-cores CPU server) for Taillard benchmarks [30] of the job shop scheduling problem, and compare them with theoretically determined values.

In the theoretical part of the chapter we consider the *Parallel Random Access Machine* (PRAM) model of the concurrent computing architecture, in which all processors have access to the common, shared memory with the constant access time $O(1)$. Additionally, we assume that we use the *Concurrent Read Exclusive Write* (CREW) model of the access to memory, in which processors can read concurrently from the same cell of memory, but it is forbidden to write concurrently to the same memory cell. Such a theoretical model is the most convenient for theoretical analysis, and it is very close to the practical hardware properties of the real GPU devices. The chapter constitutes a continuation of work presented in Bożejko et al. [5].

3.2   Job Shop Problem

Job shop scheduling problems result from many real-world cases, which means that they have good practical applications as well as industrial significance. Let us consider a set of jobs $\mathcal{J} = \{1, 2, \ldots, n\}$, a set of machines $M = \{1, 2, \ldots, m\}$ and a set of operations $\mathcal{O} = \{1, 2, \ldots, o\}$. The set $\mathcal{O}$ is decomposed into subsets connected with jobs. A job j consists of a sequence of o_j operations indexed consecutively by $(l_{j-1}+1, l_{j-1}+2, \ldots, l_j)$, which have to be executed in this order, where $l_j = \sum_{i=1}^{j} o_i$ is the total number of operations of the first j jobs, $j = 1, 2, \ldots, n$, $l_0 = 0$, $\sum_{i=1}^{n} o_i = o$. An operation i has to be executed on machine $v_i \in M$ without any idleness in time $p_i > 0$, $i \in \mathcal{O}$. Each machine can execute at most one operation at a time. A feasible solution constitutes a vector of times of the operation execution beginning $S = (S_1, S_2, \ldots, S_o)$ such that the following constraints are fulfilled

$$S_{l_{j-1}+1} \geq 0, \quad j = 1, 2, \ldots, n, \tag{3.1}$$

$$S_i + p_i \leq S_{i+1}, \quad i = l_{j-1}+1, \ l_{j-1}+2, \ldots, l_j - 1, \quad j = 1, 2, \ldots, n, \tag{3.2}$$

$$S_i + p_i \leq S_j \quad \text{or} \quad S_j + p_j \leq S_i, \quad i,j \in O, \quad v_i = v_j, \quad i \neq j. \tag{3.3}$$

Certainly, $C_j = S_j + p_j$. An appropriate criterion function has to be added to the above constraints. The most frequent are the following two criteria: minimization of the time of finishing all the jobs and minimization of the sum of job finishing times. From the formulation of the problem we obtain $\mathscr{C}_j \equiv C_{l_j}, j \in \mathscr{J}$.

The first criterion, the time of finishing all the jobs

$$C_{\max}(S) = \max_{1 \leq j \leq n} C_{l_j}, \tag{3.4}$$

corresponds to the problem denoted as $J||C_{\max}$ in the literature. The second criterion, the sum of job finishing times

$$C(S) = \sum_{j=1}^{n} C_{l_j}, \tag{3.5}$$

corresponds to the problem denoted as $J||\sum C_i$ in the literature.

Both problems described are strongly NP-hard and although they are similarly modelled, the second one is found to be harder because of the lack of some specific properties (so-called block properties, see [26]) used in optimization of execution time of solution algorithms.

3.2.1 Disjunctive Model

The disjunctive model is most commonly used, however it is very unpractical from the point of view of efficiency (and computational complexity). It is based on the notion of disjunctive graph $G = (O, U \cup V)$. This graph has a set of vertices O which represent operations, a set of conjunctive arcs (directed) which show technological order of operation execution

$$U = \bigcup_{j=1}^{n} \bigcup_{i=l_{j-1}+1}^{l_j-1} \{(i, i+1)\} \tag{3.6}$$

and the set of disjunctive arcs (non-directed) which show possible schedule of operations execution on each machine

$$V = \bigcup_{i,j \in O, i \neq j, v_i = v_j} \{(i,j),(j,i)\}. \tag{3.7}$$

A vertex $i \in O$ has a weight p_i which equals the time of execution of operation O_i. Arcs have the weight zero. A choice of exactly one arc from the set $\{(i,j),(j,i)\}$ corresponds to determining a schedule of operations execution – 'i before j' or 'j before i'. A subset $W \subset V$ consisting of exclusively directed arcs, at most one from each pair $\{(i,j),(j,i)\}$, we call a *representation* of disjunctive arcs. Such a representation is complete if all the disjunctive arcs have determined direction. A complete representation, defining a precedence relation of jobs execution on the same machine, generates one solution, not always feasible, if it includes cycles. A feasible solution is generated by a complete representation W such that the graph $G(W) = (O, U \cup W)$ is acyclic. For a feasible schedule values S_i of the vector of operations execution starting times $S = (S_1, S_2, \ldots, S_o)$ can be determined as a length of the longest path incoming to the vertex i (without p_i). As the graph $G(W)$ includes o vertices and $O(o^2)$ arcs, therefore determining the value of the cost function for a given representation W takes the time $O(o^2)$ by using Bellman algorithm of paths in graphs determination.

3.2.2 Combinatorial Model

In the case of many applications a combinatorial representation of a solution is better than a disjunctive model for the job shop problem. It is void of redundance, characteristic of the disjunctive graph, that is, the situation where many disjunctive graphs represent the same solution of the job shop problem. A set of operations O can be decomposed into subsets of operations executed on a single, determined machine $k \in M$, $M_k = \{i \in O : v_i = k\}$ and let $m_k = |M_k|$. The schedule of operations execution on a machine k is determined by a permutation $\pi_k = (\pi_k(1), \pi_k(2), \ldots, \pi_k(m_k))$ of elements of the set M_k, $k \in M$, where $\pi_k(i)$ means such an element from M_k which is in position i in π_k. Let $\Phi_n(M_k)$ be a set of all permutations of elements of M_k. A schedule of operations execution on all machines is defined as $\pi = (\pi_1, \pi_2, \ldots, \pi_m)$, where $\pi \in \Phi_n$, $\Phi_n = \Phi_n(M_1) \times \Phi_n(M_2) \times \ldots \times \Phi_n(M_m)$. For a schedule π we create a directed graph (digraph) $G(\pi) = (O, U \cup E(\pi))$ with a set of vertices O and a set of arcs $U \cup E(\pi))$, where U is a set of constant arcs representing the technological order of operations execution inside a job, and a set of arcs representing an order of operations execution on machines is defined as

$$E(\pi) = \bigcup_{k=1}^{m} \bigcup_{i=1}^{m_k-1} \{(\pi_k(i), \pi_k(i+1))\} \tag{3.8}$$

Each vertex $i \in O$ has the weight p_i, each arc has the weight zero. A schedule π is feasible if the graph $G(\pi)$ does not include a cycle. For a given π, the terms of operations' beginning can be determined in time $O(o)$ from the recurrent formula

$$S_j = \max\{S_i + p_i, S_k + p_k\}, j \in O. \tag{3.9}$$

where an operation i is a direct technological predecessor of the operation $j \in O$ and an operation k is a directed machine predecessor of the operation $j \in O$ for a fixed π. We assume $S_j = 0$ for these operations j which have not any technological or machine predecessors. For a given feasible schedule π the process of determining the cost function value requires the time $O(o)$, which is thus shorter than for the disjunctive representation.

3.3   Flexible Job Shop Problem

Flexible job shop problem constitute a generalization (hybridization) of the classic job shop problem. In this section, we discuss a flexible job shop problem in which operations have to be executed on one machine from a set of dedicated machines. Next, as a job shop problem it also belongs to the strongly NP-hard class. Although exact algorithms based on a disjunctive graph representation of the solution have been developed (see Pinedo [29]), they are not effective for instances with more than 20 jobs and 10 machines.

Many approximate algorithms, chiefly metaheuristic, have been proposed (i.e. tabu search of Dauzère-Pérès and Pauli [12] and Mastrolilli and Gambardella [24]). Many authors have proposed a method of assigning operations to machines and then determining sequence of operations on each machine. This approach was followed by Brandimarte [8] and Pauli [27]. These authors solved the assignment problem (i.e., using dispatching rules) and next applied metaheuristics to solve the job shop problem. Genetic approaches have been adopted to solve the flexible job shop problem, too. Recent works are those of Jia et al. [21], Kacem et al. [22], Pezzella et al. [23] and Bożejko et al. [7]. Gao et al. [14] proposed the hybrid genetic and variable neighborhood descent algorithm for this problem.

3.3.1   Problem Formulation

The flexible job shop problem (FJSP), also called the general job shop problem with parallel machines, can be formulated as follows. Let $\mathscr{J} = \{1, 2, \ldots, n\}$ be a set of jobs which have to be executed on machines from the set $\mathscr{M} = \{1, 2, \ldots, m\}$. There exists a partition of the set of machines into types, so-called *nests* – subsets of machines with the same functional properties. A job constitutes a sequence of some operations. Each operation has to be executed on a dedicated type of machine (from the nest) within a fixed time. The problem consists in the allocation of jobs to machines of dedicated type and in determining the schedule of jobs execution on each machine to minimize the total jobs finishing time. The following constraints have to be fulfilled:

(*i*)   each job has to be executed on only one machine of a determined type at a
        time,
(*ii*) machines cannot execute more than one job at a time,
(*iii*) there are no idle times (i.e., the job execution must not be broken),
(*iv*) the technological order has to be obeyed.

Let $\mathscr{O} = \{1, 2, \ldots, o\}$ be the set of all operations. This set can be partitioned into sequences corresponding to jobs, where the job $j \in \mathscr{J}$ is a sequence of o_j operations, which have to be executed in an order on dedicated machines (i.e., in the so-called technological order). Operations are indexed by numbers

$(l_{j-1}+1,\ldots,l_{j-1}+o_j)$ where $l_j = \sum_{i=1}^{j} o_i$ is the number of operations of the first j jobs, $j = 1,2,\ldots,n$, where $l_0 = 0$ and $o = \sum_{i=1}^{n} o_i$.

The set of machines $\mathcal{M} = \{1,2,\ldots,m\}$ can be partitioned into q subsets of the same type (*nests*) where the i-th $(i = 1,2,\ldots,q)$ type $\mathcal{M}^i$ includes m_i machines which are indexed by numbers $(t_{i-1}+1,\ldots,t_{i-1}+m_i)$, where $t_i = \sum_{j=1}^{i} m_j$ is the number of machines in the first i types, $i = 1,2,\ldots,q$, where $t_0 = 0$ and $m = \sum_{j=1}^{q} m_j$.

An operation $v \in \mathcal{O}$ has to be executed on machines of the type $\mu(v)$, i.e., on one of the machines from the set (nest) $\mathcal{M}^{\mu(v)}$ in time p_{vj}, where $j \in \mathcal{M}^{\mu(v)}$.

Let

$$\mathcal{O}^k = \{v \in \mathcal{O} : \mu(v) = k\} \tag{3.10}$$

be a set of operations executed in the k-th nest $(k = 1,2,\ldots,q)$. A sequence of operations sets

$$\mathcal{Q} = (\mathcal{Q}^1, \mathcal{Q}^2, \ldots, \mathcal{Q}^m), \tag{3.11}$$

such that for each $k = 1,2,\ldots,q$

$$\mathcal{O}^k = \bigcup_{i=t_{k-1}+1}^{t_{k-1}+m_k} \mathcal{Q}^i \text{ and } \mathcal{Q}^i \cap \mathcal{Q}^j = \emptyset, \; i \neq j, \; i,j = 1,2,\ldots,m, \tag{3.12}$$

we call an *assignment of operations from the set $\mathcal{O}$ to machines from the set $\mathcal{M}$* (or shortly, machine workload).

A sequence $(\mathcal{Q}^{t_{k-1}+1}, \mathcal{Q}^{t_{k-1}+2}, \ldots, \mathcal{Q}^{t_{k-1}+m_k})$ is an *assignment* of operations to machines in the i-th nest (shortly, an assignment in the i-th nest). In a special case a machine can execute no operations; then a set of operations assigned to be executed by this machine is an empty set.

If the assignment of operations to machines has been completed, then the optimal schedule of operations execution determination (including a sequence of operations execution on machines) leads to the classic scheduling problem solving, that is, the job shop problem (see Section 3.2 and Grabowski and Wodecki [16]).

Let $K = (K_1, K_2, \ldots, K_m)$ be a sequence of sets where $K_i \in 2^{\mathcal{O}^i}$, $i = 1,2,\ldots,m$ (in a particular case elements of this sequence can be empty sets). By $\mathcal{K}$ we denote the set of all such sequences. The number of elements of the set $\mathcal{K}$ is $2^{|\mathcal{O}^1|} \cdot 2^{|\mathcal{O}^2|} \cdot \ldots \cdot 2^{|\mathcal{O}^m|}$.

If $\mathcal{Q}$ is an assignment of operations to machines, then $\mathcal{Q} \in \mathcal{K}$ (of course, the set $\mathcal{K}$ includes also sequences which are not feasible; that is, such sequences do not constitute assignments of operations to machines).

For any sequence of sets $K = (K_1, K_2, \ldots, K_m)$ $(K \in \mathcal{K})$ by $\Pi_i(K)$ we denote the set of all permutations of elements from K_i. Thereafter, let

$$\pi(K) = (\pi_1(K), \pi_2(K), \ldots, \pi_m(K)) \tag{3.13}$$

be a concatenation of m sequences (permutations), where $\pi_i(K) \in \Pi_i(K)$. Therefore

$$\pi(K) \in \Pi(K) = \Pi_1(K) \times \Pi_2(K) \times, \ldots, \Pi_m(K). \tag{3.14}$$

It is easy to observe that, if $K = (K_1, K_2, \ldots, K_m)$ is an assignment of operations to machines, then the set $\pi_i(K)$ $(i = 1, 2, \ldots, m)$ includes all permutations (possible sequences of execution) of operations from the set K_i on the machine i. Further, let

$$\Phi = \{(K, \pi(K)) : K \in \mathcal{K} \wedge \pi(K) \in \Pi(K)\} \qquad (3.15)$$

be a set of pairs where the first element is a sequence set and the second – a concatenation of permutations of elements of these sets. Any feasible solution of the FJSP is a pair $(\mathcal{Q}, \pi(\mathcal{Q})) \in \Phi$ where $\mathcal{Q}$ is an assignment of operations to machines and $\pi(Q)$ is a concatenation of permutations determining the operations execution sequence which are assigned to each machine fulfilling constraints (i)–(iv). By Φ° we denote a set of feasible solutions for the FJSP. Of course, there is $\Phi^\circ \subset \Phi$.

3.3.2 Graph Models

Any feasible solution $\Theta = (\mathcal{Q}, \pi(\mathcal{Q})) \in \Phi^\circ$ (where $\mathcal{Q}$ is an assignment of operations to machines and $\pi(\mathcal{Q})$ determines the operations execution sequence on each machine) of the FJSP can be presented as a directed graph with weighted vertices $G(\Theta) = (\mathcal{V}, \mathcal{R} \cup \mathcal{E}(\Theta))$ where $\mathcal{V}$ is a set of vertices and $\mathcal{R} \cup \mathcal{E}(\Theta)$ is a set of arcs with:

1) $\mathcal{V} = \mathcal{O} \cup \{s, c\}$, where s and c are additional (fictitious) operations which represent 'start' and 'finish', respectively. A vertex $v \in \mathcal{V} \setminus \{s, c\}$ possesses two attributes:

   - $\lambda(v)$ – a number of machines on which an operation $v \in \mathcal{O}$ has to be executed,
   - $p_{v, \lambda(v)}$ – a weight of vertex which equals the time of operation $v \in \mathcal{O}$ execution on the assigned machine $\lambda(v)$.

   Weights of additional vertices $p_s = p_c = 0$.

2)

$$\mathcal{R} = \bigcup_{j=1}^{n} \left[\bigcup_{i=1}^{o_j - 1} \{(l_{j-1} + i, l_{j-1} + i + 1)\} \cup \{(s, l_{j-1} + 1)\} \cup \{(l_{j-1} + o_j, c)\} \right]. \qquad (3.16)$$

A set $\mathcal{R}$ includes arcs which connect successive operations of the job, arcs from vertex s to the first operation of each job and arcs from the last operation of each job to vertex c.

3)

$$\mathcal{E}(\Theta) = \bigcup_{k=1}^{m} \bigcup_{i=1}^{|\mathcal{O}^k| - 1} \{(\pi_k(i), \pi_k(i+1))\}. \qquad (3.17)$$

It is easy to notice that arcs from the set $\mathscr{E}(\Theta)$ connect operations executed on the same machine (π_k is a permutation of operations executed on the machine M_k, that is, operations from the set $\mathscr{O}^k$).

Arcs from the set $\mathscr{R}$ determine the operations execution sequence inside jobs (a technological order) and arcs from the set $\mathscr{E}(\pi)$ the operations execution sequence on each machine.

Remark 1. *A pair* $\Theta = (\mathscr{Q}, \pi(\mathscr{Q})) \in \Phi$ *is a feasible solution for the FJSP if and only if the graph* $G(\Theta)$ *does not include cycles.*

A sequence of vertices $(v_1, v_2, \ldots, v_k)$ in $G(\Theta)$ such that an arc $(v_i, v_{i+1}) \in \mathscr{R} \cup \mathscr{E}(\Theta)$ for $i = 1, 2, \ldots, k - 1$, we call a *path* from vertex v_1 to v_k. By $C(v, u)$ we denote the longest path (called a *critical path* in a Operational Research issues) in the graph $G(\Theta)$ from the vertex v to u ($v, u \in \mathscr{V}$) and by $L(v, u)$ we denote a *length* (sum of vertex weights) of this path.

It is easy to notice that the time of all operations execution $C_{\max}(\Theta)$ related with the assignment of operations $\mathscr{Q}$ and schedule $\pi(\mathscr{Q})$ equals the length $L(s, c)$ of the critical path $C(s, c)$ in the graph $G(\Theta)$. A solution of the FJSP amounts to determining a feasible solution $\Theta = (\mathscr{Q}, \pi(\mathscr{Q})) \in \Phi^\circ$ for which the graph connected with this solution $G(\Theta)$ has the shortest critical path, that is, it minimizes $L(s, c)$.

Let $C(s, c) = (s, v_1, v_2, \ldots, v_w, c)$, $v_i \in \mathscr{O}$ ($1 \leq i \leq w$) be a critical path in the graph $G(\Theta)$ from the starting vertex s to the final vertex c. This path can be divided into subsequences of vertices

$$\mathscr{B} = (B^1, B^2, \ldots, B^r), \tag{3.18}$$

called *blocks* in the permutations on the critical path $C(s, c)$ (Grabowski [15], Grabowski and Wodecki [16]) where:

(a) a block is a subsequence of verticesfrom the critical path including successive operations executed directly one after another,
(b) a block includes operations executed on the same machine,
(c) a product of any two blocks is an empty set,
(d) a block is a maximum (according to the inclusion) subset of operations from the critical path fulfilling constraints (a)–(c).

Next, only these blocks are considered for which $|B^k| > 1$, i.e., non-empty blocks. If B^k ($k = 1, 2, \ldots, r$) is a block on the machine M_i ($i = 1, 2, \ldots, m$) from the nest t ($t = 1, 2, \ldots, q$), then we shall denote it as follows

$$B^k = (\pi_i(a^k), \pi_i(a^k + 1), \ldots, \pi_i(b^k - 1), \pi_i(b^k)), \tag{3.19}$$

where $1 \leq a^k < b^k \leq |Q^i|$. Operations $\pi(a^k)$ and $\pi(b^k)$ in the block B^k are called *the first* and *the last*, respectively. In turn, a block without the first and the last operation we call an *internal block*. The definitions given are presented in Figure 3.1.

In the work of Grabowski [15] there are theorems called *elimination criteria* of blocks in the job shop problem.

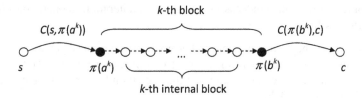

Fig. 3.1. Blocks on the critical path

Theorem 1 ([15]). *Let* $\mathscr{B} = (B^1, B^2, \ldots, B^r)$ *be a sequence of blocks of the critical path in the acyclic graph* $G(\Theta)$, $\Theta \in \Phi^\circ$. *If the graph* $G(\Omega)$ *is feasible (i.e., it represents a feasible solution) and if it is generated from* $G(\Theta)$ *by changing the order of operations execution on some machine and* $C_{\max}(\Omega) < C_{\max}(\Theta)$ *then in the* $G(\Omega)$:

(i) *at least one operation from a block* B^k, $k \in \{1, 2, \ldots, r\}$ *precedes the first element* $\pi(a^k)$ *of this block, or*
(ii) *at least one operation from a block* B^k, $k \in \{1, 2, \ldots, r\}$ *occurs after the last element* $\pi(b^k)$ *of this block.*

Changing the order of operations in any block does not generate a solution with lower value of the cost function. At least one operation from any block should be moved before the first or after the last operation of this block to generate a solution (graph) with smaller weight of the critical path. We use this property to reduce the neighborhood size, i.e., do not generate solutions with greater values (compared to the current solution) of the cost function.

3.4   Determination of the Cost Function

Both in classic job shop problem presented in the Section 3.2 and in flexible job shop problem from the Section 3.3 the method of the cost function calculation is similar because operations of machines assignment (in the flexible job shop problem) leads us to consider a number of job shop problems. Therefore, taking into consideration the constraints (3.1)–(3.3) presented in Section 3.2, it is possible to determine the time moments of job completion C_j, $j \in \mathcal{O}$ and job beginning S_j, $j \in \mathcal{O}$ in time $O(o)$ on the sequential machine using the recurrent formula

$$S_j = \max\{S_i + p_i, S_k + p_k\}, j \in \mathcal{O}. \tag{3.20}$$

where an operation i is a direct technological predecessor of the operation $j \in \mathcal{O}$ and an operation k is a directed machine predecessor of the operation $j \in \mathcal{O}$. The determination procedure of $S_j, j \in \mathcal{O}$ from the recurrent formula (3.20) should be initiated by an assignment $S_j = 0$ for those operations j which do not possess any technological or machine predecessors. Next, in each iteration an operation j has to be chosen for which:

1. the execution beginning moment S_j has not been determined yet, and
2. these moments were determined for all its direct technological and machine predecessors; for such an operation j the execution beginning moment can be determined from (3.20).

It is easy to observe that the order of determining S_j times corresponds to the index of the vertex of the graph $G(\pi)$ connected with an operation j after the topological sorting of this graph. The method mentioned above is in fact a simplistic sequential topological sort algorithm without indexing of operations (vertices of the graph). If we add to this algorithm an element of indexing vertices, for which we calculate S_j value, we obtain a sequence which is the topological order of vertices of the graph $G(\pi)$. Now, we define *layers* of the graph collecting vertices (i.e., operations) for which we can calculate S_j in parallel, as we have calculated starting times for all machine and technological predecessors of operations in the layer.

Definition 3.1. The layer of the graph $G(\pi)$ is a subsequence of the sequence of vertices ordered by the topological sort algorithm, such that there are no arcs between vertices of this subsequence.

Now we show another approach to determine cost function value, which is more time-consuming, but cost-optimal. First, we need to determine the number of layers d of the graph $G(\pi)$.

Theorem 2. *For a fixed feasible operations order π for the $J\|C_{\max}$ problem, the number of layers from Definition 3.1 of the graph $G(\pi)$ can be calculated in time $O(\log^2 o)$ on the CREW PRAMs with $O\left(\frac{o^3}{\log o}\right)$ processors.*

Proof. Here we use the graph $G^*(\pi)$ with additional vertex 0. Let $B = [b_i j]$ be an incidence matrix for the graph $G^*(\pi)$, i.e., $b_i j = 1$, if there is an arc i, j in the graph $G^*(\pi)$, otherwise $b_i j = 0$, $i, j = 1, 2, \ldots, o$. The proof is given in three steps.

1. Let us calculate the longest paths (in the sense of the number of vertices) in $G^*(\pi)$. We can use the algorithm classic parallel Bellman-Ford algorithm in this step in the time $O(\log^2 o)$ and CREW PRAMs with $O(o^3 / \log o)$ processors.
2. We sort distances from the vertex 0 to each vertex in an increasing order. Their indexes, after having been sorted, correspond to the topological order of vertices. This takes the time $O(\log o)$ and CREW PRAMs with $o + 1 = O(o)$ processors, using parallel merge sort algorithm. We obtain a sequence $Topo[i]$, $i = 0, 1, 2, \ldots, o$.
3. Let us assign each element of the sorted sequence to one processor, without the last one. If the next value of the sequence (distance from 0) $Topo[i + 1]$, $i = 0, 1, \ldots, o - 1$ is the same as $Topo[i]$ considered by the processor i, we assign $c[i] \leftarrow 1$, and $c[i] \leftarrow 0$ if $Topo[i + 1] \neq Topo[i]$. This step requires the time $O(1)$ and o processors. Next, we add all values $c[i]$, $i = 0, 1, \ldots, o - 1$. To make this

step we need the time $O(\log o)$ and CREW PRAMs with $O(o)$ processors. We get $d = 1 + \sum_{i=0}^{o-1} c[i]$ because there is an additional layer connected with exactly one vertex 0.

The most time- and processor-consuming is Step 1. We need the time $O(\log^2 o)$ and the number of processors $O\left(\frac{o^3}{\log o}\right)$ of the CREW PRAMs.   ∎

Theorem 3. *For a fixed feasible operations order π for the $J\|C_{\max}$ problem, the value of cost function can be determined in time $O(d)$ on $O(o/d)$-processor CREW PRAMs where d is the number of layers of the graph $G(\pi)$.*

Proof. Let Γ_k, $k = 1, 2, \ldots, d$ be the number of calculations of the operations' finishing moment C_i, $i = 1, 2, \ldots, o$ in the k-th layer. Certainly, $\sum_{i=1}^{d} \Gamma_i = o$. Let p be the number of processors used. The time of computations in a single layer k after having divided calculations into $\lceil \frac{\Gamma_i}{p} \rceil$ groups, each group containing (at most) p elements, is $\lceil \frac{\Gamma_i}{p} \rceil$ (the last group cannot be full). Therefore, the total computation time in all d layers equals $\sum_{i=1}^{d} \lceil \frac{\Gamma_i}{p} \rceil \leq \sum_{i=1}^{d} (\frac{\Gamma_i}{p} + 1) = \frac{o}{p} + d$. To obtain the time of computations $O(d)$ we should use $p = O(\frac{o}{d})$ processors.   ∎

This theorem provides a cost-optimal method of parallel calculation of the cost function value for the classic job shop problem with the makespan criterion. We will use it to determine the cost function value for the flexible job shop problem, after fixing of the operations-to-machines assignment.

3.5   Data Broadcasting

Here we propose a solution method to the flexible job shop problem in the distributed computing environments, such as multi-GPU clusters. Tabu search algorithm is executed in concurrent working threads, as in *multiple-walk* model of parallelization [1] (MPDS,*Multiple starting Point Different Strategies* in the Voß [31] classification of parallel tabu search metaheuristic). Additionally, MPI library is used to distribute calculations among GPUs (see Fig. 3.2). GPU devices are used for concurrent cost function calculations as it was mentioned in the Section 3.4.

Now let us consider a single cycle of the MPI data broadcasting, multi-GPU computations and batching up of the results obtained. Let us assume that the single communication procedure between two nodes of a cluster takes the time T_{comm}, the time of sequential tabu search computations is T_{seq} and the computations time of parallel tabu search is $T_{calc} = \frac{T_{seq}}{p}$ (p is the number of GPUs). Therefore, the total parallel computations time of the single cycle is

$$T_p = 2T_{comm} \log_2 p + T_{calc} = 2T_{comm} \log_2 p + \frac{T_{seq}}{p}.$$

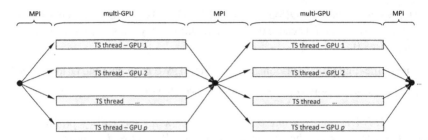

Fig. 3.2. Skeleton of the Multi-Level Tabu Search metaheuristic

Fig. 3.3 Scheme of the two
levels parallelization model.
Flexible job shop problem
is converted into a number
of classic job shop prob-
lems and distributed among
GPU devices. Results are
collected and converted
into flexible job shop problem, etc.

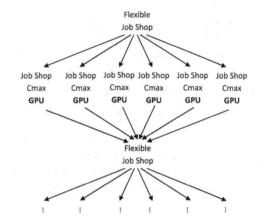

In case of using more processors, the parallel computing time $\left(\frac{T_{seq}}{p}\right)$ decreases, whereas the time of communication $(2T_{comm}\log p)$ increases. We are looking for such a number of processors p (let us call it p^*) for which T_p is minimal. By calculating $\frac{\partial T_p}{\partial p} = 0$ we obtain

$$\frac{2T_{comm}}{p\ln 2} - \frac{T_{seq}}{p^2} = 0 \tag{3.21}$$

and then

$$p = p^* = \frac{T_{seq}\ln 2}{2T_{comm}}, \tag{3.22}$$

which provides us with an optimal number of processors p^* which minimizes the value of the parallel running time T_p.

The fraction of communication time is $O(\log_2 p)$ in this tree-based data broad-casting method, therefore this is another situation than for linear-time broadcasting (discussed in [3]), for which the overall communication and calculation efficiency is much lower. On the other hand the linear-time broadcasting is similar to described by Brooks' Law [9] for project management, i.e. the expected advantage from splitting development work among n programmers is $O(n)$ but the communications cost associated with coordinating and then merging their work is $O(n^2)$.

3.6 Solution Method

Considered flexible job shop problem has been solved with a tabu search method. In each step of the tabu search method the neighbourhood of operations to machines assignment is generated (see Fig. 3.3). Each element of the neighbourhood generated in this way is a feasible solution of a classic job shop problem and its goal function is calculated in order to find the best solution from the neighbourhood. The value of cost function is determined on GPU. MPI library has been used for the communication implementation. The proposed tabu search method uses multiple point single strategy which means that MPI processes begins from different starting solutions using the same search strategy. For each MPI process one GPU is assigned for accelerating the goal function determination. For the best solution from neighbourhood TSAB algorithm is executed and solution obtained with this algorithm used as base solution in the next iteration. Also, after a number of iterations, the best solution with corresponding tabu list is broadcasted among MPI processes and each MPI process continues calculation from this solution. The best solution interchange between processors guarantees different search paths for each processor.

3.6.1 GPU Implementation Details

In our Multi-GPU tabu search metaheuristic for the flexible job shop problem we adopt the Floyd-Warshall algorithm for the longest path computation between each pair of nodes in a graph (see the proof of the theorem 2). The main idea of the algorithm can be formulated as follows. Find the longest path between nodes v_i and v_j containing the node v_k. It consist of a sub-path from v_i to v_k and a sub-path v_k to v_j. This idea can be formulated with the following formula

$$d_{ij}^{(k)} = \max\{d_{ij}^{(k-1)}, d_{ik}^{(k-1)} + d_{kj}^{(k-1)}\}, \tag{3.23}$$

where $d_{ij}^{(k)}$ is the longest path from v_i to v_j such that all intermediate nodes on the path are in set $v_1, \ldots, v_k$. Figure 3.4 shows CUDA implementation of the longest path computation between each pair of nodes. The CUDA kernel (Figure 3.5) is invoked o times, where o is the number of nodes in the graph. At the k-th iteration, the kernel computes two values for each pair of nodes in the graph: the direct distance between them and the indirect distance through the node v_k. The larger of the two distances is written back to the distance matrix. The final distance matrix reflects the lengths of the longest paths between each pair of nodes. The inputs of the CUDA kernel constitutes the number of the graph nodes, path distance matrix and the iteration (step) number.

```
1  extern "C" void FindPathsOnGPU(const int &o, int *graph)
2  {
3    int *graphDev;
4    const int dataSize = (o+1)*(o+1)*sizeof(int);
5    const int size=(int)(log((double)o)/log(2.0));
6    dim3 threads(o+1);
7    dim3 blocks(o+1);
8    cudaMalloc( (void**) &graphDev, dataSize);
9    cudaMemcpy( graphDev, graph, dataSize, cudaMemcpyHostToDevice);
10   for (int iter=1; iter <= size+1; ++iter)
11   {
12     for(int k=0; k<=o; ++k)
13     {
14       PathsKernel<<<blocks, threads>>>(o, graphDev, k);
15       cudaThreadSynchronize();
16     }
17   }
18   cudaMemcpy( graph, graphDev, dataSize, cudaMemcpyDeviceToHost);
19   cudaFree(graphDev);
20 }
```

Fig. 3.4. CUDA implementation of computing the longest path in a graph

```
1  __global__ void PathsKernel(const int o, int *graph, const int i)
2  {
3    int x = threadIdx.x;
4    int y = blockIdx.x;
5    int k = i;
6    int yXwidth = y * (o+1);
7
8    int dYtoX = graph[yXwidth + x];
9    int dYtoK = graph[yXwidth + k];
10   int dKtoX = graph[k*(o+1) + x];
11
12   int indirectDistance = dYtoK + dKtoX;
13   int max = 0;
14   int tmp = 0;
15
16   if(dYtoK !=0 and dKtoX !=0)
17   {
18     tmp = indirectDistance;
19     if(max < tmp)
20       max = tmp;
21   }
22   if(dYtoX < max)
23   {
24     graph[yXwidth + x] = max;
25   }
26 }
```

Fig. 3.5. CUDA kernel

3.6.2 Computational Experiments

Parallel multi–GPU algorithm for solving flexible job shop problem was coded in C
(CUDA) and MPI library. The proposed algorithm was run on the Tesla S2050 GPU

and tested on the benchmark problems taken from Brandimarte [8] and Hurink [19]. The GPU was installed on the server based on the Intel Core i7 CPU X980 processor working under 64-bit Linux Ubuntu 10.04 operating system. Figure 3.6 and Table 3.1 report the comparison of speedup obtained for MPI+CUDA implementation on the Tesla S2050 GPU. For Tables 3.1 and 3.2 a particular column means:

- o – number of operations in considered flexible job shop problem instance,
- s_{CPU} – speedup value calculated for sequential CPU time,
- s_{GPU} – speedup value calculated for sequential GPU time (for one GPU thread).

The speedup value is given by the following formula:

$$s = \frac{T_{seq}}{T_{par}}, \tag{3.24}$$

where T_{seq} is the calculations time of the sequential algorithm and T_{par} is the calculations time of the parallel algorithm.

The obtained results show that the usage of parallel goal function computation method for computations acceleration in metaheuristics which solves flexible job shop problem results in shorter calculation time for the number of operations greater than 120 (bdata) and 200 (edata). The use of the parallel algorithm for goal function calculation in tabu search method results in about 2.5x (bdata) and 2.9x (edata) absolute speedup (in comparison with the sequential algorithm run on CPU) and

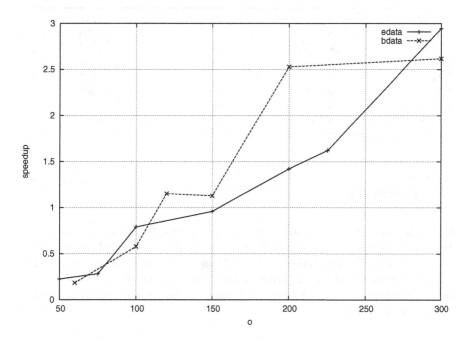

Fig. 3.6. Speedup for Brandimarte (*bdata*) and Hurink (*edata*)test instances

over 120x orthodox speedup (i.e. comparing with the sequential algorithm run on single GPU core, calculated for Mk01-Mk06 instances) for the flexible job shop problem instances with 200 and 300 operations.

Table 3.1. Speedup for MPI+CUDA implementation obtained on Tesla S2050 GPU

instance	o	s_{CPU}	s_{GPU}
Mk01	60	0.18	11.32
Mk02	60	0.19	12.53
Mk03	120	1.15	122.38
Mk04	120	0.46	34.20
Mk05	60	0.63	47.70
Mk06	150	1.12	121.33
Mk07	100	0.57	-
Mk08	200	2.52	-
Mk09	200	2.65	-
Mk10	300	2.61	-

Table 3.2. Speedup for MPI+CUDA implementation obtained on Tesla S2050 GPU

instance	$n \times m$	o	s_{CPU}
la01-05	10×5	50	0.2247
la06-10	15×5	75	0.2806
la11-15	20×5	100	0.4377
la16-20	10×10	100	0.7891
la21-25	15×10	150	0.9579
la26-30	20×10	200	1.4203
la31-35	30×10	300	2.9453
la36-40	15×15	225	1.6201

3.7 Conclusion

In this chapter we propose a framework designed to solve difficult problems of combinatorial optimization in distributed parallel architectures without shared memory, such as clusters of nodes equipped with GPU units (i.e. multiu-GPU clusters). The methodology can be especially effective for large instances of hard to solve optimization problems, such as flexible scheduling problems as well as discrete routing and assignment problems.

References

1. Alba, E.: Parallel Metaheuristics. A New Class of Algorithms. Wiley & Sons Inc. (2005)
2. Armentano, V.A., Scrich, C.R.: Tabu search for minimizing total tardiness in a job shop. International Journal of Production Economics 63(2), 131–140 (2000)
3. Bożejko, W.: A new class of parallel scheduling algorithms, pp. 1–280. Wroclaw University of Technology Publishing House (2010)
4. Bożejko, W.: On single-walk parallelization of the job shop problem solving algorithms. Computers & Operations Research 39, 2258–2264 (2012)
5. Bożejko, W., Uchroński: Distributed Tabu Search Algorithm for the Job Shop Problem. In: Proceedings of the 14th International Asia Pacific Conference on Computer Aided System Theory, Sydney, Australia, February 6-8 (2012)
6. Bożejko, W., Uchroński, M.: A Neuro-tabu Search Algorithm for the Job Shop Problem. In: Rutkowski, L., Scherer, R., Tadeusiewicz, R., Zadeh, L.A., Zurada, J.M. (eds.) ICAISC 2010, Part II. LNCS, vol. 6114, pp. 387–394. Springer, Heidelberg (2010)
7. Bożejko, W., Uchroński, M., Wodecki, M.: Parallel Meta^{2}heuristics for the Flexible Job Shop Problem. In: Rutkowski, L., Scherer, R., Tadeusiewicz, R., Zadeh, L.A., Zurada, J.M. (eds.) ICAISC 2010, Part II. LNCS, vol. 6114, pp. 395–402. Springer, Heidelberg (2010)
8. Brandimarte, P.: Routing and scheduling in a flexible job shop by tabu search. Annals of Operations Research 41, 157–183 (1993)
9. Brooks Jr., F.P.: The Mythical Man-Month, anniversary edn. Addison-Wesley, Reading (1995)
10. Bushee, D.C., Svestka, J.A.: A bi-directional scheduling approach for job shops. International Journal of Production Research 37(16), 3823–3837 (1999)
11. Crainic, T.G., Toulouse, M., Gendreau, M.: Parallel asynchronous tabu search in multicommodity locationallocation with balancing requirements. Annals of Operations Research 63, 277–299 (1995)
12. Dauzère-Pérès, S., Pauli, J.: An integrated approach for modeling and solving the general multiprocessor job shop scheduling problem using tabu search. Annals of Operations Research 70(3), 281–306 (1997)
13. Flynn, M.J.: Very highspeed computing systems. Proceedings of the IEEE 54, 1901–1909 (1966)
14. Gao, J., Sun, L., Gen, M.: A hybrid genetic and variable neighborhood descent algorithm for flexible job shop scheduling problems. Computers & Operations Research 35, 2892–2907 (2008)
15. Grabowski, J.: Generalized problems of operations sequencing in the discrete production systems. Monographs, vol. 9. Scientific Papers of the Institute of Technical Cybernetics of Wrocław Technical University (1979)
16. Grabowski, J., Wodecki, M.: A very fast tabu search algorithm for the job shop problem. In: Rego, C., Alidaee, B. (eds.) Adaptive Memory and Evolution, Tabu Search and Scatter Search. Kluwer Academic Publishers, Dordrecht (2005)
17. Hanafi, S.: On the Convergence of Tabu Search. Journal of Heuristics 7, 47–58 (2000)
18. Holthaus, O., Rajendran, C.: Efficient jobshop dispatching rules: further developments. Production Planning and Control 11, 171–178 (2000)
19. Hurink, E., Jurisch, B., Thole, M.: Tabu search for the job shop scheduling problem with Multi-purpose machine, Oper. Res. Spektrum 15, 205–215 (1994)
20. Jain, A.S., Rangaswamy, B., Meeran, S.: New and stronger job-shop neighborhoods: A focus on the method of Nowicki and Smutnicki (1996). Journal of Heuristics 6(4), 457–480 (2000)

21. Jia, H.Z., Nee, A.Y.C., Fuh, J.Y.H., Zhang, Y.F.: A modified genetic algorithm for distributed scheduling problems. International Journal of Intelligent Manufacturing 14, 351–362 (2003)
22. Kacem, I., Hammadi, S., Borne, P.: Approach by localization and multiobjective evolutionary optimization for flexible job-shop scheduling problems. IEEE Transactions on Systems, Man, and Cybernetics, Part C 32(1), 1–13 (2002)
23. Pezzella, F., Morganti, G., Ciaschetti, G.: A genetic algorithm for the Flexible Job-schop Scheduling Problem. Computers & Operations Research 35, 3202–3212 (2008)
24. Mastrolilli, M., Gambardella, L.M.: Effective neighborhood functions for the flexible job shop problem. Journal of Scheduling 3(1), 3–20 (2000)
25. Mattfeld, D.C., Bierwirth, C.: An efficient genetic algorithm for job shop scheduling with tardiness objectives. European Journal of Operational Research 155(3), 616–630 (2004)
26. Nowicki, E., Smutnicki, C.: An advanced tabu search algorithm for the job shop problem. Journal of Scheduling 8(2), 145–159 (2005)
27. Pauli, J.: A hierarchical approach for the FMS schduling problem. European Journal of Operational Research 86(1), 32–42 (1995)
28. Pezzella, F., Merelli, E.: A tabu search method guided by shifting bottleneck for the job-shop scheduling problem. European Journal of Operational Research 120, 297–310 (2000)
29. Pinedo, M.: Scheduling: theory, algorithms and systems. Prentice-Hall, Englewood Cliffs (2002)
30. Taillard, E.: Benchmarks for basic scheduling problems. European Journal of Operational Research 64, 278–285 (1993)
31. Voß, S.: Tabu search: Applications and prospects. In: Du, D.Z., Pardalos, P.M. (eds.) Network Optimization Problems, pp. 333–353. World Scientific Publishing Co., Singapore (1993)
32. Wang, T.Y., Wu, K.B.: An eficient configuration generation mechanism to solve job shop scheduling problems by the simulated annealing. International Journal of Systems Science 30(5), 527–532 (1999)

Chapter 4
Stable Scheduling with Random Processing Times

Wojciech Bożejko, Paweł Rajba, and Mieczysław Wodecki

Abstract. In this work stability of solutions determined by algorithms based on tabu search method for a certain (NP-hard) one-machine arrangement problem was examined. The times of tasks performance are deterministic and they also constitute random variables of the standard or the Erlang's schedule. The best results were obtained when as a criterion to choose an element from the neighborhood convex combinations of the first and the second moments of the random goal function were accepted. In this way determined solutions are stable, i.e. little sensitive to parameters random changes.[1]

4.1 Introduction

Research concerning problems of algorithms arrangement refers mainly to deterministic models. To solve such problems, which belong in the majority of cases to the strongly NP-hard class, rough algorithms are applied successfully. They are mainly based on local optimalization methods: simulated annealing, tabu search and a genetic algorithm. Determined by these algorithms solutions only slightly differ from best solutions. However, in practice, in the course of a process realisation (according to the fixed schedule) it appears very often that certain parameters (e.g. the task completion time) are different from the initial ones. By the lack of the solutions

Wojciech Bożejko
Institute of Computer Engineering, Control and Robotics,
Wrocław University of Technology, Janiszewskiego 11-17, 50-372 Wrocław, Poland
e-mail: wojciech.bozejko@pwr.wroc.pl

Paweł Rajba · Mieczysław Wodecki
Institute of Computer Science, University of Wrocław, Joliot-Curie 15,
50-383 Wrocław, Poland
e-mail: {pawel.rajba,mwd}@ii.uni.wroc.pl

[1] The work was supported by MNiSW Poland, within the grant No. N N514 232237.

R. Klempous et al. (eds.), *Advanced Methods and Applications in Computational Intelligence*,   61
Topics in Intelligent Engineering and Informatics 6,
DOI: 10.1007/978-3-319-01436-4_4, © Springer International Publishing Switzerland 2014

stability in the fixed schedule there may occur a big mistake, which makes such a schedule unacceptable. That's why there is a necessity to construct such models and methods of their solutions that would take into account potential changes in the course of parameters process realisation and generate stable solutions.

Problems of arrangements with uncertain data may be solved using methods based on elements of probability calculus (van den Akker and Hoogeveen [25], Vondrák [30], Dean [5], Cai [4] or fuzzy sets theories (Prade [22], Itoh and Ishii [10]). They make it possible to consider uncertainties already at the stage of the mathematical model construction or directly in algorithms being constructed.

In this work we deal with the one-machine problem of tasks arrangement with the latest completion times and the minimalisation of the costs sum of tardy tasks. Times of tasks completion are deterministic and constitute random variables of a standard or the Erlang's schedule. On the basis of this problem the resistance to a random variable of constructive solutions of parameters according to tabu search metaheuristics is examined. The paper constitutes a continuation of work presented in Bożejko et al. [2].

4.2 Problem Definition and Method of Its Solution

Algorithms based on the tabu search method have been applied successfully to solve NP-hard problems of combinatorial optimalization to date. They are simple to be implemented and comparative results in literature show that determined by these algorithms solutions only slightly differ from the best ones.

4.2.1 Single Machine Scheduling Problem

In this problem each task from the set $\mathscr{J} = \{1,2,\dots,n\}$ should be performed uninterruptedly on a machine which in any moment can do at most one task. For a task i, let p_i, d_i, w_i be the execution time, the expected completion time and the penalty for a delay of a task. Such a sequence of tasks' performance should be determined in a way that the penalty sum is minimal.

Let Π be a set of permutations of elements from $\mathscr{J}$. For any permutation $\pi \in \Pi$ by $C_{\pi(i)} = \Sigma_{j=1}^{i} p_{\pi(j)}$ we denote the completion time of i-th task] in a permutation π. Then

$$U_{\pi(i)} = \begin{cases} 0, & \text{if } C_{\pi(i)} \le d_{\pi(i)}, \\ 1, & \text{otherwise}, \end{cases} \tag{4.1}$$

we call the task delay, $w_{\pi(i)} \cdot U_{\pi(i)}$ the *penalty* for delay and

$$F(\pi) = \sum_{i=1}^{n} w_{\pi(i)} U_{\pi(i)} \tag{4.2}$$

the permutation cost.

The problem of Minimization of the Total Weighted of Late jobs (abbrev. MTWL) consist in determining such a permutation $\pi^* \in \Pi$ that

$$F(\pi^*) = \min\{\mathscr{F}(\beta) : \beta \in \Pi\}.$$

In literature (see [9]) this problem is indicated as $1||\sum w_i U_i$ and it belongs to the strongly NP-hard problems class (Karp [13]). Such problems heve been studied for quite long together with many variations, especially with polynomial computational complexity.

For the problem $1|p_i = 1|\sum w_i U_i$ (all the processing times are identical) Monma [17] has presented an algorithm with $O(n)$ complexity. Similarly, for the problem $1|w_i = c|\sum U_i$, (where the cost function factors are identical) there is the Moore algorithm [18] with $O(n \ln n)$ complexity. Lawler [15] has adapted the Moore algorithm to solve the problem $1|p_i < p_j \Rightarrow w_i \geq w_j|\sum w_i U_i$. Problems with the earliest starting times compose another group r_i. Kise et al. [12] have proven that even the problem of late tasks minimization ($1|r_i|\sum U_i$ without the cost function weight) is strongly NP-hard. They have also presented a polynomial algorithm that has computational complexity $O(n^2)$ for a particular example, the $1|r_i < r_j \Rightarrow d_i \leq d_j|\sum U_i$ problem.

If a partial order relation is given on the set of tasks, the MTWL problem is strongly NP-hard even when the task realization times are unities (Garey and Johnson [6]). Lenstra and Rinnoy Kan [16] have proven that if a partial order relation is a union of independent chains, the problem is also strongly NP-hard.

Optimal algorithms of this solution based on the dynamic programming method were presented by Lawler and Moor [14]; a pseudo-polynomial algorithm with the computational complexity $O(n \min\{\sum_j p_j, \max_j\{d_j\}\})$ and based on the branch and bound method - by Potts [20], Sourd [26] and Wodecki [27]. Exact algorithms make it possible to determine optimal solutions effectively only if the number of tasks does not exceed 50 (80 in a multi-processor neighborhood, [27]).

Therefore, in practice only rough algorithms are applied (mainly metaheuristics) constituting adaptation of algorithms of the problem solution $1||\sum w_i T_i$, ([3]). We can refer to the models and algorithms of Józefowska [11] for a detailed survey on the models and algorithms developed in this area.

4.3 Problem Description and Preliminaries

Each schedule of jobs can be represented by permutation $\pi = (\pi(1), \pi(2), \dots, \pi(n))$ on set $\mathscr{J}$. Let Π denote the set of all such permutations. The total cost of $\pi \in \Pi$ is $\sum_{i=1}^{n} w_{\pi(i)} U_{\pi(i)}$, where $C_{\pi(i)} = \sum_{j=1}^{i} p_{\pi(j)}$ is a completion time of the job $\pi(i)$.

Job $\pi(i)$ is considered as *early* one, if it is completed before its due date (i.e. $C_{\pi(i)} \leq d_{\pi(i)}$), or *tardy* if the job is completed after its due date (i.e. $C_{\pi(i)} > d_{\pi(i)}$).

Each permutation $\pi \in \Pi$ is decomposed into m ($m \le n$) subsequences $B_1, B_2, \ldots,$ B_m, called *blocks* in π, each of them contains the jobs having in common specific properties, where

1. $B_k = (\pi(f_k), \pi(f_k+1), \ldots, \pi(l_k-1)), \pi(l_k))$, $l_{k-1}+1 = f_k \le l_k$,
   $k = 1, 2, \ldots, m$, $l_0 = 0$, $l_m = n$.

2. All the jobs $j \in B_k$ satisfy the following condition:
   either

$$d_j \ge C_{\pi(l_k)}, \tag{C1}$$

   or

$$d_j < S_{\pi(f_k)} + p_j, \tag{C2}$$

   where $S_{\pi(f_k)}$ is a *starting time* of the job $\pi(f_k)$, i.e. $S_{\pi(f_k)} = C_{\pi(f_k)} - p_{\pi(f_k)}$. Clearly, each job $j \in B_k$ satisfying Condition C1 (or C2) is a early (or tardy) one in π.

3. B_k is maximal subsequence of π in which all the jobs satisfy either Condition C1 or Condition C2.

Jobs $\pi(f_k)$ and $\pi(l_k)$ in B_k are the *first* and *last* ones, respectively. Note that a block can contain only one job, i.e. $|B_k| = 1$, and then $f_k = l_k$. Note that all the blocks are connected "in series" in permutation π. By definition, there exist two type of blocks implied by either C1 or C2. To distinguish them, we will use the *E-block* and *T-block* notions (or alternatively B_k^E and B_k^T), respectively (see Figure 4.1).

Example 1. Let us consider the $n=10$ jobs' instance that is specified in Table 4.1.

Table 4.1. Data for the instance

i	1	2	3	4	5	6	7	8	9	10
p_i	2	3	1	2	3	2	3	3	2	4
d_i	12	19	12	9	5	1	17	24	19	3
w_i	3	1	2	5	3	3	4	2	4	5

Let $\pi = (1, 2, 3, 4, 5, 6, 7, 8, 9, 10)$. Permutation π contains five blocks, (i.e. $m = 5$), $l_1 = 1$, $f_1 = 4$, $l_2 = 5$, $f_2 = 6$, $l_3 = f_3 = 7$, $l_4 = f_4 = 8$, $l_5 = 9$ and $f_5 = 10$. Blocks $B_1 = (1, 2, 3, 4)$, $B_2 = (5, 6)$, $B_3 = (7)$, $B_4 = (8)$ and $B_5 = (9, 10)$. There are: three E-blocks: B_1, B_3, B_4, and two T-blocks: B_2 and B_5. These blocks are shown on Figure 4.1.

Let

$$F_k(\pi) = \sum_{j \in B_k} w_j U_j,$$

be a *partial value* of the objective associated with the block B_k in π. It is clear that by the definition of blocks in π, we have

$$F(\pi) = \sum_{k=1}^{m} F_k(\pi).$$

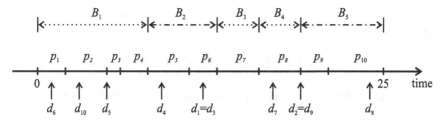

Fig. 4.1. Blocks in permutation π

It is evident that by Condition $C1$, for any permutation of jobs within an E-block B_k^E of π, (i.e. in the positions $f_k, f_k + 1, ..., l_k - 1, l_k$ of B_k^E) , we have

$$F_k(\pi) = \sum_{j \in B_k^E} w_j U_j = 0. \tag{4.3}$$

With respect to T-blocks B_k^T in π, it should be noticed that by Condition $C2$, for any permutation of jobs within B_k^T (i.e. in the positions $f_k, f_k + 1, ..., l_k - 1, l_k$ of B_k^T), all the jobs are tardy. Therefore, an optimal sequence of the jobs within B_k^T of π can be obtained, using well-known Weighted Shortest Processing Time (WSPT) rule, proposed by Smith [24]. The WSPT rule creates an optimal sequence of the jobs in the non-increasing order of the ratios w_j/p_j.

Let $\overrightarrow{B}_k^T$ denote a T-block of the jobs from B_k^T of π ordered by the WSPT rule, then we have

$$F_k(\pi) = \sum_{j \in B_k^T} w_j U_j \geq \sum_{j \in \overrightarrow{B}_k^T} w_j U_j = \overrightarrow{F}_k(\pi). \tag{4.4}$$

where the jobs from B_k^T are ordered in any permutation.

Fundamental Block Properties of the MTWL problem are derived from the following Theorem.

Theorem 4.1. *Let $\pi \in \Pi$ be any permutation with blocks $B_1, B_2, ..., B_m$, and let the jobs of each T-block of π to be ordered according to the WSPT rule. If the permutation β has been obtained from π by an interchange of jobs that $F(\beta) < F(\pi)$, then in β*

   (i) *at least one job from B_k precedes at least one job from blocks $B_1, B_2, ..., B_{k-1}$, for some $k \in \{2, 3, ..., m\}$, or*
   (ii) *at least one job from B_k succeeds at least one job from blocks $B_{k+1}, B_{k+2}, ..., B_m$ for some $k \in \{1, 2, ..., m-1\}$.*

Proof. Without loss of generality and for the simplicity of denotation one can assume that $\pi(i) = i$. Thus π takes the form

$$\pi = (1, 2, ..., l_1, f_2, f_2 + 1, ..., l_2, ..., f_k, f_k + 1, ..., l_k, ..., f_m, f_m + 1, ..., n),$$

where

$$B_k = (f_k, f_k + 1, ..., l_k), \qquad\qquad k = 1, 2, ..., m,$$

and

$$F(\pi) = \sum_{k=1}^{m} F_k(\pi) = \sum_{k=1}^{m} \sum_{j=f_k}^{l_k} w_j U_j.$$

Suppose that the theorem is false, and let β be an arbitrary permutation of the following form

$$\beta = (x_1^1, x_2^1, ..., x_{t_1}^1, x_1^2, x_2^2, ..., x_{t_2}^2, ..., x_1^k, x_2^k, ..., x_{t_k}^k, ..., x_1^m, x_2^m, ..., x_{t_m}^m), \qquad (4.5)$$

where $X_k = (x_1^k, x_2^k, ..., x_{t_k}^k)$, $k = 1, 2, ..., m$, is any permutation of $(f_k, f_k + 1, ..., l_k)$, i.e. any permutation of the jobs from block B_k of π, then it is easy to verify that the thesis of the theorem does not hold for β.

Now, for permutation β, we have

$$F(\beta) = \sum_{k=1}^{m} F_k(\beta) = \sum_{k=1}^{m} \sum_{j \in X_k} w_j U_j.$$

Since $\{x_1^k, x_2^k, ..., x_{t_k}^k\} = \{f_k, f_k + 1, ..., l_k\}$ for $k = 1, 2, ..., m$, then

$$F_k(\pi) = F_k(\beta)$$

whenever the permutations of jobs within both X_k and B_k are the same. Hence, since the jobs from X_k of β are ordered in any permutation, then, for each T-block $\overrightarrow{B}_k^T$ of π, using (4.4), we get

$$F_k(\beta) = \sum_{j \in X_k} w_j U_j \geq \sum_{j \in \overrightarrow{B}_k^T} w_j U_j = \overrightarrow{F}_k(\pi),$$

where the jobs from $\overrightarrow{B}_k^T$ are ordered by the WSPT rule, according to the assumption of the theorem.

Further, for each E-blocks B_k^E of π, using (4.3), we have

$$F_k(\beta) = F_k(\pi) = 0.$$

Therefore, for each block B_k of π, we get

$$F_k(\beta) \geq F_k(\pi).$$

Hence, for any permutation β given by (4.5), we get

$$F(\beta) = \sum_{k=1}^{m} F_k(\beta) \geq \sum_{k=1}^{m} F_k(\pi) = F(\pi),$$

which contradicts the assumption of the theorem. $\square$

Note that Theorem 4.1 provides the necessary condition to obtain a permutation β from π such that $F(\beta) < F(\pi)$.

In any permutation of jobs from T-block B^E, each job is early in permutation π. Using this property we present the algorithm of determining the first E-block in π.

Algorithm AE-block

> **Input:** permutation $\pi = (\pi(1), \pi(1), \ldots, \pi(n))$;
> **Output:** subpermutation (E-block) $B^E = (\pi(l), \pi(l+1), \ldots, \pi(k-1), \pi(k))$;
> > Let $\pi(l)$ be the first job in π such, that $C_{\pi(l)} \leq d_{\pi(l)}$;
> > $B^E \leftarrow \pi(l); \; k \leftarrow l$;
> > **while** $\left| B^E \right| = k - l + 1$ **and** $k < n$ **do**
> > **begin**
> > > **if** $(C_{\pi(k+1)} \leq d_{\pi(k+1)})$ **and**
> > > **(if (for all** $\pi(i) \in B^E$, $C_{\pi(i)} \leq d_{\pi(k+1)}$)) **then** $B^E \leftarrow B^E \cup \{\pi(k+1)\}$;
> > > $k \leftarrow k + 1$
> > **end.**

Computational complexity of this algorithm is $O(n)$.

In any permutation of jobs from T-block B^T, each job in permutation π is late. Similarly, like for T-block, on the basis of the above definition, we announce the algorithm of determining the first T-block in the permutation π.

Algorithm AT-block

> **Input:** permutation $\pi = (\pi(1), \pi(1), \ldots, \pi(n))$;
> **Output:** subpermutation (T-block) $B^T = (\pi(v), \pi(v+1), \ldots, \pi(r-1), \pi(r))$;
> > Let $\pi(t)$ be the first job in π such, that $C_{\pi(v)} > d_{\pi(v)}$.
> > $B^T \leftarrow \pi(v); \quad P_{first} \leftarrow C_{\pi(v)} - p_{\pi(v)}; \quad r \leftarrow t$;
> > **while** $\left| B^T \right| = r - v + 1$ **and** $r < n$ **do**
> > **begin**
> > > **if** $P_{first} + p_{\pi(r+1)} > d_{\pi(r+1)}$ **then** $B^T \leftarrow B^T \cup \{\pi(r+1)\}$;
> > > $r \leftarrow r + 1$
> > **end.**

Computational complexity of the above algorithm is $O(n)$.

Considering in turn, jobs in permutation π (beginning from $\pi(1)$) and applying respectively algorithm AE-block or AT-block, we will break π into E and T blocks. Computational complexity of this break is $O(n)$.

4.3.1   The Tabu Search Method

Rough algorithms are used mainly to solve NP-hard problems of discrete optimization. Solutions determined by these algorithms are found to be fully satisfactory (very often they differ from the best known solutions approximately less than a few percent). One of realizations of constructive methods of these algorithms is tabu search, whose basic elements are

- *movement* – a function which transforms one task into another,
- *neighborhood* – a subset of acceptable solutions set,
- *tabu list* – a list which contains attributes of a number of examined solutions.

Let $\pi \in \Pi$ be a starting permutation, L_{TS} a tabu list and π^* the best solution found so far.

Algorithm Tabu Search (*TS*)
```
1   repeat
2       Determine the neighborhood N (π) of permutation π;
3       Delete from N (π) permutations forbidden by the list L_TS;
4       Determine a permutation δ ∈ N (π), such that
5           F(δ) = min{F(β): β ∈ N (π)};
6       if ( F(δ) < F(π*) ) then
7           π* := δ;
8       Place attributes δ on the list L_TS;
9       π := δ
10  until (the completion condition).
```

4.3.2   Movement and Neighborhood

Let us notice that Theorem 1 provides the necessary condition to obtain a permutation β from π such that $F(\beta) < F(\pi)$.

Let $\mathscr{B} = [B_1, B_2, \dots, B_v]$ be an ordered partition of the permutation $\pi \in \Pi$ into blocks. If a job $\pi(j) \in B_i$ ($B_i \in \mathscr{B}$), therefore existang moves, which can improve goal function value, consist in reordering a job $\pi(j)$ before the first or after the last job of this block. Let $\mathscr{M}_j^{bf}$ and $\mathscr{M}_j^{af}$ be sets of such moves (obviously $\mathscr{M}_1^{bf} = \mathscr{M}_v^{af} = \oslash$). Therefore, the neighborhood of the permutation $\pi \in \Pi$ has the form of

$$\mathscr{M}(\pi) = \bigcup_{j=1}^{n} \mathscr{M}_j^{bf} \cup \bigcup_{j=1}^{n} \mathscr{M}_j^{af}. \tag{4.6}$$

The *neighborhood* of the π is a set of permutations

$$\mathscr{N}(\pi) = \{m(\pi): \ m \in \mathscr{M}(\pi)\}. \tag{4.7}$$

To prevent from arising cycle too quickly (returning to the same permutation after some small number of iterations of the algorithm), some attributes of each move are saved on so-called tabu list (list of the prohibited moves). This list is served as a FIFO queue.

By implementing an algorithm from the neighborhood permutations whose attributes are on the tabu list L_{TS} are removed.

Generally, in our algorithm, for the given initial permutation, we identify the blocks (if there is more than one partition of the permutation into blocks, any of them can be used), and order the jobs of each T-block according to the WSPT rule. Then, for the resulting (basic) permutation π, we calculate $F(\pi)$, create the set of moves ME, compound move $\hat{v}$, and the permutation $\pi_{\hat{v}}$. Next, the search process of algorithm is repeated for the new initial permutation $\pi_{\hat{v}}$ until a given number of iterations is reached. According to the philosophy of tabu search, the compound move cannot contain the single moves with a status tabu; these moves are not allowed.

4.3.3 The Tabu Moves List

To prevent a cycle from arising some attributes of each movement are put on the list of tabu moves.

In our algorithm we use the cyclic tabu list defined as a finite list (set) L_{TS} with length $LengthT$ containing ordered triplets. The list L_{TS} is a realization of the short-term search memory. If a move $v = (x, y)$ is performed on permutation π, then, a triplet $(\pi(x), y, F(\pi_v))$ is added to L_{TS}. If the compound move $\hat{v}$ is performed, then the triplet corresponding to each move from $\hat{v}$ is added to the tabu list. Each time before adding a new element to L_{TS}, we must remove the oldest one. With respect to a permutation π, a move $v = (x, y)$ is forbidden i.e. it has *tabu* status, if there is a triplet (r, s, ϕ) in L_{TS} such that $\pi(x) = r$, $y = s$, and $F(\pi_v) \geq \phi$.

As mentioned above, our algorithm uses a tabu list with dynamic length. This length is changed, as the current iteration number *iter* of algorithm increases, using a "pick" that can be treated as a specific disturbance (diversification).

This kind of tabu list was employed on those very fast tabu search algorithms proposed by Grabowski and Wodecki, where it was successfully applied to the classical flow shop and job shop problems [7, 8]. Here, we extend this component of algorithm in the original form [7], to the problem considered. In this tabu list, length $LengthT$ is a cyclic function shown in Figure 4.2, and defined by the expression

$$LengthT = \begin{cases} LTS, & \text{if } W(l) < iter \leq W(l) + h(l), \\ LTS + \psi, & \text{if } W(l) + h(l) < iter \leq W(l) + h(l) + H, \end{cases}$$

where $l = 1, 2, \ldots$ is the number of the cycle, $W(l) = \sum_{s=1}^{l} h(s-1) + (l-1) \times H$ (here $h(0) = 0$). Further, H is the width of the pick equal to ψ, and $h(l)$ is the interval between the neighbour pick equal to $3 \times LTS$. If $LengthT$ decreases, then a suitable number of the oldest elements of tabu list L_{TS} is deleted and the search process is continued. The LTS and ψ are tuning parameters which are to be chosen experimentally.

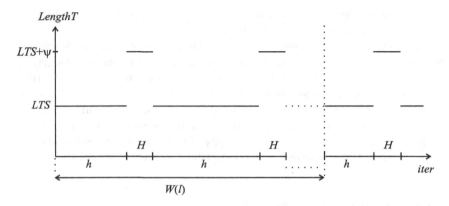

Fig. 4.2. Dynamic tabu list

4.4 Stochastic Processing Times

In literature there has been explored a problem of arrangement with random times tasks, mainly with a normal or a uniform distribution (van den Akker Hoogeveen [25]) and an exponential one (Pinedo [19]).

We consider a set of n jobs $J = \{1, 2, \ldots, n\}$ to be processed on a single machine. The job processing times $\widetilde{p}_i$ ($i = 1, 2, \ldots, n$) are independent random variables. Then for a determined sequence of occurring tasks in a permutation π the completion time of a task performance $\widetilde{C}_{\pi(i)} = \sum_{j=1}^{i} \widetilde{p}_{\pi(j)}$ is a random variable. Random variables are delays $\widetilde{U}_{\pi(i)} = 0$ when $\widetilde{C}_{\pi(i)} \leq d_{\pi(i)}$ and $\widetilde{U}_{\pi(i)} = 1$ when $\widetilde{C}_{\pi(i)} > d_{\pi(i)}$ as well as the goal function

$$\widetilde{\mathscr{F}}(\pi) = \sum_{i=1}^{n} w_{\pi(i)} \widetilde{U}_{\pi(i)}. \tag{4.8}$$

In the tabu search (TS) algorithm selecting the best element from an environment (instruction 5) is comparable with the goal function values. As (4.8) is a random variable we replace it with a convex combination of the expected value and the standard deviation

$$\mathscr{W}(\pi) = c \cdot E(\widetilde{\mathscr{F}}(\pi)) + (1 - c) \cdot D(\widetilde{\mathscr{F}}(\pi)) \quad (c \in [0, 1]). \tag{4.9}$$

In the probabilistic version of an algorithm in a place of a goal function $\mathscr{F}$ (instructions 5 and 6) there should be placed a function W defined in (4.9).

4.4.1   Normal Distribution

Let's assume that the times of tasks performance $\tilde{p}_i$, $i \in J$ are independent random variables with a normal distribution with an average m_i and a standard deviation σ_i i.e. $\tilde{p}_i \sim N(m_i, \sigma_i)$. Then, the time of completion of a task performance $\tilde{C}_i$ (performed as i-th in a sequence) is a random variable with a normal distribution with an average $m^{(i)} = \sum_{j=1}^i m_j$ and a standard deviation $\sigma^{(i)} = \sqrt{\sum_{j=1}^i \sigma_j^2}$.

The delay of a task $\tilde{U}_i$ (according to (4.6)) is also a random variable whose expected value is:

$$E(\tilde{U}_i) = \sum_x x * P(\tilde{U}_i = x) = 0 * P(\tilde{C}_i \le d_i) + 1 * P(\tilde{C}_i > d_i) = 1 - \Phi(\delta^i),$$

where Φ is a cumulative distribution function of a normal distribution $N(0,1)$, and a parameter $\delta^i = \frac{d_i - m^{(i)}}{\sigma^{(i)}}$.

In that case, the expected value of a goal function (4.8) equals

$$E(\widetilde{\mathscr{F}}(\pi)) = E(\sum_{i=1}^n w_i * \tilde{U}_i) = \sum_{i=1}^n w_i * E(\tilde{U}_i) = \sum_{i=1}^n w_i * (1 - \Phi(\delta^i)). \qquad (4.10)$$

It's easy to notice that $E(\tilde{U}_i^2) = 1 - \Phi(\delta^i)$, therefore

$$D^2(\tilde{U}_i) = E(\tilde{U}_i^2) - (E(\tilde{U}_i)^2 = \Phi(\delta^i)(1 - \Phi(\delta^i)). \qquad (4.11)$$

Making use of (4.8) and (4.11) we get

$$D^2(\widetilde{\mathscr{F}}(\pi)) = \sum_{i=1}^n w_i \Phi(\delta^i)(1 - \Phi(\delta^i)) + 2 \sum_{i<j} w_i w_j \mathrm{cov}(\tilde{U}_i, \tilde{U}_j).$$

In line with the covariance definition

$$\mathrm{cov}\left(\tilde{U}_i, \tilde{U}_j\right) = E\left(\tilde{U}_i, \tilde{U}_j\right) - E\left(\tilde{U}_i\right) E\left(\tilde{U}_j\right) =$$

$$E\left(\tilde{U}_i, \tilde{U}_j\right) - \left(1 - \Phi\left(\delta^i\right)\right)\left(1 - \Phi\left(\delta^j\right)\right).$$

It's possible to prove easily that the expected value

$$E(\tilde{U}_i, \tilde{U}_j) = (1 - \Phi(\delta^i))(1 - \Phi(\gamma^j)),$$

where $\gamma^j = \frac{1}{\sqrt{1 - \rho^2}}\left(\frac{d_j - m^{(j)}}{\sigma^{(j)}} - \rho \frac{d_i - m^{(i)}}{\sigma^{(i)}}\right)$ and $\rho^2 = \frac{(\sigma^{(i)})^2}{(\sigma^{(j)})^2}$. Thus,

$$\mathrm{cov}\left(\tilde{U}_i, \tilde{U}_j\right) = \left(1 - \Phi\left(\delta^i\right)\right)\left(1 - \Phi\left(\gamma^j\right)\right) - \left(1 - \Phi\left(\delta^i\right)\right)\left(1 - \Phi\left(\delta^i\right)\right) =$$

$$\left(1 - \Phi\left(\delta^i\right)\right)\left(\Phi\left(\delta^i\right) - \Phi\left(\gamma^j\right)\right).$$

Summarizing the goal function variance

$$D^2\left(\mathscr{F}(\pi)\right) = \sum_{i=1}^{n} w_i \Phi\left(\delta^i\right)\left(1 - \Phi\left(\delta^i\right)\right) + 2\sum_{i<j} w_i w_j \text{cov}\left(\widetilde{U}_i, \widetilde{U}_j\right) =$$

$$\sum_{i=1}^{n} w_i \Phi\left(\delta^i\right)\left(1 - \Phi\left(\delta^i\right)\right) + 2\sum_{i<j} w_i w_j [(1 - \Phi\left(\delta^i\right))\left(\Phi\left(\delta^i\right) - \Phi\left(\gamma^j\right)\right)]. \quad (4.12)$$

Therefore, to calculate the value $\mathscr{W}(\pi)$ determined in (4.9), (4.10) and (4.12) should be used.

4.4.2 The Erlang's Distribution

Let's assume that the time of tasks' performance has the Erlang's distribution $\widetilde{p}_i \sim \mathscr{E}(\alpha_i, \lambda)$, $i \in J$. Then, the time of completing the task execution (in a permutation $\pi = (1, 2, \ldots, n)$) $\widetilde{C}_i = \sum_{j=1}^{i} \widetilde{p}_j \sim \mathscr{E}(\alpha_1 + \ldots + \alpha_i, \lambda)$.

Let $F_i(x) = F_{\widetilde{p}_1 + \ldots + \widetilde{p}_i}(x)$ be the cumulative distribution function of the time of completion of the i-th task $\widetilde{C}_i$ execution. The expected value

$$E(\widetilde{U}_i) = 0 \cdot P(\widetilde{C}_i \le d_i) + 1 \cdot P(\widetilde{C}_i > d_i) = 1 - F_i(d_i)$$

and

$$E(\widetilde{\mathscr{F}}(\pi)) = E\left(\sum_{i=1}^{n} w_i \widetilde{U}_i\right) = \sum_{i=1}^{n} w_i E\left(\widetilde{U}_i\right) = \sum_{i=1}^{n} w_i(1 - F_i(d_i)). \quad (4.13)$$

It's easy to observe that $E(\widetilde{U}_i^2) = 1 - F_i(d_i)$, that's why a variance

$$D^2(\widetilde{U}_i) = D^2(\sum_{i=1}^{n} w_i \widetilde{U}_i) = E(\widetilde{U}_i^2) - (E(\widetilde{U}_i))^2 = F_i(d_i)(1 - F_i(d_i)).$$

Thus,

$$D^2(\widetilde{\mathscr{F}}(\pi)) = \sum_{i=1}^{n} w_i\left(F_i(d_i)(1 - F_i(d_i))\right) + 2\sum_{i<j} w_i w_j \text{cov}(\widetilde{U}_i, \widetilde{U}_j),$$

The covariance $\text{cov}(\widetilde{U}_i, \widetilde{U}_j)$ between variables $\widetilde{U}_i$ and $\widetilde{U}_j$ is calculated according to the formulae

$$\text{cov}(\widetilde{U}_i, \widetilde{U}_j) = E(\widetilde{U}_i \widetilde{U}_j) - E(\widetilde{U}_i)E(\widetilde{U}_j).$$

Finally,

$$D^2(\widetilde{\mathscr{F}}(\pi)) = \sum_{i=1}^{n} w_i(F_i(d_i)(1 - F_i(d_i))) +$$

$$2\sum_{i<j} w_i w_j (FI + SI - (1 - F_i(d_i))(1 - F_j(d_j))), \tag{4.14}$$

where

$$FI = \int_{d_i}^{d_j} \int_{d_j-x}^{\infty} f_i(x) f_j(y) dy dx, \quad \text{and} \quad SI = \int_{d_j}^{\infty} \int_{0}^{\infty} f_i(x) f_j(y) dy dx.$$

As a result, in order to calculate the value of a function $\mathscr{W}(\pi)$ determined in (4.9) the formulae (4.13) and (4.14) should be applied.

4.5   The Algorithms' Stability

In this section we shall introduce a certain measure which let us examine the influence of the change of tasks' parameters on the goal function value (4.2) i.e. the solution stability.

Let $\delta = ((p_1, w_1, d_1), \dots, (p_n, w_n, d_n))$ be an example of data (deterministic) for the MTWL problem. By $D(\delta)$ we denote a set of data generated from δ by a disturbance of times of tasks performance. A disturbance consists in changing these times on random determined values. Disturbed data $\gamma \in D(\delta)$ take the form of $\gamma = ((p_1', w_1, d_1), \dots, (p_n', w_n, d_n))$, where the time of execution p_i' $(i = 1, \dots, n)$ is a realization of a random variable $\tilde{p}_i$ in the Erlang's distribution $\mathscr{E}(\lambda, \alpha_i)$ (see Section 4.4.2), and the expected value is $E\tilde{p}_i = p_i$.

Let $\mathscr{A} = \{\mathscr{A}\mathscr{D}, \widetilde{\mathscr{A}\mathscr{N}}\}$ where $\mathscr{A}\mathscr{D}$ i $\widetilde{\mathscr{A}\mathscr{N}}$ is the deterministic and the probabilistic algorithm respectively (i.e. solving examples with deterministic or random times of tasks' performance) for the MTWL problem. By π_δ we denote a solution (a permutation) determined by the algorithm $\mathscr{A}$ for a data δ. Then, let $\mathscr{F}(\mathscr{A}, \pi_\delta, \varphi)$ be the cost of tasks' execution (4.2) for the example φ in a sequence determined by a solution (a permutation) π_δ determined by the algorithm $\mathscr{A}$ for data δ. Then,

$$\Delta(\mathscr{A}, \delta, D(\delta)) = \frac{1}{|D(\delta)|} \sum_{\varphi \in D(\delta)} \frac{\mathscr{F}(\mathscr{A}, \pi_\delta, \varphi) - \mathscr{F}(\mathscr{A}\mathscr{D}, \pi_\varphi, \varphi)}{\mathscr{F}(\mathscr{A}\mathscr{D}, \pi_\varphi, \varphi)},$$

we call *the solution stability* π_δ (of an example δ) determined by the algorithm $\mathscr{A}$ on the set of disturbed data $D(\delta)$.

Because in our studies on the π_φ determining as the starting solution in the $\mathscr{A}$ algorithm (tabu search) we have adopted π_δ, we have

$$\mathscr{F}(\mathscr{A}, \pi_\delta, \varphi) - \mathscr{F}(\mathscr{A}, \pi_\varphi, \varphi) \geq 0.$$

Let Ω be a set of deterministic examples for the problem of tasks' arrangement. The *stability* rate of the algorithm $\mathscr{A}$ on the set Ω is defined in the following way:

$$S(\mathscr{A}, \Omega) = \frac{1}{|\Omega|} \sum_{\delta \in \Omega} \Delta(\mathscr{A}, \delta, D(\delta)). \tag{4.15}$$

In the following section we will present numerical experiments that allow comparisons of the deterministic stability coefficient $S(\mathscr{A}\mathscr{D},\Omega)$ with the probabilistic stability coefficient $S(\widetilde{\mathscr{A}\mathscr{N}},\Omega)$.

4.6   The Calculation Experiments

Presented in this work algorithms were examined on many examples. The deterministic data were generated randomly (for a problem $1||\sum w_iT_i$ (see [27])) For a fixed number of tasks n ($n = 40, 50, 100, 150, \ldots, 500, 1000$) we have determined n tuples (p_i, w_i, d_i), $i = 1, 2, \ldots, n$, where the processing time p_i and the cost w_i are the realization of a random variable with a uniform distribution, respectively from the range [1,100] and [1,10]. Similarly, the critic lines are drawn from the range $[P(1 - TF - RDD/2, P(1 - TF + RDD/2]$ depending on the parameters RDD (relative range of due dates) and TF (average tardiness factor) from the set $\{0.2, 0.4, 0.6, 0.8, 1.0\}$, whereas $P = \sum_{i=1}^{n} p_i$. For every couple of parameters RDD, TF (there are 25 such couples) 5 examples have been generated. The whole deterministic data set Ω contains 1500 examples (125 for every n). On the basis of each example of deterministic data (p_i, w_i, d_i), $i = 1, 2, \ldots, n$, an example of probabilistic data $(\widetilde{p}_i, w_i, d_i)$ was determined, where $\widetilde{p}_i$ is a random variable (with the normal/standard or the Erlang's schedule) representing the time of task performance (detailed description in Section 4.4.2). The set we denote by $\widetilde{\Omega}$.

By initiating of each algorithm the starting permutation was the natural permutation $\pi = (1, 2, \ldots, n)$. In addition, the following parameters were used:

- the length of list of tabu moves: $\ln n$,
- the maximal number of iterations of an algorithm (*the completion condition*): $n/2$ or n,
- in the formula (4.9) parameter $c = 0.8$.
- dynamic parameters of tabu list: $h = \lceil n/4 \rceil$, $H = \lceil n/10 \rceil$, $LTS = \lceil \sqrt{(n)} \rceil$ and $\psi = \lceil \sqrt{(n/4)} \rceil$.

Calculations of the deterministic algorithm $\mathscr{A}\mathscr{D}$ were made on the examples from the set Ω, whereas the probabilistic algorithm $\widetilde{\mathscr{A}\mathscr{N}}$ (normal distribution) and $\widetilde{\mathscr{A}\mathscr{E}}$ (Erlang's distribution) on examples from the set $\widetilde{\Omega}$.

First of all, the quality designated by the algorithms solutions was examined. In order to do that, for every solution of the example of deterministic data designated by algorithm $A = \{\mathscr{A}\mathscr{D}, \widetilde{\mathscr{A}\mathscr{N}}, \widetilde{\mathscr{A}\mathscr{E}}\}$ relative error was calculated

$$\varepsilon(A) = \frac{\mathscr{F}_A - \mathscr{F}^*}{\mathscr{F}^*}, \qquad (4.16)$$

where $\mathscr{F}_A$ is a value of a solution set by the algorithm A, and $\mathscr{F}^*$ a value of the solution set by a very good algoritm presented in the work [28]. The average relative error (for every set of data) was presented in Table 4.2. According to the expectations, the best appeared to be a deterministic algorithm $\mathscr{A}\mathscr{D}$ (for every group of data the mean error equals 6%)). The error for the two left $\widetilde{\mathscr{A}\mathscr{N}}$ and $\widetilde{\mathscr{A}\mathscr{E}}$ algorithms is similar and equals 11% and 12% respectively.

In order to examine the algorithms stability (i.e. the determination of parameter's value (4.15)) for each example of deterministic data from the set Ω, 100 examples of disturbed data were generated (the way of generating is described in Section 4.5). Each of these examples was solved by the algorithm $\mathscr{A}\mathscr{D}$. On the basis of these calculations the stability rates of all three algorithms were determined. The comparable results are presented in Table 4.2. After completing n iterations the average stability rate of an algorithm $S(\mathscr{A}\mathscr{D}, \Omega) = 0.36$, and the probabilistic coefficient(for times with the Erlang's schedule) $S(\widetilde{\mathscr{A}\mathscr{E}}, \Omega) = 0.07$.

It means that the disturbance of a solution determined by an algorithm $\mathscr{A}\mathscr{D}$ causes worsening of the goal function value on average by about 36%. In case of the algorithm $\mathscr{A}\mathscr{D}$ the worsening equals on average only 7%. As a result, we can state that the average deviation of the deterministic algorithm is about 5 bigger than the average deviation of the probabilistic algorithm. For the algorithm $\mathscr{A}\mathscr{N}$ (with times with the normal distribution) the rate $S(\widetilde{\mathscr{A}\mathscr{N}}, \Omega) = 0.24$. It means it's smaller than the rate of the deterministic algorithm, however, nearly 4 times bigger than the probabilistic algorithm $\widetilde{\mathscr{A}\mathscr{E}}$. The maximal mistake of an algorithm $\widetilde{\mathscr{A}\mathscr{E}}$ does not exceed 41%, and the deterministic algorithm $\mathscr{A}\mathscr{D}$ equals over 124%. Calculations for a bigger number of iterations were made as well. The stability rate of algorithms improved slightly. The n number of iterations of the algorithm based on the tabu

Table 4.2. Algorithms' deviation $\varepsilon(A)$ and stability $S(\mathscr{A}, \Omega)$)

n	Algorithm $\mathscr{A}\mathscr{D}$ (deterministic)		Algorithm $\widetilde{\mathscr{A}\mathscr{N}}$ (normal distribution)		Algorithm $\widetilde{\mathscr{A}\mathscr{E}}$ (Erlang's distribution)	
	Deviation	Stability	Deviation	Stability	Deviation	Stability
40	0.03	0.62	0.08	0.26	0.05	0.08
50	0.01	0.34	0.07	0.30	0.011	0.07
100	0.04	0.53	0.09	0.19	0.08	0.06
150	0.06	0.31	0.11	0.17	0.09	0.09
200	0.05	0.28	0.07	0.28	0.12	0.09
250	0.03	0.30	0.06	0.11	0.09	0.06
300	0.08	0.36	0.14	0.26	0.28	0.07
350	0.06	0.27	0.09	0.18	0.08	0.02
400	0.05	0.32	0.16	0.25	0.15	0.08
450	0.07	0.25	0.05	0.21	0.06	0.09
500	0.11	0.17	0.16	0.26	0.16	0.07
1000	0.14	0.48	0.16	0.36	0.17	0.09
Average	0.06	0.36	0.11	0.24	0.12	0.15

search method is very small. Due to this fact, the average time of calculations of one example, on a personal computer with a processor Pentium 2.6 GHz, does not exceed one second.

The calculational experiments proved explicitly that solutions determined by the probabilistic algorithm with times of tasks performance with the Erlang's distribution are the most stable.

4.7 Conclusion

In this work we presented a method of modelling uncertain data by means of random variables with the normal and the Erlang's distribution. The Erlang's distribution, as much better than other ones, represents times of tasks performance which in the course of realization are exposed to change (extending). Algorithms based on the tabu search method were presented for solving a certain single machine problem of tasks arrangement. The calculations experiments proved that an algorithm in which times of tasks performance are random variables with the Erlang's distribution is very stable. The relative average deviation for disturbed data by a small number of iterations and a short calculation time does not exceed 7%.

References

1. Beasley, J.E.: OR-Library: distributing test problems by electronic qmail. Journal of the Operational Research Society 41, 1069–1072 (1990),
   http://people.brunel.ac.uk/~mastjjb/jeb/info.html
2. Bożejko, W., Rajba, P., Wodecki, M.: Arrangement Algorithms Stability with Probabilistic Parameters of Tasks. In: Proceedings of the 14th International Asia Pacific Conference on Computer Aided System Theory, Sydney, Australia, February 6-8 (2012)
3. Bożejko, W., Grabowski, J., Wodecki, M.: Block approach-tabu search algorithm for single machine total weighted tardiness problem. Computers & Industrial Engineering 50, 1–14 (2006)
4. Cai, X., Zhou, X.: Single machine scheduling with expotential processing times and general stochastic cost functions. Journal of Global Optimization 31, 317–332 (2005)
5. Dean, B.C.: Approximation algorithms for stochastic scheduling problems. PhD thesis, MIT (2005)
6. Garey, M.R., Johnson, D.S.: Scheduling tasks with nonuniform deadlines on two processor. Journal of ACM 23, 461–467 (1976)
7. Grabowski, J., Wodecki, M.: A very fast tabu search algorithm for the permutation flow shop problem with makespan criterion. Computers and Operations Research 31, 1891–1909 (2004)
8. Grabowski, J., Wodecki, M.: A very fast tabu search algorithm the job shop problem. In: Rego, C., Alidaee, A. (eds.) Metaheuristic Optimization via Memory and Evolution; Tabu Search and Scatter Search, pp. 117–144. Kluwer Academic Publishers, Boston (2005)

9. Graham, R.L., Lawler, E.L., Lenstra, J.K., Rinnooy Kan, A.H.G.: Optimization and approximation in deterministic sequencing and scheduling. Annals of Discrete Mathematics 5, 287–326 (1979)
10. Itoh, T., Ishii, H.: Fuzzy due-date scheduling problem with fuzzy processing times. Int. Trans. Oper. Res. 6, 639–647 (1999)
11. Józefowska, J.: Just-in-time scheduling. Springer, Berlin (2007)
12. Kise, H., Ibaraki, T., Mine, H.: A solvable case of the one-machine scheduling problem with ready times and due times. Operations Research 26, 121–126 (1978)
13. Karp, R.M.: Reducibility among Combinatorial Problems. In: Millerand, R.E., Thatcher, J.W. (eds.) Complexity of Computations, pp. 85–103. Plenum Press, NY (1972)
14. Lawler, E.L., Moore, J.M.: A Functional Equation and its Applications to Resource Allocation and Sequencing Problems. Management Science 16, 77–84 (1969)
15. Lawler, E.L.: A "pseudopolinomial" algorithm for sequencing jobsto minimize total tardiness. Annals of Discrete Mathematics 1, 331–342 (1977)
16. Lenstra, J.K., Rinnoy Kan, A.H.G.: Complexity results for scheduling chains on a single machine. European Journal of Operational Research 4, 270–275 (1980)
17. Monma, C.I.: Linear-time algorithms for scheduling on parallel processor. Operations Research 30, 116–124 (1982)
18. Moore, J.M.: An n-job, one machine sequencig algorithm for minimizing the number of late jobs. Menagement Science 15, 102–109 (1968)
19. Pinedo, M.: Stochastic scheduling with release dates. Operation Research 31, 559–572 (1983)
20. Potts, C.N., Van Wassenhove, L.N.: A Branch and Bound Algorithm for the Total Weighted Tardiness Problem. Operations Research 33, 177–181 (1985)
21. Potts, C.N., Van Wassenhove, L.N.: Algorithms for Scheduling a Single Machine to Minimize the Weighted Number of Late Jobs. Management Science 34(7), 843–858 (1988)
22. Prade, H.: Using fuzzy set theory in a scheduling problem. Fuzzy Sets and Systems 2, 153–165 (1979)
23. Sahni, S.K.: Algorithms for Scheduling Independent Jobs. J. Assoc. Comput. Match. 23, 116–127 (1976)
24. Smith, W.E.: Various Optimizers for Single-Stage Production. Naval Research Logist Quartely 3, 59–66 (1956)
25. Van den Akker, M., Hoogeveen, H.: Minimizing the number of late jobs in a stochastic setting usinga chance constraint. Journal of Scheduling 11, 59–69 (2008)
26. Sourd, F., Kedad-Sidhoum, S.A.: A faster branch-and-bound algorithm for the earliness-tardiness scheduling problem. Journal of Scheduling 11, 49–58 (2008)
27. Wodecki, M.: A Branch-and-Bound Parallel Algorithm for Single-Machine Total Weighted Tardiness Problem. J. Adv. Manuf. Tech. 37(9-10), 996–1004 (2008)
28. Wodecki, M.: A block approach to earliness-tardiness scheduling problems. Advanced Manufacturing Technology 40, 797–807 (2009)
29. Villareal, F.J., Bulfin, R.L.: Scheduling a Single Machine to Minimize the Weighted Number of Tardy Jobs. IEE Trans. 15, 337–343 (1983)
30. Vondrák, J.: Probabilistic methods in combinatorial and stochastic optimization. PhD thesis, MIT (2005)

Chapter 5
Neural Networks Based Feature Selection in Biological Data Analysis

Witold Jacak, Karin Pröll, and Stephan Winkler

Abstract. In this chapter we present a novel method for scoring function specification and feature selection by combining unsupervised learning with supervised cross validation. Various clustering algorithms such as one dimensional Kohonen SOM, k-means, fuzzy c-means and hierarchical clustering procedures are used to perform a clustering of object-data for a chosen subset of input features and a given number of clusters. The resulting object clusters are compared with the predefined target classes and a matching factor (score) is calculated. This score is used as criterion function for heuristic sequential and cross feature selection.

5.1 Introduction

Classification of biological data means to develop a model that will divide biological observations into a set of predetermined classes N. Typically a biological data set is composed of many variables (features) that represent measures of biological attributes in biological experiments. A common aspect of biological data is its high dimensionality, that means the data dimension is high, but the sample size is relatively small. This phenomenon is called high dimensionality-small sample problem [12, 5, 14, 7, 1, 18, 10, 23, 26]. The smaller the sample, the less accurate are the results of classification and the amount of error increases.

The biological objective presented in this chapter is to study the influence of genetically modified (GM) foods on health and to identify biomarkers (measured immunologic parameters), which could be used to predict potential harmful GMO effects for post market surveillance. In experiments mice, rats, fish and pigs were tested for the effect of Bt-maize on a myriad of immunologic and general health

Witold Jacak · Karin Pröll · Stephan Winkler
Dept. of Software Engineering at Hagenberg,
Upper Austrian University of Applied Sciences, Softwarepark 11, A 4232 Hagenberg, Austria
e-mail: {jacak,proell,winkler}@fh-hagenberg.at

R. Klempous et al. (eds.), *Advanced Methods and Applications in Computational Intelligence,*   79
Topics in Intelligent Engineering and Informatics 6,
DOI: 10.1007/978-3-319-01436-4_5, © Springer International Publishing Switzerland 2014

parameters [28]. Identification of potential biomarkers in the experimental datasets can be transformed into a feature selection and classification problem.

Traditional statistical classification procedures such as discriminant analysis are built on the Bayesian decision theory [3, 2, 9, 10]. In these procedures, a probability model must be assumed in order to calculate the posterior probability upon which the classification decision is performed. One major limitation of statistical models is that they work well only when the underlying assumptions are satisfied. Users must have a good knowledge of both data properties and model capabilities before the models can be successfully applied [3, 10, 23].

Recent research activities in classification problem have shown that neural networks are a promising alternative to various conventional classification methods [18, 2, 7, 6, 9, 26, 27]. Neural networks are data driven self-adaptive systems in a way that they can adjust themselves to the data without any explicit specification of a functional or distributional form for the underlying model. Neural networks are able to estimate the posterior probabilities, which provide the basis for establishing classification rules and performing statistical analysis [26, 27]. The effectiveness of neural network classification has been tested empirically, it depends on the quality of input data.

Mining biological data may involve hundreds of input variables that can potentially be included in the model. Using the whole set of input variables to create the model does not necessarily give the best performance (peaking phenomenon) [7, 18]. Feature can be used to narrow down the number of features identifying the variables that are most important for a given analysis. Generally feature selection deals with choosing those input variables from the measurement space (all input variables) that are most predictive for a given target and constitute the feature space. The main objective of this process is to retain the optimum number of input variables necessary for the target recognition and to reduce the dimensionality of the measurement space so that an effective and easily computable model can be created for efficient data classification. Feature selection is a major issue in all supervised and unsupervised neural network learning tasks including classification, regression, time-series prediction, and clustering [14, 7, 9, 23]. A model with less features (variables) is faster to construct and easier to interpret, especially in biological data mining where a domain expert should interpret and validate such a model. Using classical supervised clustering and classification methods could lead - especially in case of small sample sets - to a faster overfitting of a model during the training phase and to worse prediction performance. Performing feature selection prior to training phase could avoid overfitting and improve the predictive capability of the model.

Feature selection methods can be grouped in four categories [2, 7, 26].

- Filter techniques select the features by looking only at the intrinsic property of input data and ignore feature dependencies. These techniques are independent of the classification model [2].
- Wrapper methods use the hypothesis model for the search of a feature subset. Hypothesis models are usually constructed in a supervised way during the training phase (e.g. Bayes model or multi-layer perceptron neural network (MLP))

but they increase the risk of overfitting especially in case of a small sample sizes [13].

- Embedded techniques include the search for an optimal subset of features in classifier construction [10, 6, 19].
- Feature selection after classification uses discriminant analysis to find sensitive variables [2, 18].

These techniques can be applied in exhaustive or heuristic search. Exhaustive search methods try to find the best subset among 2^m candidate subsets, where m is the total number of features. Exhaustive search can be very time consuming and is practically unacceptable, even for a medium-sized feature set.

Heuristic search methods use a learning approach which reduces computational complexity [12, 16]. There are two kinds of learning methods: supervised learning and unsupervised learning. In case of supervised learning the incorporation of prior knowledge about target classes into the training data is the key element to substantially improve classification performance. But full knowledge increases the risk of overfitting during the training phase of the model and as a consequence prediction performance for novel data decreases. Unsupervised learning uses no prior knowledge of target classes and tries to find a hidden structure in unlabeled training data [2, 7, 23, 27].

In this chapter we present a novel method for scoring function specification and feature selection by combining unsupervised learning methods with supervised cross validation. A one dimensional Kohonen SOM (Self-Organizing Map) network, k-means, fuzzy c-means and hierarchical agglomerative clustering (HAC) algorithms [11, 15, 17, 3, 6, 27] are used to perform clustering of sample data for a chosen subset of input features and a given number of clusters.

The resulting clusters are compared with the target classes and a matching factor (score) is calculated. This score is used as criterion (objective) function in heuristic sequential feature selection [16, 13, 2, 5]. Additionally the importance of an individual feature for recognition of target classes or composed groups of target classes is calculated using this matching factor. The results are compared and aggregated with the result of heuristic sequential feature selection to determine a final sensitive feature space. The final result is a reduced model with only few of the original features which are used to train a MLP network model [3, 26] for object classification.

5.2   Unsupervised Clustering and Matching Factor

For chosen input variables (selected features) from all input variables F and a given number N of clusters a Kohonen SOM-SL neural network with single layer is constructed [17, 3, 14, 7, 6, 23, 27]. Usually the output neurons are arranged in a bi-dimensional array, but we use an implementation of the network where the output neurons are arranged along a single layer (SL configuration). In the SL configuration the target classes are given by the number of output neurons. In this case there is no topological similarity between output neurons since adjacent output neurons do not represent necessarily similar classes. Additionally, we apply k-means, fuzzy c-means and hierarchical agglomerative (HAC) clustering algorithms on the same subset of features [11, 20, 3, 7, 27]. The resulting four sets of clusters are compared among one another and with the target classes. Based on comparison and validation with target classes the matching factors (score) p_f is calculated. The matching factor calculation module of the whole system is presented in Fig. 5.1.

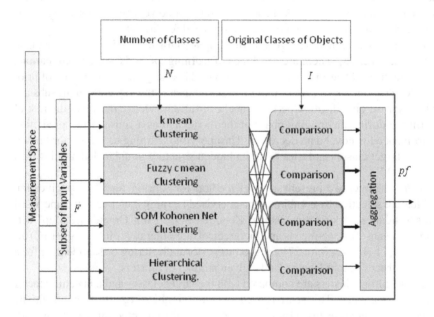

Fig. 5.1. Calculation of global matching score

5.2.1 *Matching Factor Calculation*

Let $I = \{1, ..., N\}$ be a set of indexes of target classes L and $J = \{1, ..., N\}$ be a set of indexes of cluster set K generated from input data of objects with unsupervised learning using the clustering methods.

Let $L_i \in L$ be a subset of objects belonging to target class $i \in I$ and $K_j \in K$ be a subset of objects belonging to cluster $j \in J$ created by one of the clustering methods. For each $i \in I$ and $j \in J$ we define the Jaccard matching coefficient [21, 4, 8, 25] c_{ij} as follows:

$$c_{ij} = \frac{|L_i \cap K_j|}{|L_i \cup K_j|} \tag{5.1}$$

Now, the task is to find the best allotment of generated classes K to the target classes L, that the number of hits reaches the maximal value. We use the c_{ij} coefficient to calculate the global matching factor p_f between the generated cluster and the target classes in an iterative manner [17].

For $n = 1, .., N$ we determine

$$p_{ij}^n = \begin{cases} c_{ij} \text{ if } c_{ij} = max\{c_{lk}|l \in I_{n-1}, k \in J_{n-1}\} \\ \quad \wedge s_i = \Sigma(c_{ik}|k \in J_{n-1}) \Rightarrow min \\ \quad \wedge s_j = \Sigma(c_{lj}|l \in I_{n-1}) \Rightarrow min \\ 0 \quad other \end{cases} \tag{5.2}$$

where $I_0 = I$, $J_0 = J$, and $J_n = J_{n-1} - \{j\}$, $I_n = I_{n-1} - \{i\}$ and $p_{ij}^{n-1} > 0$. Calculation stops when J_n and I_n are empty sets. If more than one pair (i, j) meets conditions (2) we randomly select one of those pairs and set all others p to 0.

It is easy to observe that the $N \times N$ sized matrix $p = [p_{ij}]$ has only N elements greater zero and represent the best allotment of the generated clusters to the original classes. The global matching factor p_f can be defined as mean value of non-zero elements of p:

$$p_f = avg(p_{ij}) \quad \text{where } p_{ij} > 0 \tag{5.3}$$

The matching factor describes the score of recognizing the target classes by unsupervised clustering which is simply based on input data without prior knowledge of classes. This factor uses posterior knowledge to calculate the score for validating the chosen set of individual features or feature groups. The choice of features should maximize the matching factor.

5.2.2   Aggregation

The matching factor can also be calculated between the results of two different clustering methods. The final score takes into account only these results of clustering methods, which have pairwise high matching factor and similar high matching factor between its clusters and the target classes. The final aggregated score is calculated as mean value of matching factors of all clustering procedures.

Let $Class = \{Cl_{k-mean}, Cl_{SOM}, Cl_{c-mean}, Cl_{hierarchic}, Cl_{target}\}$ be a set of clusters generated by different methods and target clusters. At first we look for such a subset C of $Class$ which has maximal matching factor between all elements in the subset. The subset C always includes the original target clusters Cl_{target}. This task can be formulated as follows:

Find such subset C^k (k number of elements in C^k) of $Class$, where C^k always includes Cl_{target} and has more than $k > 1$ elements, that the minimal matching factor p_f inside the subset C^k is maximal.

$$min\{p_f(Cl_i, Cl_j)|Cl_i, Cl_j \in C^k\} \Rightarrow max \tag{5.4}$$

In the second step the aggregated matching factor is calculated as mean value between the non target elements set C^k and Cl_{target}.

$$p_f^{agv} = avg\{p_f(Cl_i, Cl_{target}) | Cl_i \neq Cl_{target} \wedge Cl_i \in C^k\} \tag{5.5}$$

The choice of features should maximize the matching factor p_f^{avg}. Parameter k can be used to increase objectivity of unsupervised clustering when $(k > 2)$ methods give similar results.

5.3   Feature Selection System

The matching factor p_f is used as score for finding a sensitive group of input features. The feature selection system contains four modules:

- module for single feature filtering based on nonparametric Kruskal-Wallis test for significance of single variable,
- module for single feature filtering based matching factor test for significance of single variable
- module for sequential feature selection techniques (forward search) with matching factor as quality criterion (objective) function,
- module for sequential feature selection techniques (forward search) with naive Bayes classifier as quality criterion (objective) function

The system is presented in Fig.5.2. Each module calculates a candidate feature set. The filtering modules in the system reject a feature, when it's recognition performance is less (with a threshold of $tr\%$) than the maximum value of the objective function for all input data. Additionally, a novel method for feature selection, so called cross feature selection, is introduced. This method can only be applied on input data having more than two target classes, because it combines single target classes into larger hyper-classes (see next section).

To compare the recognition performance of all candidate sets the MLP-networks are used [9, 26], which are individually trained for selected candidates sets using target class labels. As a result confusion matrices indicating the recognition performance of a candidate feature set are calculated. The final result is the candidate set with best recognition performance.

5.3.1   Cross Feature Selection Method

When target classes L are known then it is possible to construct different subsets of target classes. The subset of target classes $\{L_i, L_k, \cdots, L_m\} \subset L$ is grouped together into one new class $LC_i = \cup\{L_i, L_k, \cdots, L_m\}$ called hyper-class, which contains all

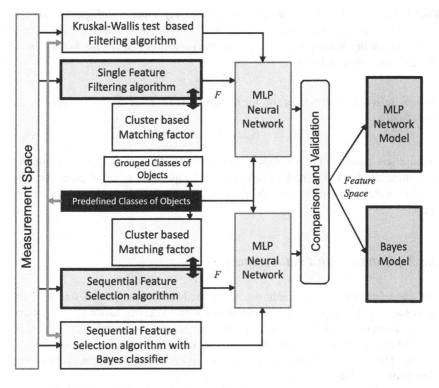

Fig. 5.2. Mixed feature selection system based on unsupervised learning

objects data from its components. The new hyper-classes set LC covers a set of target classes, i.e.

$$\cup LC_i = \cup L. \tag{5.6}$$

The new set of target hyper-classes determines the new number of classes. This number $|LC| = K$ is used for unsupervised clustering of the previously grouped input data from measurement space with Kohonen SOM-SL, k-means, fuzzy c-means and hierarchical HAC procedures and to calculate the matching factor between generated clusters and the new hyper-classes set LC.

In this approach we can search for feature sets, which discriminate the subsets of original classes. Usually the grouping of original classes into larger new hyper-classes is not accidental and is performed based on domain specific knowledge concerning properties of objects being classified.

For each group of target classes and also for each individual target class it is possible to test the significance of every single variable due to its predictability of the target classes (feature filtering). In filtering process, after the calculation of the matching factor individually for each variable, only variables with matching factor greater than a given threshold value (threshold $tr\%$) are chosen as candidates for feature space.

Fact 1. *Let LC^k $(k = 1, \cdots, M)$ denote the k-th experiment with chosen subsets (hyper-classes) of original target classes and $F^k \subset F$ the set of selected sensitive features recognizing best the hyper-classes in this experiment. When the hyper-classes sets LC^k meet the condition*

$$\cap LC^k = L \tag{5.7}$$

then the selected sensitive features for recognition of the full original target L could be defined as

$$F_{select} = \cup F^k. \tag{5.8}$$

Feature sets based on cross comparison of subsets of target classes are compared to feature sets found by the forward/backward sequential selection procedure. On these candidates exhaustive search can be applied to find the optimal final feature space.

5.3.2 Classifier in Feature Space

The candidates of sensitive feature sets obtained from cross feature selection and from sequential feature selection techniques with matching factor as quality criterion function are used to train a multi-layer perceptron neural network (MLP) to build the model and to evaluate the performance of target recognition and prediction. This training is performed based on full knowledge of target classification (supervised learning).

The candidate set with the maximal recognition ratio creates the final feature space. As a final classifier a multi-layer perceptron neural network (MLP) and a naive Bayes classifier are constructed (see Fig. 5.2).

5.4 Case Study

The aim of this study is to test the quality of the feature selection method based on a combination of unsupervised learning and posterior validation in comparison to standard statistic algorithms. A collection of biological data for 288 objects (Norwegian farmed salmon) is used in our experiment. The objects belong to four predefined classes (target classes) $L = \{A, B, C, D\}$. The pairs of classes (A, C) and (B, D) have property A in common and pairs (A, B) and (C, D) have Property B in common [28].

Each object is described with 89 input variables (parameters: organ weights, body composition, intestinal morphology and clinical biochemistry [28]). The measurement space - full feature set - is therefore $F = \{f_1, \cdots, f_{89}\}$. Missing values were replaced by the mean value of the parameter into consideration from the respective class.

At first the filtering for single sensitive features based on Kruskal-Wallis test and matching factor is performed. The Kruskal-Wallis test based filtering found only three sensitive parameters $\{f_{50}, f_{56}, f_{75}\}$ and filtering based on matching factor selects a set of seven sensitive parameters $\{f_{30}, f_{50}, f_{56}, f_{60}, f_{72}, f_{75}, f_{89}\}$, which include the Kruskal-Wallis test result. The comparison between performance of class recognition of two features f_{50} chosen by Kruskal-Wallis test and f_{66} selected by matching factor presents Fig. 5.3.

For validation a neural network (MLP) was trained with 70% data of whole datasets and evaluated with 30% of data.

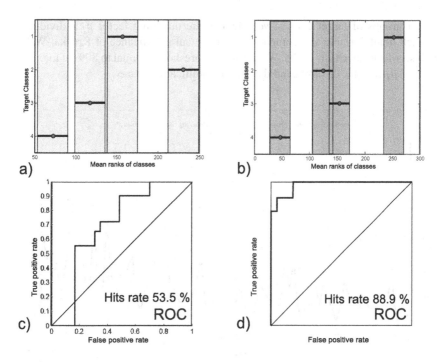

Fig. 5.3. Comparison of class recognition performance between two features f_{50} (left side of figure) and f_{66} (right side of figure) based on MLP classifier. Figures (a) and (b) show the Kruskal-Wallis test separation and figures (c) and (d) show ROC of trained MLP for both parameters respectively. The recognition ratio for f_{50} is only 54% and for f_{66} is 89%.

The general task is to find a subset of input variables $F_i \subset F$ with a minimal number of sensitive features, which recognize the four target classes without loss of classification quality. This subset will be used to develop a model for object classification. Additionally feature sets, which recognize two subsets of the original target classes were identified.

In the first step the classes A, B and C, D are combined together into one hyper-class i.e. $LC^1 = \{\{A \cup B\}, \{C \cup D\}\}$ and in the second step the classes A, C and B, D into one hyper-class i.e. $LC^2 = \{\{A \cup C\}, \{B \cup D\}\}$.

Since $LC^1 \cap LC^2 = L$, i.e. the condition of the previously defined fact comes true, then the union of sensible features for recognition of LC^1 and LC^2 could be used as a discriminant feature set for the whole target classes set.

5.4.1 Case Study

For all groups of target classes we calculate the matching factor p_f individually for each input variable and compare it to statistical significance of Kruskal-Wallis analysis of variance test results. We consider a threshold tr equal to 80% of maximal value of p_f to filter the first candidate set of significant features.

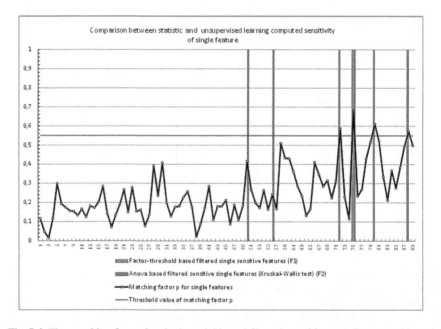

Fig. 5.4. The matching factor for single variable and filtered set of features for recognition of four original classes. Matching factor based filtered single features $F1 = \{66, 72, 75, 80, 88\}$, Kruskal-Wallis test based filtered single features $F2 = \{50, 56, 75\}$.

The matching factor for single variable, the inverse of the p-value of Kruskal-Wallis test and filtered set of features for recognition of four target classes is presented in Fig. 5.4.

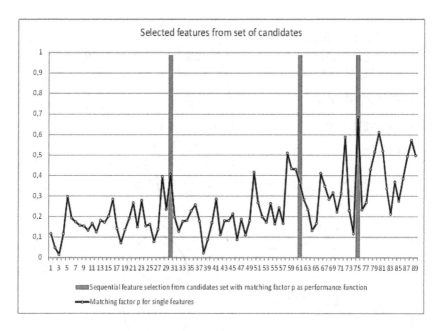

Fig. 5.5. The matching factor and selected set of features for recognition of four target classes with sequential selection procedure ($F3 = \{30, 61, 75\}$)

The same data are used for selecting the features subset with sequential forward selection algorithm using matching factor as quality criterion function. The resulting subset of features is presented in Fig. 5.5. It can be observed that not all candidates selected individually are present in the solution set of sequential search. Both methods applied for previously described hyper-classes LC^1 and LC^2 found only one single sensitive feature in each of both cases, f_{61} and f_{71} respectively. The cross selection gives therefore the new feature candidates set $F5 = \{61, 71\}$. The union of these is used as the new candidate subset for feature space.

The complete collection of data was used as input feature space for a feature selection method based on matching factor p_f as performance criterion and additionally for sequential feature selection based on naive Bayes classifier as performance function. The selected features of all approaches are presented in Fig. 5.6.

After the feature selection phase a multi-layer perceptron network is used to classify the training and validation dataset in the reduced feature space. The classification results are compared with the classes indicated in the original dataset. The results of feature selection were validated with a two-layer perceptron network (MLP) with four neurons in the hidden layer. The data set was divided into training set 70% of samples and test set 30% of samples. The correct classification rate to target class in percent is presented in Fig. 5.7.

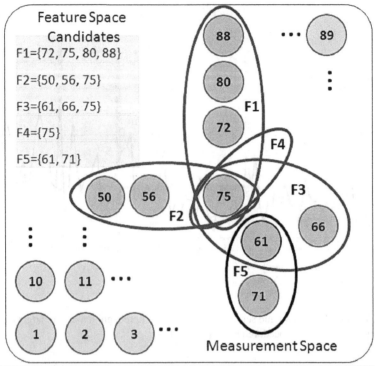

Method	Selected Features	Prediction performance
Factor-threshold based filtered single sensitive features (F1)	F1={72, 75, 80, 88}	87,8%
Anova based filtered sensitive single features (Kruskal-Wallis test) (F2)	F2={50, 56, 75}	78,3%
Sequential feature selection with matching factor p as performance function (F3)	F3={61, 66, 75}	95,7%
Sequential feature selection with Bayes classifier as performance function (F4)	F4={75}	73,9%
Features selected as union of discriminants of subsets of target classes (F5)	F5={61, 71}	99,2%

Fig. 5.6. Features selected with different algorithms for recognition of four original object classes

The feature spaces found using the combined unsupervised SOM, k-means, fuzzy c-means and hierarchical clustering and the posterior validation for matching factor calculation show a better classification rate for target classes as features selected with classical statistic methods (Fig. 5.6). It demonstrates the feasibility and effectiveness of the proposed novel method for score calculation, which is independent of the classification model.

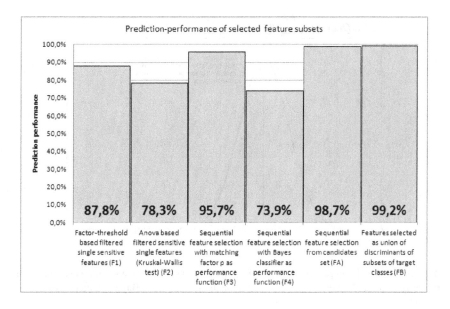

Fig. 5.7. Comparison of classification-performance of selected features subsets

5.5   Results and Conclusion

In this chapterr a feature selection method independent of the underlying classification model and based on clustering algorithms with posterior validation has been proposed. The analysis was performed on biological data. Mice, rats, fish and pigs have been tested for the effect of Bt-maize on general health as well as immune and allergic parameters. 12 different experiments were performed. All experiments data contain data-sets with samples sizes between 20 and 288, and numbers of features between 56 and 93. The numbers of target classes were between 2 and 7 [28].

Empirical distribution of p_m/p_f ratio , where p_m denotes the recognition quality of matching factor based selection and p_f is the recognition quality based on other objecive functions (MLP, naïïœeve Bayes, standard cross validation) is presented in Fig. 5.8. The recognition quality is determined as ratio of number of correct recognized test samples to number of all samples. The expectation value of empirical distribution is 1.19, thus feature selection with matching factor score gives about 20% better recognition accuracy.

Similarly we can compare the quality of class recognition between selected features with matching factor and selected features with other objective functions. For each experiment we perform feature selection using four different objective functions (matching factor, MLP, standard cross validation algorithm and naive Bayes) and the same heuristic feature selection method (forward heuristic search). Ratio $p_i(F_{selected})/p_i(F_{all-feature})$ for different objective functions and feature sets (p_i recognition quality of selection based on i-th objective function) is shown in Fig. 5.9.

Ratio $Quality(F_{matching})/Quality(F_{standard})$

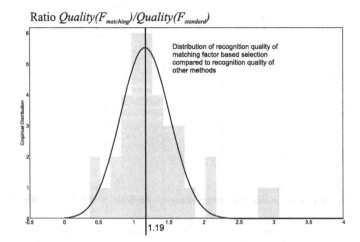

Fig. 5.8. Empirical comparison of recognition performance of selected features subsets with matching factor and other objective functions such as MLP, naive Bayes, standard cross validation

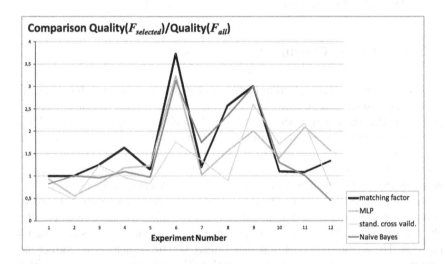

Fig. 5.9. Comparison of recognition performance of selected features subsets in relation to the whole features set obtainig by using matching factor MLP, naive Bayes, standard cross validation functions

Matching factor based selection improves recognition quality in approximately 60% of experiments. Recognition quality of matching factor based feature selection is about 66% better than recognition quality based on full feature set (robustness to noisy data).

The presented method can be classified as a heuristic method, which obtains an effective feature reduction with almost the same correct classification rate. Heuristics is designed on an unsupervised learning approach. The optimal number of feature space dimensions is therefore difficult to determine. Similar results for feature selection can be expected using evolutionary computing methods presented in [24]. Future works is to further investigate the distance measures to assess the impact on classification performance.

References

1. Bezdek, J.C., Ehrlich, R., Full, W.: FCM: The fuzzy c-means clustering algorithm. Computers and Geosciences 10(23) (1984)
2. Chen, Y., Abraham, A., Yang, B.: Feature selection and classification using flexible neural tree. Neurocomputing 70 (2006)
3. Campello, R.J.G.B.: A fuzzy extension of the Rand index and other related indexes for clustering and classification assessment. Pattern Recognition Letters 28 (2007)
4. Faro, D., Maiorana, F.: Discovering complex regularities by adaptive Self Organizing classification. Proceedings of WASET 4 (2005)
5. Halkidi, M., Batistakis, Y., Vazirgiannis, M.: On Clustering Validation Techniques. Journal of Intelligent Information Systems 17, 2–3 (2001)
6. Höppner, F.: Fuzzy cluster analysis: methods for classification, data analysis, and image recognition. Wiley Publ. (1999)
7. Jacak, W., Pröll, K.: Unsupervised Neural Networks based Scoring and Feature Selection in Biological Data Analysis. In: Proc. 14th International Asia Pacific Conference on Computer Aided System Theory, Sydney (2012)
8. Jain, A.K., Dubes, R.C.: Algorithms for clustering data. Prentice-Hall, Inc., Upper Saddle River (1988)
9. Jain, A.K., Zongker, D.: Feature Selection: Evaluation, Application and Small Sample Performance. IEEE Trans. PAMI 19(2) (1997)
10. Kohavi, R., John, G.H.: Wrappers for feature subset selection. Artificial Intelligence 97(1-2) (1997)
11. Kwak, N., Choi, C.: Input Feature Selection for Classification Problems. IEEE Transactions on Neural Networks 13(1) (2002)
12. Law, M.H.C., Figueiredo, M.A.T., Jain, A.K.: Simultaneous feature selection and clustering using mixture models. IEEE Trans. PAMI 26(9) (2004)
13. Legny, C., Juhsz, S., Babos, A.: Cluster validity measurement techniques. World Scientific and Engineering Academy and Society, WSEAS (2006)
14. Lee, K., Booth, D., State, K., Alam, P.: Backpropagation and Kohonen Self-Organizing Feature Map in Bankruptcy Prediction. In: Zhang, G.P. (ed.) Neural Networks in Business Forecasting. Idea Group Inc. (2004)
15. Maiorana, F.: Feature Selection with Kohonen Self Organizing Classification Algorithm, World Academy of Science, Engineering and Technology. Proceedings of WASET 45 (2008)

16. Su, M.-C., Chang, H.-T.: Fast Self-Organizing Feature Map Algorithm. IEEE Transactions on Neural Networks 11(3) (2000)
17. Su, M.-C., Liu, T.-K., Chang, H.-T.: Improving the Self-Organizing Feature Map Algorithm Using an Efficient Initialization Scheme. Journal of Science and Engineering 5(1) (2002)
18. Pal, S.K., De, R.K., Basak, J.: Unsupervised Feature Evaluation: A Neuro-Fuzzy Approach. IEEE Transactions on Neural Networks 11(2) (2000)
19. Pal, N.R., Bezdek, J.C.: On cluster validity for the fuzzy c-means model. IEEE Transactions on Fuzzy Systems (3) (1995)
20. Raghavendra, B.K., Simha, J.B.: Evaluation of Feature Selection Methods for Predictive Modeling Using Neural Networks in Credits Scoring. Int. J. Advanced Networking and Applications 2(3) (2010)
21. Silva, B., Marques, N.: Feature clustering with self-organizing maps and an application to financial time-series for portfolio selection. In: Proceedings of Intern. Conf. of Neural Computation, ICNC (2010)
22. Thangavel, K., Shen, Q., Pethalakshmi, A.: Application of Clustering for Feature Selection Based on Rough Set Theory Approach. AIML Journal 6(1) (January 2006)
23. Törmä, M.: Self-organizing neural networks in feature extraction. Intern. Arch. of Photogrammetry and Remote Sensing XXXI, Part 2 (1996)
24. Winkler, S., Affenzeller, M., Kronberger, G., Kommenda, M., Wagner, S., Jacak, W., Stekel, H.: Feature selection in the analysis of tumor marker data using evolutionary algorithms. In: Proceedings of the 7th International Mediterranean and Latin American Modelling Multiconference, EMSS 2010 (2010)
25. Vendramin, L., Campello, R.J.G.B., Hruschka, E.R.: Relative clustering validity criteria: A comparative overview. Statistical Analysis and Data Mining 3(4) (2010)
26. Zhang, G.P.: Neural Networks for Classification: A Survey. IEEE Transactions on Systems, Man, and Cybernetics Part C: Applications and Reviews 30(4) (2000)
27. Ye, H., Liu, H.: A SOM-based method for feature selection. In: Proceedings of the 9th International Conference on Neural Information Processing (ICONIP 2002), vol. 3 (2002)
28. Epstein, M. (ed.): Proc. of GMSAFOOD Conference, Vienna (2012)

Chapter 6
On the Identification of Virtual Tumor Markers and Tumor Diagnosis Predictors Using Evolutionary Algorithms

Stephan M. Winkler, Michael Affenzeller, Gabriel K. Kronberger, Michael Kommenda, Stefan Wagner, Witold Jacak, and Herbert Stekel

Abstract. In this chapter we present results of empirical research work done on the data based identification of estimation models for tumor markers and cancer diagnoses: Based on patients' data records including standard blood parameters, tumor markers, and information about the diagnosis of tumors we have trained mathematical models that represent virtual tumor markers and predictors for cancer diagnoses, respectively. We have used a medical database compiled at the Central Laboratory of the General Hospital Linz, Austria, and applied several data based modeling approaches for identifying mathematical models for estimating selected tumor marker values on the basis of routinely available blood values; in detail, estimators for the tumor markers AFP, CA-125, CA15-3, CEA, CYFRA, and PSA have been identified and are discussed here. Furthermore, several data based modeling approaches implemented in HeuristicLab have been applied for identifying estimators for selected cancer diagnoses: Linear regression, k-nearest neighbor learning, artificial neural networks, and support vector machines (all optimized using evolutionary algorithms) as well as genetic programming. The investigated diagnoses of breast cancer, melanoma, and respiratory system cancer can be estimated correctly in up to 81%, 74%, and 91% of the analyzed test cases, respectively; without tumor markers up to 75%, 74%, and 87% of the test samples are correctly estimated, respectively.

Stephan M. Winkler · Michael Affenzeller · Gabriel K. Kronberger · Michael Kommenda · Stefan Wagner · Witold Jacak
Heuristic and Evolutionary Algorithms Laboratory,
University of Applied Sciences Upper Austria, School of Informatics,
Communication and Media, Softwarepark 11, 4232 Hagenberg, Austria

Herbert Stekel
Central Laboratory, General Hospital Linz, Krankenhausstraße 9, 4021 Linz, Austria

R. Klempous et al. (eds.), *Advanced Methods and Applications in Computational Intelligence*,   95
Topics in Intelligent Engineering and Informatics 6,
DOI: 10.1007/978-3-319-01436-4_6, © Springer International Publishing Switzerland 2014

6.1   Introduction and Research Goals

In this chapter we present research results achieved within the research center
Heureka![1]: Data of thousands of patients of the General Hospital (AKH) Linz, Austria, have been analyzed in order to identify mathematical models for tumor markers
and tumor diagnoses. We have used a medical database compiled at the blood laboratory of the General Hospital Linz, Austria, in the years 2005 – 2008: 28 routinely
measured blood values of thousands of patients are available as well as several tumor
markers; not all values are measured for all patients, especially tumor marker values
are determined and documented if there are indications for the presence of cancer.

In Figure 6.1 the main modeling tasks addressed in this research work are illustrated: Tumor markers are modeled using standard blood parameters and tumor
marker data; tumor diagnosis models are trained using standard blood values, tumor
marker data, and diagnosis information, and alternatively we also train diagnosis
estimation models only using standard blood parameters and diagnosis information.

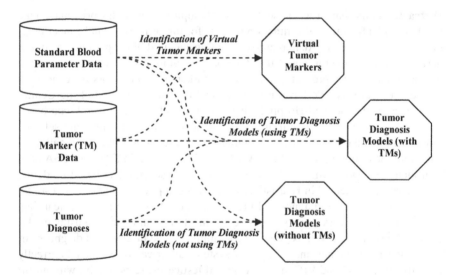

Fig. 6.1. Modeling tasks investigated in this research work: Tumor markers are modeled, and
tumor diagnosis models are trained using standard blood values, diagnosis information, and
optionally tumor marker data

6.1.1   Identification of Virtual Tumor Markers

In general, tumor markers are substances found in humans (especially blood and /
or body tissues) that can be used as indicators for certain types of cancer. There are
several different tumor markers which are used in oncology to help detect the presence of cancer; elevated tumor marker values can indicate the presence of cancer,

---

[1] Josef Ressel Center for Heuristic Optimization; http://heureka.heuristiclab.com/

but there can also be other causes. As a matter of fact, elevated tumor marker values themselves are not diagnostic, but rather only suggestive; tumor markers can be used to monitor the result of a treatment (as for example chemotherapy). Literature discussing tumor markers, their identification, their use, and the application of data mining methods for describing the relationship between markers and the diagnosis of certain cancer types can be found for example in [26] (where an overview of clinical laboratory tests is given and different kinds of such test application scenarios as well as the reason of their production are described), [10], [40], [56], and [57].

We have used data based modeling approaches (including enhanced genetic programming as well as other established data mining methods) for identifying mathematical models for estimating selected tumor marker values on the basis of routinely available blood values; in detail, estimators for the tumor markers *AFP*, *CA-125*, *CA15-3*, *CEA*, *CYFRA*, and *PSA* have been identified and are analyzed in this chapter. These tumor marker estimation models are also referred to as *virtual tumor markers*.

The documented tumor marker values are classified as "normal" (class 0), "slightly elevated" (class 1), "highly elevated" (class 2), and "beyond plausible" (class 3); this classification is done according to medical aspects using classification rules based on medical knowledge. In principle, our goal is to design classifiers for classifying samples into one of these classes. Still, in the context of the research work summarized here we have decided to produce classifiers that classify samples as "normal (belonging to class 0)" or "elevated (belonging to class 1, 2, or 3)"; i.e., we here document results for a simplified 2-class-classification problem.

6.1.2   Identification of Tumor Diagnosis Estimators

In addition, information about cancer diagnoses is also available in the AKH database: If a patient is diagnosed with any kind of cancer, then this is also stored in the database.

Our goal in the research work described here is to identify estimation models for the presence of the following types of cancer: Malignant neoplasms in the respiratory system (RSC, cancer classes C30–C39 according to the International Statistical Classification of Diseases and Related Health Problems 10th Revision (ICD-10)), melanoma and malignant neoplasms on the skin (Mel, C43–C44), and breast cancer (BC, C50).

Tumor markers are used optionally - on the one hand information about tumor markers values increases the accuracy of diagnosis estimations, on the other hand their acquisition is considered expensive and they are therefore not available by default.

We have applied two modeling methods for identifying estimation models for tumor markers and cancer diagnoses:

- Several machine learning methods (implemented in HeuristicLab [47]) have been used for producing classifiers, namely linear regression, k-nearest neighbor classification, neural networks, and support vector machines.

- Evolutionary algorithms have been applied for parameter optimization and feature selection. Feature selection is often considered an essential step in data based modeling; it is used to reduce the dimensionality of the datasets and often leads to better analyses. Additionally, each data based modeling method (except plain linear regression) has several parameters that have to be set before starting the modeling process. We have used evolutionary algorithms for finding optimal feature sets as well as optimal modeling parameters for models for tumor markers; details can be found in [54] and [52], e.g.
- Alternatively, we have applied genetic programming (GP, [28]) using a structure identification framework described in [4] and [50]. Genetic programming has been repeatedly used successfully for building formulas that describe the behavior of systems from measured data, see for example [4], [28], [32], or [50].

6.1.3   Organization of This Chapter

This chapter is structured in the following way: In Section 6.2 we give details about the data basis investigated in the research summarized here, in Section 6.3 we describe the modeling methods applied for identifying classification models for tumor data, and in Sections 6.4 and 6.5 we summarize empirical results achieved modeling tumor markers and tumor diagnoses using machine learning and evolutionary algorithms. This chapter is completed by a conclusion given in Section 6.6.

6.2   Data Basis

The blood data measured at the AKH in the years 2005–2008 have been compiled in a database storing each set of measurements (belonging to one patient): Each sample in this database contains an unique ID number of the respective patient, the date of the measurement series, the ID number of the measurement, and a set of parameters summarized in the Tables 6.1 and 6.2; standard blood parameters are stored as well as tumor marker values. Patients personal data (e.g. name, date of birth, etc.) where at no time available to the authors except the head of the laboratory.

In total, information about 20,819 patients is stored in 48,580 samples. Please note that of course not all values are available in all samples; there are very many missing values simply because not all blood values are measured during each examination.

Information about the blood parameters stored in this database and listed in Table 6.1 can for example be found in [6], [30], [34], [43], and [49].

In Table 6.2 we list those tumor markers that are available in the AKH database and have been used within the research work described here; in detail, we have analyzed data of the following tumor markers:

Table 6.1. Patient data collected at AKH Linz in the years 2005 – 2008: Blood parameters and general patient information

Parameter Name	Description	Unit	Plausible Range	Number of Available Values
ALT	Alanine transaminase, a transaminase enzyme; also called glutamic pyruvic transaminase (GPT)	U/l	[1; 225]	29,202
AST	Aspartate transaminase, an enzyme also called glutamic oxaloacetic transaminase (GOT)	U/l	[1; 175]	29,201
BSG1	Erythrocyte sedimentation rate; the rate at which red blood cells settle / precipitate within 1 hour	mm	[0; 50]	10,201
BUN	Blood urea nitrogen; measures the amount of nitrogen in the blood (caused by urea)	mg/dl	[1; 150]	28,995
CBAA	Basophil granulocytes; type of leukocytes	G/l	[0.0; 0.2]	21,184
CEOA	Eosinophil granulocytes; type of leukocytes	G/l	[0.0; 0.4]	21,184
CH37	Cholinesterase, an enzyme	kU/l	[2; 23]	7,266
CHOL	Cholesterol, a component of cell membranes	mg/dl	[40; 550]	14,981
CLYA	Lymphocytes; type of leukocytes	G/l	[1; 4]	21,188
CMOA	Monocytes; type of leukocytes	G/l	[0.2; 0.8]	21,184
CNEA	Neutrophils; most abundant type of leukocytes	G/l	[1.8; 7.7]	21,184
CRP	C-reactive protein, a protein; inflammations cause the rise of CRP	mg/dl	[0; 20]	22,560
FE	Iron	ug/dl	[30; 210]	6,792
FER	Ferritin, a protein that stores and transports iron in a safe form	ng/ml	[10; 550]	2,428
GT37	γ-glutamyltransferase, an enzyme	U/l	[1; 290]	29,173
HB	Hemoglobin; a protein that contains iron and transports oxygen	g/dl	[6; 18]	29,574
HDL	High-density lipoprotein; this protein enables the transport of lipids with blood	mg/dl	[25; 120]	7,998
HKT	Hematocrit; the proportion of red blood cells within the blood volume	%	[25; 65]	29,579
HS	Uric acid, also called urate	mg/dl	[1; 12]	24,330
KREA	Creatinine; a chemical by-product produced in muscles	mg/dl	[0.2; 5.0]	29,033
LD37	Lactate dehydrogenase (LDH); an enzyme that can be used as a general marker of injuries to cells	U/l	[5; 744]	28,356
MCV	Mean corpuscular / cell volume; the average size (i.e., volume) of red blood cells	fl ($=\mu m^3$)	[69; 115]	29,576
PLT	Thrombocytes, also called platelets; irregularly-shaped cells that do not have a nucleus	G/l	[25; 1,000]	29,579
RBC	Erythrocytes, red blood cells; the most abundant type of blood cells that transport oxygen	T/l	[2.2; 8.0]	29,576
TBIL	Bilirubin; the yellow product of the heme catabolism	mg/dl	[0; 5]	28,565
TF	Transferrin; a protein, delivers iron	mg/dl	[100; 500]	2,017
WBC	Leukocytes, also called white blood cells (WBCs); cells that help the body fight infections or foreign materials	G/l	[1.5; 50]	29,585
AGE	The patient's age	years	[0; 120]	48,580
SEX	The patient's sex	f/m		48,580

Table 6.2. Patient data collected at AKH Linz in the years 2005 – 2008: Selected tumor markers

Marker Name	Unit	Normal Range	Elevated Range	Plausible Range	Number of Available Values
AFP	IU/ml	[0.0; 5.8]	]5.8; 28]	[0.0; 90]	5,415
CA 125	U/ml	[0.0; 35.0]	]35.0; 80]	[0.0; 150]	3,661
CA 15-3	U/ml	[0.0; 25.0]	]25.0; 50.0]	[0.0; 100.0]	6,944
CEA	ng/ml	[0.0; 3.4]	]3.4; 12.0]	[0.0; 50.0]	12,981
CYFRA	ng/ml	[0.0; 3.3]	]3.3; 5.0]	[0.0; 10.0]	2,861
PSA	ng/ml	[0.0; 2.5] (age ≤ 50) [0.0; 2.5] (age 51–60) [0.0; 2.5] (age 61–70) [0.0; 2.5] (age ≥ 71)	]2.5; 10.0] (age ≤ 50)]2.5; 10.0] (age 51–60)]2.5; 10.0] (age 61–70)]2.5; 10.0] (age ≥ 71)	[0.0; 20.0]	23,130

- **AFP:** Alpha-fetoprotein (AFP, [36]) is a protein found in the blood plasma; during fetal life it is produced by the yolk sac and the liver. In humans, maximum AFP levels are seen at birth; after birth, AFP levels decrease gradually until adult levels are reached after 8 to 12 months. Adult AFP levels are detectable, but usually rather low.

  For example, AFP values of pregnant women can be used in screening tests for developmental abnormalities as increased values might for example indicate open neural tube defects, decreased values might indicate Down syndrome. AFP is also often measured and used as a marker for a set of tumors, especially endodermal sinus tumors (yolk sac carcinoma), neuroblastoma, hepatocellular carcinoma, and germ cell tumors [17]. In general, the level of AFP measured in patients often correlates with the size / volume of the tumor.

- **CA 125:** Cancer antigen 125 (CA 125) ([55]), also called carbohydrate antigen 125 or mucin 16 (MUC16), is a protein that is often used as a tumor marker that may be elevated in the presence of specific types of cancers, especially recurring ovarian cancer [39]. Still, its use in the detection of ovarian cancer is controversial, mainly because its sensitivity is rather low (as documented in [41], only 79% of all ovarian cancers are positive for CA 125) and it is not possible to detect early stages of cancer using CA 125.

  Even though CA 125 is best known as a marker for ovarian cancer, it may also be elevated in the presence of other types of cancers; for example, increased values are seen in the context of cancer in fallopian tubes, lungs, the endometrium, breast and gastrointestinal tract.

- **CA 15-3:** Mucin 1 (MUC1), also known as cancer antigen 15-3 (CA 15-3), is a protein found in humans; it is used as a tumor marker in the context of monitoring certain cancers [38], especially breast cancer. Elevated values of CA 15-3 have been reported in the context of an increased chance of early recurrence in breast cancer [25].

- **CEA:** Carcinoembryonic antigen (CEA; [22], [23]) is a protein that is in humans normally produced during fetal development. As the production of CEA usually is stopped before birth, it is usually not present in the blood of healthy adults. Elevated levels are seen in the blood or tissues of heavy smokers; persons with pancreatic carcinoma, colorectal carcinoma, lung carcinoma, gastric carcinoma, or breast carcinoma, often have elevated CEA levels. When used as a tumor marker, CEA is mainly used to identify recurrences of cancer after surgical resections.

- **CYFRA:** Fragments of cytokeratin 19, a protein found in the cytoskeleton, are found in many places of the human body; especially in the lung and in malign lung tumors high concentrations of these fragments, which are also called CYFRA 21-1, are found. Due to elevated values in the presence of lung cancer CYFRA is often used for detecting and monitoring malign lung tumors. Elevated CYFRA values have already been reported for several different kinds of tumors, especially for example in stomach, colon, breast, and ovaries. The use of CYFRA 21-1 as a tumor marker has for example been discussed in [31].
- **PSA:** Prostate-specific antigen (PSA; [7], [45]) is a protein produced in the prostate gland; PSA blood tests are widely considered the most effective test currently available for the early detection of prostate cancer since PSA is often elevated in the presence of prostate cancer and in other prostate disorders. Still, the effectiveness of these tests has also been considered questionable since PSA is prone to both false positive and false negative indications: According to [45], 70 out of 100 men with elevated PSA values do not have prostate cancer, and 25 out of 100 men suffering from prostate cancer do not have significantly elevated PSA.

As already mentioned, information about cancer diagnoses is also available in the AKH database: If a patient is diagnosed with any kind of cancer, then this is also stored in the database. All cancer diagnoses are classified according to the International Statistical Classification of Diseases and Related Health Problems 10th Revision (ICD-10) system.

In this research work we concentrate on diagnoses regarding the following types of cancer: Malignant neoplasms in the respiratory system (RSC, cancer classes C30–C39 according to ICD-10), melanoma and malignant neoplasms on the skin (Mel, C43–C44), and breast cancer (BC, C50).

6.3   Modeling Approaches

In this section we describe the modeling methods applied for identifying estimation models for cancer diagnosis: On the one hand we apply hybrid modeling using machine learning algorithms (linear regression, neural networks, the k-nearest neighbor method, support vector machines) and evolutionary algorithms for parameter optimization and feature selection (as described in Section 6.3.5), on the other hand apply use genetic programming (as described in Section 6.3.6).

All these machine learning methods have been implemented using the HeuristicLab framework [47], a framework for prototyping and analyzing optimization techniques for which both generic concepts of evolutionary algorithms and many functions to evaluate and analyze them are available. HeuristicLab is developed by the Heuristic and Evolutionary Algorithm Laboratory[2] and can be downloaded from the HeuristicLab homepage[3]. HeuristicLab is licensed under the GNU General Public License[4].

6.3.1 Linear Modeling

Given a data collection including m input features storing the information about N samples, a linear model is defined by the vector of coefficients $\theta_{1...m}$. For calculating the vector of modeled values e using the given input values matrix $u_{1...m}$, these input values are multiplied with the corresponding coefficients and added: $e = u_{1...m} * \theta$. The coefficients vector can be computed by simply applying matrix division. For conducting the test series documented here we have used an implementation of the matrix division function: $\theta = InputValues \backslash TargetValues$. Additionally, a constant additive factor is also included into the model; i.e., a constant offset is added to the coefficients vector. Theoretical background of this approach can be found in [33].

6.3.2 kNN Classification

Unlike other data based modeling methods, k-nearest neighbor classification [16] works without creating any explicit models. During the training phase, the samples are simply collected; when it comes to classifying a new, unknown sample x_{new}, the sample-wise distance between x_{new} and all other training samples x_{train} is calculated and the classification is done on the basis of those k training samples (x_{NN}) showing the smallest distances from x_{new}.

In the context of classification, the numbers of instances (of the k nearest neighbors) are counted for each given class and the algorithm automatically predicts that class that is represented by the highest number of instances (included in x_{NN}). In the test series documented in this chapter we have applied weighting to kNN classification: The distance between x_{new} and x_{NN} is relevant for the classification statement, the weight of "nearer" samples is higher than that of samples that are "further away" from x_{new}. In this research work we have varied k between 1 and 10.

---

[2] http://heal.heuristiclab.com/

[3] http://dev.heuristiclab.com/

[4] http://www.gnu.org/licenses/gpl.txt

6.3.3   Artificial Neural Networks

For training artificial neural network (ANN) models, three-layer feed-forward neural networks with one linear output neuron were created applying backpropagation (using gradient descent optimization); theoretical background and details can for example be found in [37] (Chapter 11, "Neural Networks"). In the tests documented in this chapter the number of hidden (sigmoidal) nodes hn has been varied from 5 to 100; the learning rate as well as the momentum were also varied, the range of these parameters was set to [0.01 - 0.5]. We have applied ANN training algorithms that use internal validation sets, i.e., training algorithms use 30% of the given training data as validation data and eventually return those network structures that perform best on these internal validation samples.

6.3.4   Support Vector Machines

Support vector machines (SVMs) are a widely used approach in machine learning based on statistical learning theory [46]. The most important aspect of SVMs is that it is possible to give bounds on the generalization error of the models produced, and to select the corresponding best model from a set of models following the principle of structural risk minimization [46].

In this work we have used the LIBSVM implementation described in [12], which is used in the respective SVM interface implemented for HeuristicLab; here we have used Gaussian radial basis function kernels with varying values for the cost parameters c ($c \in [0, 512]$) and the γ parameter of the SVM's kernel function ($\gamma \in [0, 1]$).

6.3.5   Hybrid Modeling Using Machine Learning Algorithms and Evolutionary Algorithms for Parameter Optimization and Feature Selection

An essential step in data mining and machine learning is (especially when there are very many available features / variables) the selection of subsets of variables that are used for learning models. On the one hand, simpler models (i.e., models that use fewer variables and have simpler structures) are preferred over more complex ones following Occam's law of parsimony [8] that states that simpler theories are in general more favorable; on the other hand, simpler models are less likely to be prone to overfitting ([4], [29]).

So-called forward approaches iteratively add variables that are essentially important for improving the quality of the achievable models, while backward elimination methods initially use all variables and iteratively eliminate those that show the least statistical significance. Early variable selection algorithms were published several decades ago (such as, e.g., [18]). Since then, numerous variable selection algorithms have been developed, many of them relying on the concept of mutual information ([11], [13], [14], [20], [21], [42], [44]). An overview of well-established variable selection methods can be found for example in [15], the use of variable selection methods in cancer classification in [35].

Unlike these variable selection methods, we here use an evolutionary algorithm that is able to simultaneously optimize variable selections and modeling parameters with respect to specific machine learning algorithms. The main advantages of this approach are on the one hand that variable selection is not necessary as a separate step in the data mining process, on the other hand variable selections and modeling parameter settings are automatically optimized for the modeling algorithm at hand. Parsimony pressure is realized by incorporating the size of sets of selected variables into the fitness function that is used for evaluating solution candidates.

Given a set of n features $F = \{f_1, f_2, \ldots, f_n\}$, our goal here is to find a subset $F' \subseteq F$ that is on the one hand as small as possible and on the other hand allows modeling methods to identify models that estimate given target values as well as possible. Additionally, each data based modeling method (except plain linear regression) has several parameters that have to be set before starting the modeling process.

The fitness of feature selection F' and training parameters with respect to the chosen modeling method is calculated in the following way: We use a machine learning algorithm m (with parameters p) for estimating predicted target values $est(F', m, p)$ and compare those to the original target values $orig$; the coefficient of determination (R^2) function is used for calculating the quality of the estimated values. Additionally, we also calculate the ratio of selected features $|F'|/|F|$. Finally, using a weighting factor α, we calculate the fitness of the set of features F' using m and p as

$$fitness(F', m, p) = \alpha * |F'|/|F| + (1 - \alpha) * (1 - R^2(est(F', m, p), orig)). \quad (6.1)$$

As an alternative to the coefficient of determination function we can also use a classification specific function that calculates the ratio of correctly classified samples, either in total or as the average of all classification accuracies of the given classes (as for example described in [50], Section 8.2): For all samples that are to be considered we know the original classifications $origCl$, and using (predefined or dynamically chosen) thresholds we get estimated classifications $estCl(F', m, p)$ for estimated target values $est(F', m, p)$. The total classification accuracy $ca_k(F', m, p)$ is calculated as

$$ca(F', m, p) = \frac{|\{j : estCl(F', m, p)[j] = origCl[j]\}|}{|estCl|} \quad (6.2)$$

Class-wise classification accuracies $cwca$ are calculated as the average of all classification accuracies for each given class $c \in C$ separately:

$$ca(F', m, p)_c = \frac{|\{j : estCl(F', m, p)[j] = origCl[j] = c\}|}{|\{j : origCl[j] = c\}|} \quad (6.3)$$

$$cwca(F', m, p) = \frac{\sum_{c \in C} ca(F', m, p)_c}{|C|} \qquad (6.4)$$

We can now define the classification specific fitness of feature selection F' using m and p as

$$fitness_{ca}(F', m, p) = \alpha * |F'|/|F| + (1 - \alpha) * (1 - ca(F', m, p)) \qquad (6.5)$$

or

$$fitness_{cwca}(F', m, p) = \alpha * |F'|/|F| + (1 - \alpha) * (1 - cwca(F', m, p)). \qquad (6.6)$$

In [5], for example, the use of evolutionary algorithms for feature selection optimization is discussed in detail in the context of gene selection in cancer classification; in [53] we have analyzed the sets of features identified as relevant in the modeling of tumor markers AFP and CA15-3.

We have now used evolutionary algorithms for finding optimal feature sets as well as optimal modeling parameters for models for tumor diagnosis; this approach is schematically shown in Figure 6.2. A solution candidate is here represented as $[s_{1,...,n} \ p_{1,...,q}]$ where s_i is a bit denoting whether feature F_i is selected or not and p_j is the value for parameter j of the chosen modeling method m. This rather simple definition of solution candidates enables the use of standard concepts for genetic operators for crossover and mutation of bit vectors and real valued vectors: We use uniform, single point, and 2-point crossover operators for binary vectors and bit flip mutation that flips each of the given bits with a given probability. Explanations of these operators can for example be found in [19] and [24].

In the test series described later in Section 6.5 we have used strict offspring selection [1] which means that individuals are accepted to become members of the next generation if they are evaluated better than both parents. Standard fitness evaluation as given in Equation 6.1 has been used during the execution of the evolutionary processes, and classification specific fitness evaluation as given in Equation 6.6 has been used for selecting the solution candidate eventually returned as the algorithm's result.

6.3.6   Genetic Programming

We have also applied a classification algorithm based on genetic programming (GP) [28] using a structure identification framework described in [4] and [50], in combination with an enhanced, hybrid selection scheme called offspring selection ([1], [2], [3]). In the left part of Figure 6.3 we show the overall GP workflow including offspring selection, in the right part the here used strict version of OS is depicted; we have used the GP implementation in HeuristicLab.

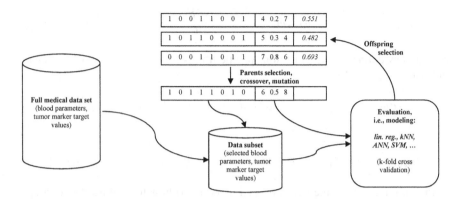

Fig. 6.2. A hybrid evolutionary algorithm for feature selection and parameter optimization in data based modeling

In addition to splitting the given data into training and test data, the GP based training algorithm implemented in HeuristicLab has been designed in such a way that a part of the given training data is not used for training models and serves as validation set; in the end, when it comes to returning classifiers, the algorithm returns those models that perform best on validation data. This approach has been chosen because it is assumed to help to cope with over-fitting; it is also applied in other GP based machine learning algorithms as for example described in [9].

We have used the following parameter settings for our GP test series: The mutation rate was set to 20%, gender specific parents selection [48] (combining random and roulette selection) was applied as well as strict offspring selection [1] (OS, with success ratio as well as comparison factor set to 1.0). The functions set described in [50] (including arithmetic as well as logical ones) was used for building composite function expressions.

The following parameter settings have been used in our GP test series:

- Single population approach; the population size was set to 700
- Mutation rate: 15%
- Varying maximum formula tree complexity
- Parents selection: Gender specific [48], random & roulette
- Offspring selection [1]: Strict offspring selection (success ratio as well as comparison factor set to 1.0)
- One-elitism
- Termination criteria:

    - Maximum number of generations: 1000; this criterion was not reached in the tests documented here, all executions were terminated via the
    - Maximum selection pressure [1]: 555

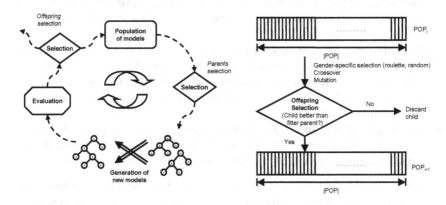

Fig. 6.3. The GP cycle [32] including offspring selection. Flowchart for embedding a simplified version of offspring selection into the GP based machine learning process.

- Function set: All functions (including arithmetic as well as logical ones) as described in [50]

In addition to splitting the given data into training and test data, the GP based training algorithm implemented in HeuristicLab has been implemented in such a way that a part of the given training data is not used for training models and serves as validation set; in the end, when it comes to returning classifiers, the algorithm returns those models that perform best on validation data. This approach has been chosen because it is assumed to help to cope with over-fitting; it is also applied in other GP based machine learning algorithms as for example described in [9].

6.4 Empirical Study: Identification of Models for Tumor Markers

In this section we summarize empirical results previously described in [51].

6.4.1 Data Preprocessing

Before analyzing the data and using them for training classifiers for tumor markers we have preprocessed the data available in the AKH data base:

- All variables have been linearly scaled to the interval [0;1]: For each variable v_i, the minimum value min_i is subtracted from all contained values and the result divided by the difference between min_i and the maximum plausible value $maxplau_i$; all values greater than the given maximum plausible value are replaced by 1.0.

- All samples belonging to the same patient with not more than one day difference with respect to the measurement data have been merged. This has been done in order to decrease the number of missing values in the data matrix. In rare cases, more than one value might thus be available for a certain variable; in such a case, the first value is used.

- Additionally, all measurements have been sample-wise re-arranged and clustered according to the patients' IDs. This has been done in order to prevent data of certain patients being included in the training as well as in the test data.

Before modeling algorithms can be used for training classifiers, we have compiled separate data sets for each analyzed target tumor marker tm_i: First, all samples containing measured values for tm_i are extracted. Second, all variables are removed from the resulting data set that contain values in less than 80% of the remaining samples. Third, all samples are removed that still contain missing values. This procedure results in a specialized data set $dstm_i$ for each tumor marker tm_i. In Table 6.3 we summarize statistical information about all resulting data sets for the markers analyzed here[5]; the numbers of samples belonging to each of the defined classes are also given for each resulting data set.

6.4.2   Test Series and Results

All machine learning mentioned in Section 6.3 have been applied using several different parametrizations; for each modeling method we also give relevant parameter settings, namely the number of neighbors k for kNN learning ($k \in \{1, 3, 5, 10\}$), the number of nodes n in the hidden layer of ANNs ($n \in \{10, 25, 50, 100\}$), the γ value for SVMs ($\gamma \in \{0.001, 0.01, 0.05, 0.1, 0.5, 1, 8\}$), and the maximum tree size s (i.e., the number of nodes in formula trees) for GP ($s \in \{25, 50, 100, 150\}$). Five-fold cross-validation [27] training / test series have been executed; this means that the available data are separated in five (approximately) equally sized, complementary subsets, and in each training / test cycle one data subset is chosen as used as test and the rest of the data as training samples. The classifiers learned using training data are evaluated on training as well as on test data; in the following Tables 6.4 – 6.9 we give statistics about the quality on the produced classifiers, namely the average classification accuracies (as the average of correct sample classifications) and their standard deviations on training as well as on test data.

---

[5] Please note that the number of total samples and the number of samples in class 0 for the PSA data set differ from the numbers stated in [51]; the here given numbers are the correct ones.

Table 6.3. Overview of the data sets compiled for selected tumor markers

Marker Name	Input Variables	Total Samples	Distribution of Samples			
			Class 0	Class 1	Class 2	Class 3
AFP	AGE, SEX, ALT, AST, BUN, CH37, GT37, HB, HKT, KREA, LD37, MCV, PLT, RBC, TBIL, WBC	2,755	2146 (77.9%)	454 (16.5%)	64 (2.32%)	91 (3.3%)
CA 125	AGE, SEX, ALT, AST, BUN, CRP, GT37, HB, HKT, HS, KREA, LD37, MCV, PLT, RBC, TBIL, WBC	1,053	532 (50.5%)	143 (13.6%)	84 (8.0%)	294 (27.9%)
CA 15-3	AGE, SEX, ALT, AST, BUN, CBAA, CEOA, CLYA, CMOA, CNEA, CRP, GT37, HB, HKT, HS, KREA, LD37, MCV, PLT, RBC, TBIL, WBC	4,918	3,159 (64.2%)	1,011 (20.6%)	353 (7.2%)	395 (8.0%)
CEA	AGE, SEX, ALT, AST, BUN, CBAA, CEOA, CLYA, CMOA, CNEA, CRP, GT37, HB, HKT, HS, KREA, LD37, MCV, PLT, RBC, TBIL, WBC	5,567	3,133 (56.3%)	1,443 (25.9%)	492 (8.8%)	499 (9.0%)
CYFRA	AGE, SEX, ALT, AST, BUN, CH37, CHOL, CRP, CYFS, GT37, HB, HKT, HS, KREA, MCV, PLT, RBC, TBIL, WBC	419	296 (70.6%)	37 (8.8%)	36 (8.6%)	50 (11.9%)
PSA	AGE, SEX, ALT, AST, BUN, CBAA, CEOA, CHOL, CLYA, CMOA, CNEA, CRP, GT37, HB, HKT, HS, KREA, LD37, MCV, PLT, RBC, TBIL, WBC	2,366	1,145 (48.4%)	779 (32.9%)	249 (10.5%)	193 (8.2%)

Table 6.4. Classification results for AFP

Modeling method		Classification accuracy ($\mu \pm \sigma$)	
		Training	Test
LinReg		0.8356 ($\pm$ 0.009)	0.8221 ($\pm$ 0.018)
kNN	1	1.0000 ($\pm$ 0.000)	0.7659 ($\pm$ 0.026)
	3	1.0000 ($\pm$ 0.000)	0.7790 ($\pm$ 0.031)
	5	1.0000 ($\pm$ 0.000)	0.7974 ($\pm$ 0.037)
	10	1.0000 ($\pm$ 0.000)	0.8061 ($\pm$ 0.041)
ANN	10	0.8625 ($\pm$ 0.014)	0.8215 ($\pm$ 0.045)
	25	0.8608 ($\pm$ 0.029)	0.8239 ($\pm$ 0.049)
	50	0.8564 ($\pm$ 0.017)	**0.8548 ($\pm$ 0.020)**
	100	0.8642 ($\pm$ 0.014)	0.8217 ($\pm$ 0.049)
SVM	0.01	0.7846 ($\pm$ 0.012)	0.7826 ($\pm$ 0.067)
	0.05	0.8301 ($\pm$ 0.015)	0.8137 ($\pm$ 0.048)
	0.1	0.8490 ($\pm$ 0.009)	0.8107 ($\pm$ 0.054)
	0.5	0.9057 ($\pm$ 0.011)	0.8075 ($\pm$ 0.043)
	1	0.9426 ($\pm$ 0.009)	0.7998 ($\pm$ 0.039)
	8	1.0000 ($\pm$ 0.000)	0.7773 ($\pm$ 0.045)
GP	25	0.771 ($\pm$ 0.014)	0.7735 ($\pm$ 0.028)
	50	0.7884 ($\pm$ 0.014)	0.7681 ($\pm$ 0.014)
	100	0.7985 ($\pm$ 0.026)	0.7739 ($\pm$ 0.037)

Table 6.5. Classification results for CA125

Modeling method		Classification accuracy ($\mu \pm \sigma$)	
		Training	Test
LinReg		0.7243 ($\pm$ 0.022)	0.5913 ($\pm$ 0.071)
kNN	1	1.0000 ($\pm$ 0.000)	0.5611 ($\pm$ 0.070)
	3	1.0000 ($\pm$ 0.000)	0.5449 ($\pm$ 0.074)
	5	1.0000 ($\pm$ 0.000)	0.5560 ($\pm$ 0.062)
	10	1.0000 ($\pm$ 0.000)	0.5680 ($\pm$ 0.084)
ANN	10	0.7941 ($\pm$ 0.049)	0.5342 ($\pm$ 0.073)
	25	0.7707 ($\pm$ 0.184)	0.6275 ($\pm$ 0.071)
	50	0.7601 ($\pm$ 0.034)	0.5851 ($\pm$ 0.059)
	100	0.7661 ($\pm$ 0.048)	0.5993 ($\pm$ 0.047)
SVM	0.01	0.7140 ($\pm$ 0.024)	0.5594 ($\pm$ 0.081)
	0.05	0.7663 ($\pm$ 0.023)	0.5503 ($\pm$ 0.080)
	0.1	0.8054 ($\pm$ 0.015)	0.5239 ($\pm$ 0.080)
	0.5	0.9180 ($\pm$ 0.017)	0.5357 ($\pm$ 0.068)
	1	0.9905 ($\pm$ 0.003)	0.5607 ($\pm$ 0.063)
	8	1.0000 ($\pm$ 0.000)	0.5548 ($\pm$ 0.103)
GP	**25**	0.7194 ($\pm$ 0.017)	**0.6810 ($\pm$ 0.081)**
	50	0.7474 ($\pm$ 0.022)	0.6677 ($\pm$ 0.067)
	100	0.7628 ($\pm$ 0.014)	0.6344 ($\pm$ 0.048)

Table 6.6. Classification results for CA 15-3

Modeling method		Classification accuracy ($\mu \pm \sigma$)	
		Training	Test
LinReg		0.7533 ($\pm$ 0.017)	0.7069 ($\pm$ 0.020)
kNN	1	1.0000 ($\pm$ 0.000)	0.6495 ($\pm$ 0.030)
	3	1.0000 ($\pm$ 0.000)	0.6767 ($\pm$ 0.029)
	5	1.0000 ($\pm$ 0.000)	0.6905 ($\pm$ 0.032)
	10	1.0000 ($\pm$ 0.000)	0.7028 ($\pm$ 0.029)
ANN	10	0.7920 ($\pm$ 0.008)	0.7127 ($\pm$ 0.022)
	25	0.8000 ($\pm$ 0.027)	0.7007 ($\pm$ 0.028)
	50	0.7878 ($\pm$ 0.008)	0.6997 ($\pm$ 0.012)
	100	0.7866 ($\pm$ 0.040)	0.6974 ($\pm$ 0.009)
SVM	0.01	0.7252 ($\pm$ 0.017)	0.7077 ($\pm$ 0.039)
	0.05	0.7700 ($\pm$ 0.010)	0.7236 ($\pm$ 0.034)
	0.1	0.8089 ($\pm$ 0.006)	0.7181 ($\pm$ 0.034)
	0.5	0.9405 ($\pm$ 0.001)	0.6843 ($\pm$ 0.026)
	1	0.9874 ($\pm$ 0.001)	0.7135 ($\pm$ 0.078)
	8	1.0000 ($\pm$ 0.000)	0.6433 ($\pm$ 0.031)
GP	25	0.7348 ($\pm$ 0.012)	0.7167 ($\pm$ 0.036)
	50	0.7327 ($\pm$ 0.009)	0.7163 ($\pm$ 0.039)
	100	0.7504 ($\pm$ 0.011)	**0.7302** ($\pm$ 0.030)

Table 6.7. Classification results for CEA

Modeling method		Classification accuracy ($\mu \pm \sigma$)	
		Training	Test
LinReg		0.6847 ($\pm$ 0.020)	0.6566 ($\pm$ 0.053)
kNN	1	1.0000 ($\pm$ 0.000)	0.5877 ($\pm$ 0.009)
	3	1.0000 ($\pm$ 0.000)	0.5979 ($\pm$ 0.017)
	5	1.0000 ($\pm$ 0.000)	0.6006 ($\pm$ 0.022)
	10	1.0000 ($\pm$ 0.000)	0.6115 ($\pm$ 0.034)
ANN	10	0.7526 ($\pm$ 0.025)	0.6556 ($\pm$ 0.030)
	25	0.7681 ($\pm$ 0.026)	0.6393 ($\pm$ 0.039)
	50	0.7486 ($\pm$ 0.010)	0.6555 ($\pm$ 0.044)
	100	0.7459 ($\pm$ 0.005)	0.6625 ($\pm$ 0.039)
SVM	0.001	0.7277 ($\pm$ 0.012)	0.6461 ($\pm$ 0.054)
	0.01	0.6758 ($\pm$ 0.014)	0.6470 ($\pm$ 0.061)
	0.05	0.7432 ($\pm$ 0.008)	0.6417 ($\pm$ 0.048)
	0.1	0.7904 ($\pm$ 0.003)	0.6373 ($\pm$ 0.043)
	0.5	0.9321 ($\pm$ 0.011)	0.6030 ($\pm$ 0.034)
	1	0.9788 ($\pm$ 0.008)	0.5813 ($\pm$ 0.026)
	8	1.0000 ($\pm$ 0.000)	0.5508 ($\pm$ 0.014)
GP	25	0.6791 ($\pm$ 0.034)	0.6532 ($\pm$ 0.007)
	50	0.6854 ($\pm$ 0.009)	0.6686 ($\pm$ 0.014)
	100	0.6874 ($\pm$ 0.005)	0.6772 ($\pm$ 0.008)
	150	0.6822 ($\pm$ 0.005)	**0.6828** ($\pm$ 0.011)

Table 6.8. Classification results for CYFRA

Modeling method		Classification accuracy ($\mu \pm \sigma$)	
		Training	Test
LinReg		0.7957 ($\pm$ 0.020)	0.7061 ($\pm$ 0.076)
kNN	1	1.0000 ($\pm$ 0.000)	0.6372 ($\pm$ 0.048)
	3	1.0000 ($\pm$ 0.000)	0.6445 ($\pm$ 0.111)
	5	1.0000 ($\pm$ 0.000)	0.6586 ($\pm$ 0.119)
	10	1.0000 ($\pm$ 0.000)	0.6965 ($\pm$ 0.106)
ANN	10	0.7590 ($\pm$ 0.014)	0.7139 ($\pm$ 0.073)
	25	0.7639 ($\pm$ 0.037)	0.7134 ($\pm$ 0.080)
	50	0.8304 ($\pm$ 0.045)	0.7087 ($\pm$ 0.079)
	100	0.7768 ($\pm$ 0.035)	**0.7303** ($\pm$ 0.048)
	200	0.7969 ($\pm$ 0.051)	0.7160 ($\pm$ 0.055)
SVM 0.001		0.7063 ($\pm$ 0.039)	0.6727 ($\pm$ 0.058)
0.01		0.7195 ($\pm$ 0.025)	0.7041 ($\pm$ 0.071)
0.05		0.8269 ($\pm$ 0.015)	0.7206 ($\pm$ 0.073)
0.1		0.8585 ($\pm$ 0.069)	0.7229 ($\pm$ 0.056)
0.5		0.9988 ($\pm$ 0.002)	0.6985 ($\pm$ 0.059)
1		1.0000 ($\pm$ 0.000)	0.6634 ($\pm$ 0.092)
8		1.0000 ($\pm$ 0.000)	0.7088 ($\pm$ 0.086)
GP	25	0.7565 ($\pm$ 0.033)	0.6763 ($\pm$ 0.069)
	50	0.7708 ($\pm$ 0.025)	0.6520 ($\pm$ 0.056)
	100	0.7865 ($\pm$ 0.039)	0.6563 ($\pm$ 0.050)

Table 6.9. Classification results for PSA

Modeling method		Classification accuracy ($\mu \pm \sigma$)	
		Training	Test
LinReg		0.6403 ($\pm$ 0.014)	0.5858 ($\pm$ 0.026)
kNN	1	1.0000 ($\pm$ 0.000)	0.5365 ($\pm$ 0.037)
	3	1.0000 ($\pm$ 0.000)	0.5420 ($\pm$ 0.029)
	5	1.0000 ($\pm$ 0.000)	0.5517 ($\pm$ 0.018)
	10	1.0000 ($\pm$ 0.000)	0.5622 ($\pm$ 0.021)
ANN	10	0.6507 ($\pm$ 0.043)	0.5840 ($\pm$ 0.018)
	25	0.6327 ($\pm$ 0.014)	0.5828 ($\pm$ 0.014)
	50	0.6456 ($\pm$ 0.016)	0.5718 ($\pm$ 0.013)
	100	0.6462 ($\pm$ 0.017)	0.5786 ($\pm$ 0.018)
	200	0.6661 ($\pm$ 0.016)	0.5808 ($\pm$ 0.016)
SVM 0.001		0.5882 ($\pm$ 0.027)	0.5536 ($\pm$ 0.052)
0.01		0.6403 ($\pm$ 0.016)	0.5913 ($\pm$ 0.022)
0.05		0.6813 ($\pm$ 0.011)	0.5672 ($\pm$ 0.018)
0.1		0.7331 ($\pm$ 0.005)	0.5740 ($\pm$ 0.015)
0.5		0.9286 ($\pm$ 0.012)	0.5600 ($\pm$ 0.026)
1		0.9903 ($\pm$ 0.003)	0.5499 ($\pm$ 0.021)
8		0.9996 ($\pm$ 0.001)	0.5351 ($\pm$ 0.024)
GP	25	0.6857 ($\pm$ 0.090)	0.6527 ($\pm$ 0.077)
	50	0.7176 ($\pm$ 0.080)	**0.6746** ($\pm$ 0.080)
	100	0.7071 ($\pm$ 0.029)	0.6730 ($\pm$ 0.039)

6.5   Empirical Study: Identification of Models for Tumor Diagnoses

In this section we summarize empirical results previously described in [52].

6.5.1   Data Preprocessing

Before analyzing the data and using them for training data-based tumor diagnosis estimators we have preprocessed the available data:

- All variables have been linearly scaled to the interval [0;1]: For each variable v_i, the minimum value min_i is subtracted from all contained values and the result divided by the difference between min_i and the maximum plausible value $maxplau_i$; all values greater than the given maximum plausible value are replaced by 1.0.
- All samples belonging to the same patient with not more than one day difference with respect to the measurement data have been merged. This has been done in order to decrease the number of missing values in the data matrix. In rare cases, more than one value might thus be available for a certain variable; in such a case, the first value is used.
- Additionally, all measurements have been sample-wise re-arranged and clustered according to the patients' IDs. This has been done in order to prevent data of certain patients being included in the training as well as in the test data.

Before starting the modeling algorithms for training classifiers we had to compile separate data sets for each analyzed target tumor t_i: First, blood parameter measurements were joined with diagnosis results; only measurements and diagnoses with a time delta less than a month were considered. Second, all samples containing measured values for t_i are extracted. Third, all samples are removed that contain less than 15 valid values. Finally, variables with less than 10% valid values are removed from the data base.

This procedure results in a specialized data set dst_i for each tumor marker t_i. In Table 6.10 we summarize statistical information about all resulting data sets for the markers analyzed here; the numbers of samples belonging to each of the defined classes are also given for each resulting data set.

6.5.2   Test Series and Results

Five-fold cross-validation [27] training / test series have been executed; this means that the available data are separated in five (approximately) equally sized, complementary subsets, and in each training / test cycle one data subset is chosen is used as test and the rest of the data as training samples.

In this section we document test accuracies (μ, σ) for the investigated cancer types; we here summarize test results for modeling cancer diagnoses using tumor markers (TMs) as well as for modeling without using tumor markers. Linear modeling, kNN modeling, ANNs, and SVMs have been applied for identifying estimation models for the selected tumor types, genetic algorithms with strict OS have been applied for optimizing variable selections and modeling parameters; standard fitness calculation as given in Equation 6.1 has been used by the evolutionary process, the classification specific one as given in Equation 6.6 has been used for selecting the eventually returned model. The probability of selecting a variable initially was set to 30%. Additionally, we have also applied simple linear regression using all available variables. Finally, genetic programming with strict offspring selection (OSGP) has also been applied.

Table 6.10. Overview of the data sets compiled for selected cancer types

Cancer Type	Input Variables	Total Samples	Samples in Class 0	Class 1	Missing Values
Breast Cancer	AGE, SEX, AFP, ALT, AST, BSG1, BUN, C125, C153, C199, C724, CBAA, CEA,	706	324 (45.9%)	382 (54.1%)	46.67%
Melanoma	CEOA, CH37, CHOL, CLYA, CMOA, CNEA, CRP, CYFS,	905	485 (53.6%)	420 (46.4%)	47.79%
Respira-tory System Cancer	FE, FER, FPSA, GT37, HB, HDL, HKT, HS, KREA, LD37, MCV, NSE, PLT, PSA, PSAQ, RBC, S100, SCC, TBIL, TF, TPS, WBC	2,363	1,367 (57.9%)	996 (42.1%)	44.76%

In all test series the maximum selection pressure [1] was set to 100, i.e., the algorithms were terminated as soon as the selection pressure reached 100. The population size for genetic algorithms optimizing variable selections and modeling parameters was set to 10, for GP the population size was set to 700. In all modeling cases except kNN modeling regression models have been trained, the threshold for classification decisions was in all cases set to 0.5 (since the absence of the specific tumor is represented by 0.0 in the data and its presence by 1.0).

Table 6.11. Modeling results for breast cancer diagnosis

Modeling Method	Using TMs Test accuracies		Not using TMs Test accuracies	
	μ	σ	μ	σ
LR, full features set	79.32%	1.06	70.63%	1.28
OSGA + LR, $\alpha = 0.0$	81.78%	0.21	73.13%	0.36
OSGA + LR, $\alpha = 0.1$	81.49%	1.18	72.66%	0.14
OSGA + LR, $\alpha = 0.2$	81.44%	0.37	71.40%	0.57
OSGA + kNN, $\alpha = 0.0$	79.21%	0.78	74.22%	2.98
OSGA + kNN, $\alpha = 0.1$	78.99%	0.57	75.55%	0.87
OSGA + kNN, $\alpha = 0.2$	78.33%	1.04	74.50%	0.20
OSGA + ANN, $\alpha = 0.0$	81.41%	1.14	75.60%	2.47
OSGA + ANN, $\alpha = 0.1$	80.19%	1.68	72.38%	6.08
OSGA + ANN, $\alpha = 0.2$	79.37%	1.17	70.54%	6.10
OSGA + SVM, $\alpha = 0.0$	81.23%	1.10	73.90%	2.36
OSGA + SVM, $\alpha = 0.1$	80.46%	1.80	72.19%	0.94
OSGA + SVM, $\alpha = 0.2$	77.43%	3.55	71.89%	0.70
OSGP, $ms = 50$	79.72%	1.80	75.32%	0.45
OSGP, $ms = 100$	75.50%	4.95	71.63%	2.75
OSGP, $ms = 150$	79.20%	6.60	75.75%	2.16

Table 6.12. Modeling results for melanoma diagnosis

Modeling Method	Using TMs Test accuracies		Not using TMs Test accuracies	
	μ	σ	μ	σ
LR, full features set	73.81%	3.39	71.09%	4.14
OSGA + LR, $\alpha = 0.0$	72.45%	4.69	72.36%	2.30
OSGA + LR, $\alpha = 0.1$	74.73%	2.35	72.09%	4.01
OSGA + LR, $\alpha = 0.2$	73.85%	2.54	72.70%	2.02
OSGA + kNN, $\alpha = 0.0$	68.77%	2.38	71.00%	1.97
OSGA + kNN, $\alpha = 0.1$	71.33%	0.27	70.21%	3.41
OSGA + kNN, $\alpha = 0.2$	67.33%	0.31	69.65%	3.14
OSGA + ANN, $\alpha = 0.0$	74.78%	1.63	69.17%	2.97
OSGA + ANN, $\alpha = 0.1$	73.81%	2.23	71.82%	0.61
OSGA + ANN, $\alpha = 0.2$	74.12%	1.03	71.40%	0.49
OSGA + SVM, $\alpha = 0.0$	69.72%	7.57	68.87%	4.78
OSGA + SVM, $\alpha = 0.1$	71.75%	4.88	68.22%	1.88
OSGA + SVM, $\alpha = 0.2$	61.48%	3.99	63.20%	2.09
OSGP, $ms = 50$	71.24%	9.54	74.89%	3.66
OSGP, $ms = 100$	69.91%	5.20	65.16%	13.06
OSGP, $ms = 150$	71.79%	4.31	70.13%	3.60

Table 6.13. Modeling results for respiratory system cancer diagnosis

Modeling Method	Using TMs Test accuracies		Not using TMs Test accuracies	
	μ	σ	μ	σ
LR, full features set	91.32%	0.37	85.97%	0.27
OSGA + LR, $\alpha = 0.0$	91.57%	0.46	86.41%	0.36
OSGA + LR, $\alpha = 0.1$	91.16%	1.18	85.80%	0.45
OSGA + LR, $\alpha = 0.2$	89.45%	0.37	85.02%	0.15
OSGA + kNN, $\alpha = 0.0$	90.98%	0.84	87.09%	0.46
OSGA + kNN, $\alpha = 0.1$	90.01%	2.63	87.01%	0.83
OSGA + kNN, $\alpha = 0.2$	90.16%	0.74	86.92%	0.81
OSGA + ANN, $\alpha = 0.0$	90.28%	1.63	85.97%	4.07
OSGA + ANN, $\alpha = 0.1$	90.99%	1.97	85.82%	4.52
OSGA + ANN, $\alpha = 0.2$	88.64%	1.87	87.24%	1.91
OSGA + SVM, $\alpha = 0.0$	89.03%	1.38	83.12%	3.79
OSGA + SVM, $\alpha = 0.1$	89.91%	1.58	86.25%	0.79
OSGA + SVM, $\alpha = 0.2$	88.33%	1.94	84.66%	2.06
OSGP, $ms = 50$	89.58%	2.75	85.98%	5.74
OSGP, $ms = 100$	90.44%	3.02	86.54%	6.02
OSGP, $ms = 150$	89.58%	3.75	87.97%	5.57

In Table 6.14 we summarize the effort of the modeling approaches applied in this research work: For the combination of GAs and machine learning methods we document the number of modeling executions, and for GP we give the number of evaluated solutions (i.e., models).

For the combination of genetic algorithms with linear regression, kNN modeling, ANNs, and SVMs (with varying variable ratio (vr) weighting factors) as well as GP with varying maximum tree sizes ms we give the sizes of selected variable sets, and (where applicable) also k, hn, c, and γ. Obviously there are different variations in the parameters identified as optimal by the evolutionary process: The numbers of variables used as well as the neural networks' hidden nodes vary to a relatively small extent, e.g., whereas especially the SVMs' parameters (especially the c factors) vary very strongly.

Table 6.14. Effort in terms of executed modeling runs and evaluated model structures

Modeling Method	Modeling executions					
	$vrfw = 0.0$		$vrfw = 0.1$		$vrfw = 0.2$	
	μ	σ	μ	σ	μ	σ
LR	3260.4	717.8	2339.6	222.6	2465.2	459.2
kNN	2955.3	791.8	3046.0	362.4	3791.3	775.9
ANN	3734.0	855.9	3305.0	582.6	3297.0	475.9
SVM	2950.0	794.8	2846.0	391.4	3496.7	859.8

Modeling Method	Evaluated solutions (models)		
	$ms = 50$ μ, σ	$ms = 100$ μ, σ	$ms = 150$ μ, σ
OSGP	1483865.0, 674026.2	1999913.3, 198289.1	2238496.7, 410123.6

Table 6.15. Optimized parameters for linear regression

Problem Instance, vr weighting		Variables used	
		μ	σ
BC, TM	$\alpha = 0.0$	16.6	2.10
	$\alpha = 0.1$	11.8	1.50
	$\alpha = 0.2$	6.4	0.60
BC, no TM	$\alpha = 0.0$	9.6	1.15
	$\alpha = 0.1$	8.8	0.58
	$\alpha = 0.2$	6.4	1.20
Mel, TM	$\alpha = 0.0$	16.6	0.55
	$\alpha = 0.1$	12.2	0.84
	$\alpha = 0.2$	9.2	4.09
Mel, no TM	$\alpha = 0.0$	10.8	1.79
	$\alpha = 0.1$	8.8	2.28
	$\alpha = 0.2$	8.2	1.92
RSC, TM	$\alpha = 0.0$	17.2	2.95
	$\alpha = 0.1$	13.4	2.51
	$\alpha = 0.2$	9.0	2.55
RSC, no TM	$\alpha = 0.0$	16.0	4.64
	$\alpha = 0.1$	9.6	0.89
	$\alpha = 0.2$	8.6	3.21

Table 6.16. Optimized parameters for kNN modeling

Problem Instance, vr weighting		Variables used		k	
		μ	σ	μ	σ
BC, TM	$\alpha = 0.0$	18.2	2.20	9.8	2.10
	$\alpha = 0.1$	14.0	3.60	12.6	4.60
	$\alpha = 0.2$	11.0	1.80	11.2	3.00
BC, no TM	$\alpha = 0.0$	14.4	1.67	11.2	1.64
	$\alpha = 0.1$	14.0	2.45	13.8	3.11
	$\alpha = 0.2$	11.8	0.84	18.8	1.10
Mel, TM	$\alpha = 0.0$	15.6	1.82	17.8	2.86
	$\alpha = 0.1$	16.4	1.34	14.4	5.90
	$\alpha = 0.2$	13.6	1.67	19.4	1.34
Mel, no TM	$\alpha = 0.0$	15.0	1.58	14.2	1.10
	$\alpha = 0.1$	10.4	1.52	18.2	1.64
	$\alpha = 0.2$	9.6	1.14	16.8	2.05
RSC, TM	$\alpha = 0.0$	14.6	1.67	20.0	0.00
	$\alpha = 0.1$	13.6	1.67	16.8	6.06
	$\alpha = 0.2$	10.4	1.52	12.8	3.90
RSC, no TM	$\alpha = 0.0$	15.6	2.79	15.2	1.64
	$\alpha = 0.1$	12.2	1.10	10.6	1.82
	$\alpha = 0.2$	10.2	2.95	13.2	2.95

Table 6.17. Optimized parameters for ANNs

Problem Instance, vr weighting		Variables used		hn	
		μ	σ	μ	σ
BC, TM	$\alpha = 0.0$	17.0	1.20	75.6	20.80
	$\alpha = 0.1$	14.8	3.40	51.0	5.80
	$\alpha = 0.2$	11.0	0.80	35.8	13.00
BC, no TM	$\alpha = 0.0$	12.6	1.41	82.4	23.46
	$\alpha = 0.1$	12.2	0.89	70.8	26.40
	$\alpha = 0.2$	11.2	1.10	68.2	14.58
Mel, TM	$\alpha = 0.0$	19.6	2.19	56.8	13.31
	$\alpha = 0.1$	12.8	2.28	61.0	2.55
	$\alpha = 0.2$	15.6	5.18	51.2	6.98
Mel, no TM	$\alpha = 0.0$	15.4	2.51	68.6	14.24
	$\alpha = 0.1$	8.2	1.64	59.8	3.83
	$\alpha = 0.2$	8.0	1.00	58.6	5.81
RSC, TM	$\alpha = 0.0$	13.4	3.44	64.6	10.97
	$\alpha = 0.1$	11.2	2.28	68.2	6.69
	$\alpha = 0.2$	8.2	1.64	60.2	13.92
RSC, no TM	$\alpha = 0.0$	13.2	2.28	71.2	12.38
	$\alpha = 0.1$	12.2	2.05	70.6	12.99
	$\alpha = 0.2$	11.6	2.19	64.4	14.24

Table 6.18. Optimized parameters for SVMs

Problem Instance, vr weighting		Variables used μ	σ	C μ	σ	γ μ	σ
BC, TM	$\alpha = 0.0$	21.6	3.50	101.50	92.30	0.05	0.06
	$\alpha = 0.1$	18.8	3.50	12.44	13.85	0.09	0.01
	$\alpha = 0.2$	16.0	2.00	64.79	67.59	0.04	0.01
BC, no TM	$\alpha = 0.0$	15.6	1.83	47.16	12.63	0.05	0.05
	$\alpha = 0.1$	15.4	1.10	22.50	25.88	0.07	0.04
	$\alpha = 0.2$	13.0	2.65	8.14	10.09	0.07	0.04
Mel, TM	$\alpha = 0.0$	13.0	4.53	166.23	236.61	0.27	0.25
	$\alpha = 0.1$	10.8	3.42	204.74	210.43	0.18	0.19
	$\alpha = 0.2$	4.2	2.95	123.08	44.14	0.26	0.20
Mel, no TM	$\alpha = 0.0$	21.4	6.95	116.21	196.73	0.41	0.30
	$\alpha = 0.1$	19.8	1.64	492.73	8.10	0.48	0.41
	$\alpha = 0.2$	14.4	3.29	310.17	208.60	0.36	0.35
RSC, TM	$\alpha = 0.0$	21.2	8.50	183.54	95.38	0.27	0.26
	$\alpha = 0.1$	14.6	1.14	74.56	67.98	0.09	0.10
	$\alpha = 0.2$	11.2	3.83	37.55	68.31	0.45	0.35
RSC, no TM	$\alpha = 0.0$	13.4	4.10	23.14	31.91	0.35	0.25
	$\alpha = 0.1$	12.4	3.21	144.73	96.79	0.19	0.08
	$\alpha = 0.2$	12.4	3.21	376.66	206.84	0.09	0.10

Table 6.19. Number of variables used by models returned by OSGP

Problem Instance, maximum tree size ms		Variables used by returned model μ	σ
BC, TM	$ms = 50$	9.0	2.74
	$ms = 100$	9.6	1.34
	$ms = 150$	17.8	0.45
BC, no TM	$ms = 50$	10.5	0.71
	$ms = 100$	10.0	1.41
	$ms = 150$	11.5	0.71
Mel, TM	$ms = 50$	10.2	2.05
	$ms = 100$	10.0	2.55
	$ms = 150$	12.0	2.00
Mel, no TM	$ms = 50$	8.0	1.58
	$ms = 100$	8.8	0.84
	$ms = 150$	11.4	3.36
RSC, TM	$ms = 50$	7.8	2.05
	$ms = 100$	12.0	2.35
	$ms = 150$	12.0	1.22
RSC, no TM	$ms = 50$	9.4	3.91
	$ms = 100$	12.2	2.17
	$ms = 150$	13.6	3.13

6.6 Conclusion

We have described the data based identification of mathematical models for the tumor markers AFP, CA-125, CA15-3, CEA, CYFRA, and PSA as well as selected tumors, namely breast cancer, melanoma, and respiratory system cancer. Data collected at the General Hospital Linz, Austria have been used for creating models that predict tumor marker values and tumor diagnoses; several different techniques of applied statistics and computational intelligence have been applied, namely linear regression, kNN learning, artificial neural networks, support vector machines (all optimized using evolutionary algorithms), and genetic programming.

On the one hand, it seems that none of the methods used here produced the best results for all modeling tasks; in two cases (AFP and CYFRA) ANNs produced models that perform best on test data, in all remaining four cases (CA-125, CA15-3,

CEA, and PSA) extended genetic programming has produced best results. Additionally, we here see that in those modeling cases, for which GP found best results, these best results (produced by GP) are significantly better than those produced by other methods; for the other two modeling tasks, results found by linear regression were almost quite as good as the best ones (trained used ANNs). I.e., for those medical modeling tasks described here, GP performs best among those techniques that are able to identify nonlinearities (ANNs, SVMs, GP). Furthermore, we also see that GP results show less overfitting than those produced using other methods.

On the other hand, the investigated diagnoses of breast cancer, melanoma, and respiratory system cancer can be estimated correctly in up to 81%, 74%, and 91% of the analyzed test cases, respectively; without tumor markers up to 75%, 74%, and 88% of the test samples are correctly estimated, respectively. Linear modeling performs well in all modeling tasks, feature selection using genetic algorithms and nonlinear modeling yield even better results for all analyzed modeling tasks. No modeling method performs best for all diagnosis prediction tasks.

Acknowledgements. The work described in this chapter was done within the Josef Ressel-Centre *Heureka!* for Heuristic Optimization sponsored by the Austrian Research Promotion Agency (FFG).

References

1. Affenzeller, M., Wagner, S.: SASEGASA: A new generic parallel evolutionary algorithm for achieving highest quality results. Journal of Heuristics - Special Issue on New Advances on Parallel Meta-Heuristics for Complex Problems 10, 239–263 (2004)
2. Affenzeller, M., Wagner, S.: Offspring selection: A new self-adaptive selection scheme for genetic algorithms. In: Ribeiro, B., Albrecht, R.F., Dobnikar, A., Pearson, D.W., Steele, N.C. (eds.) Adaptive and Natural Computing Algorithms, Springer Computer Science, pp. 218–221. Springer (2005)
3. Affenzeller, M., Wagner, S., Winkler, S.: Goal-oriented preservation of essential genetic information by offspring selection. In: Proceedings of the Genetic and Evolutionary Computation Conference (GECCO), vol. 2, pp. 1595–1596. Association for Computing Machinery, ACM (2005)
4. Affenzeller, M., Winkler, S., Wagner, S., Beham, A.: Genetic Algorithms and Genetic Programming - Modern Concepts and Practical Applications. Chapman & Hall / CRC (2009)
5. Alba, E., Garca-Nieto, J., Jourdan, L., Talbi, E.G.: Gene selection in cancer classification using PSO/SVM and GA/SVM hybrid algorithms. In: IEEE Congress on Evolutionary Computation 2007, pp. 284–290 (2007)
6. Alberts, B.: Leukocyte functions and percentage breakdown. In: Molecular Biology of the Cell. NCBI Bookshelf (2005)
7. Andriole, G.L., Crawford, E.D., Grubband, R.L., Buys, S.S., Chia, D., Church, T.R., et al.: Mortality results from a randomized prostate-cancer screening trial. New England Journal of Medicine 360(13), 1310–1319 (2009)
8. Ariew, R.: Ockham's Razor: A Historical and Philosophical Analysis of Ockham's Principle of Parsimony. University of Illinois, Champaign-Urbana (1976)

9. Banzhaf, W., Lasarczyk, C.: Genetic programming of an algorithmic chemistry. In: O'Reilly, U., Yu, T., Riolo, R., Worzel, B. (eds.) Genetic Programming Theory and Practice II, pp. 175–190. Ann Arbor (2004)

10. Bitterlich, N., Schneider, J.: Cut-off-independent tumour marker evaluation using ROC approximation. Anticancer Research 27, 4305–4310 (2007)

11. Brown, G.: A new perspective for information theoretic feature selection. In: International Conference on Artificial Intelligence and Statistics, pp. 49–56 (2009)

12. Chang, C.C., Lin, C.J.: LIBSVM: a library for support vector machines (2001), Software available at http://www.csie.ntu.edu.tw/~cjlin/libsvm

13. Cheng, H., Qin, Z., Feng, C., Wang, Y., Li, F.: Conditional mutual information-based feature selection analyzing for synergy and redundancy. Electronics and Telecommunications Research Institute (ETRI) Journal 33(2) (2011)

14. Cover, T.M., Thomas, J.A.: Elements of information theory. Wiley-Interscience, New York (1991)

15. Duch, W.: Feature Extraction: Foundations and Applications. Springer (2006)

16. Duda, R.O., Hart, P.E., Stork, D.G.: Pattern Classification, 2nd edn. Wiley Interscience (2000)

17. Duffy, M.J., Crown, J.: A personalized approach to cancer treatment: how biomarkers can help. Clinical Chemistry 54(11), 1770–1779 (2008)

18. Efroymson, M.A.: Multiple regression analysis. Mathematical Methods for Digital Computers. Wiley (1960)

19. Eiben, A., Smith, J.: Introduction to Evolutionary Computation. Natural Computing Series. Springer, Heidelberg (2003)

20. El Akadi, A., El Ouardighi, A., Aboutajdine, D.: A powerful feature selection approach based on mutual information. International Journal of Computer Science and Network Security 8(4), 116–121 (2008)

21. Fleuret, F.: Fast binary feature selection with conditional mutual information. The Journal of Machine Learning Research 5, 1531–1555 (2004), http://dl.acm.org/citation.cfm?id=1005332.1044711

22. Gold, P., Freedman, S.O.: Demonstration of tumor-specific antigens in human colonic carcinomata by immunological tolerance and absorption techniques. The Journal of Experimental Medicine 121, 439–462 (1965)

23. Hammarstrom, S.: The carcinoembryonic antigen (cea) family: structures, suggested functions and expression in normal and malignant tissues. Seminars in Cancer Biology 9, 67–81 (1999)

24. Holland, J.H.: Adaption in Natural and Artifical Systems. University of Michigan Press (1975)

25. Keshaviah, A., Dellapasqua, S., Rotmensz, N., Lindtner, J., Crivellari, D., et al.: Ca15-3 and alkaline phosphatase as predictors for breast cancer recurrence: a combined analysis of seven international breast cancer study group trials. Annals of Oncology 18(4), 701–708 (2007)

26. Koepke, J.A.: Molecular marker test standardization. Cancer 69, 1578–1581 (1992)

27. Kohavi, R.: A study of cross-validation and bootstrap for accuracy estimation and model selection. In: Proceedings of the 14th International Joint Conference on Artificial Intelligence, vol. 2, pp. 1137–1143. Morgan Kaufmann (1995)

28. Koza, J.R.: Genetic Programming: On the Programming of Computers by Means of Natural Selection. The MIT Press (1992)

29. Kronberger, G.K.: Symbolic regression for knowledge discovery - bloat, overfitting, and variable interaction networks. Ph.D. thesis, Institute for Formal Models and Verification, Johannes Kepler University Linz (2010)

30. LaFleur-Brooks, M.: Exploring Medical Language: A Student-Directed Approach, 7th edn. Mosby Elsevier, St. Louis (2008)
31. Lai, R.S., Chen, C.C., Lee, P.C., Lu, J.Y.: Evaluation of cytokeratin 19 fragment (cyfra 21-1) as a tumor marker in malignant pleural effusion. Japanese Journal of Clinical Oncology 29(9), 421–424 (1999)
32. Langdon, W.B., Poli, R.: Foundations of Genetic Programming. Springer, Heidelberg (2002)
33. Ljung, L.: System Identification – Theory For the User, 2nd edn. PTR Prentice Hall, Upper Saddle River (1999)
34. Maton, A., Hopkins, J., McLaughlin, C.W., Johnson, S., Warner, M.Q., LaHart, D., Wright, J.D.: Human Biology and Health. Prentice Hall, Englewood Cliffs (1993)
35. Meyer, P., Bontempi, G.: On the use of variable complementarity for feature selection in cancer classification. In: Evolutionary Computation and Machine Learning in Bioinformatics, pp. 91–102 (2006)
36. Mizejewski, G.J.: Alpha-fetoprotein structure and function: relevance to isoforms, epitopes, and conformational variants. Experimental Biology and Medicine 226(5), 377–408 (2001)
37. Nelles, O.: Nonlinear System Identification. Springer, Heidelberg (2001)
38. Niv, Y.: Muc1 and colorectal cancer pathophysiology considerations. World Journal of Gastroenterology 14(14), 2139–2141 (2008)
39. Osman, N., O'Leary, N., Mulcahy, E., Barrett, N., Wallis, F., Hickey, K., Gupta, R.: Correlation of serum ca125 with stage, grade and survival of patients with epithelial ovarian cancer at a single centre. Irish Medical Journal 101(8), 245–247 (2008)
40. Rai, A.J., Zhang, Z., Rosenzweig, J., Ming Shih, I., Pham, T., Fung, E.T., Sokoll, L.J., Chan, D.W.: Proteomic approaches to tumor marker discovery. Archives of Pathology & Laboratory Medicine 126(12), 1518–1526 (2002)
41. Rosen, D.G., Wang, L., Atkinson, J.N., Yu, Y., Lu, K.H., Diamandis, E.P., Hellstrom, I., Mok, S.C., Liu, J., Bast, R.C.: Potential markers that complement expression of ca125 in epithelial ovarian cancer. Gynecologic Oncology 99(2), 267–277 (2005)
42. Shannon, C.E.: A mathematical theory of communication. The Bell Systems Technical Journal 27, 379–423 (1948)
43. Tallitsch, R.B., Martini, F., Timmons, M.J.: Human anatomy, 5th edn. Pearson/Benjamin Cummings, San Francisco (2006)
44. Tesmer, M., Estevez, P.A.: Amifs: Adaptive feature selection by using mutual information. In: IEEE International Joint Conference on Neural Networks, vol. 1 (2004)
45. Thompson, I.M., Pauler, D.K., Goodman, P.J., Tangen, C.M., et al.: Prevalence of prostate cancer among men with a prostate-specific antigen level $<= 4.0$ ng per milliliter. New England Journal of Medicine 350(22), 2239–2246 (2004)
46. Vapnik, V.: Statistical Learning Theory. Wiley, New York (1998)
47. Wagner, S.: Heuristic optimization software systems – modeling of heuristic optimization algorithms in the heuristiclab software environment. Ph.D. thesis, Johannes Kepler University Linz (2009)
48. Wagner, S., Affenzeller, M.: SexualGA: Gender-specific selection for genetic algorithms. In: Callaos, N., Lesso, W., Hansen, E. (eds.) Proceedings of the 9th World Multi-Conference on Systemics, Cybernetics and Informatics (WMSCI 2005). International Institute of Informatics and Systemics, vol. 4, pp. 76–81 (2005)
49. Williams, P.W., Gray, H.D.: Gray's anatomy, 37th edn. C. Livingstone, New York (1989)
50. Winkler, S.: Evolutionary system identification - modern concepts and practical applications. Ph.D. thesis, Institute for Formal Models and Verification, Johannes Kepler University Linz (2008)

51. Winkler, S., Affenzeller, M., Jacak, W., Stekel, H.: Classification of tumor marker values using heuristic data mining methods. In: Proceedings of the GECCO 2010 Workshop on Medical Applications of Genetic and Evolutionary Computation, MedGEC 2010 (2010)
52. Winkler, S., Affenzeller, M., Jacak, W., Stekel, H.: Identification of cancer diagnosis estimation models using evolutionary algorithms - a case study for breast cancer, melanoma, and cancer in the respiratory system. In: Proceedings of the Genetic and Evolutionary Computation Conference, GECCO 2010 (2011)
53. Winkler, S., Affenzeller, M., Kronberger, G., Kommenda, M., Wagner, S., Jacak, W., Stekel, H.: Feature selection in the analysis of tumor marker data using evolutionary algorithms. In: Proceedings of the 7th International Mediterranean and Latin American Modelling Multiconference, pp. 1–6 (2010)
54. Winkler, S., Affenzeller, M., Kronberger, G., Kommenda, M., Wagner, S., Jacak, W., Stekel, H.: On the use of estimated tumor marker classifications in tumor diagnosis prediction - a case study for breast cancer. In: Proceedings of 23rd IEEE European Modeling & Simulation Symposium, EMSS 2011 (2011)
55. Yin, B.W., Dnistrian, A., Lloyd, K.O.: Ovarian cancer antigen CA125 is encoded by the MUC16 mucin gene. International Journal of Cancer 98(5), 737–740 (2002)
56. Yonemori, K., Ando, M., Taro, T.S., Katsumata, N., Matsumoto, K., Yamanaka, Y., Kouno, T., Shimizu, C., Fujiwara, Y.: Tumor-marker analysis and verification of prognostic models in patients with cancer of unknown primary, receiving platinum-based combination chemotherapy. Journal of Cancer Research and Clinical Oncology 132(10), 635–642 (2006)
57. Zhong, L., Zhou, X., Wei, K., Yang, X., Ma, C., Zhang, C., Zhang, Z.: Application of serum tumor markers and support vector machine in the diagnosis of oral squamous cell carcinoma. Shanghai Kou Qiang Yi Xue (Shanghai Journal of Stomatology) 17(5), 457–460 (2008)

Chapter 7
Affinity Based Slotting in Warehouses with Dynamic Order Patterns

Monika Kofler, Andreas Beham, Stefan Wagner, and Michael Affenzeller

Abstract. There has been a wealth of research on warehouse optimization since the 1960s, and in particular on increasing order picking efficiency, which is one of the most labor intensive processes in many logistics centers. In the last ten years, affinity based slotting strategies, which place materials that are frequently ordered/picked together close to each other, have started to emerge. However, the effects of changing customer demand patterns on warehousing efficiency have not been investigated in detail. The aim of this chapter is to extend the classic storage location assignment problem (SLAP) to a multi-period formulation (M-SLAP) and to test and compare how various allocation rules, and in particular an affinity based policy, perform in such dynamic scenarios. A first benchmark instance for the M-SLAP is presented.

7.1 Introduction

The storage location assignment problem (SLAP) was first formulated by Hausman, Schwarz and Graves [8], who had identified a need for research on the design and scheduling of automated warehousing systems. Automated warehousing systems were introduced in the 1950s but became increasingly pervasive in the 1970s. Hausman et al. [8] also developed a widely-used taxonomy for assignment policies, distinguishing between dedicated storage, randomized storage and class-based storage, which are described in detail in Section 7.2.

Frazelle [4] has since shown that the SLAP is NP-Hard. Previous studies mainly focused on finding optimal solutions for instances of this combinatorial optimization problem. However, a lack of commonly shared benchmark instances makes comparisons between research results very challenging. Researchers either rely on randomly generated instances or use proprietary data from a real-world warehouse that can not

Monika Kofler · Andreas Beham · Stefan Wagner · Michael Affenzeller
University of Applied Sciences Upper Austria, School of Informatics,
Communication and Media, Softwarepark 11, 4232 Hagenberg, Austria
e-mail: {mkofler,abeham,swagner,maffenze}@heuristiclab.com

R. Klempous et al. (eds.), *Advanced Methods and Applications in Computational Intelligence*,   123
Topics in Intelligent Engineering and Informatics 6,
DOI: 10.1007/978-3-319-01436-4_7, © Springer International Publishing Switzerland 2014

be released to the public. Despite this, the main research efforts in this field are still dominated by the search for best possible solutions. Common SLAP optimization approaches therefore result in assignments that completely change the - usually randomly slotted - initial assignment. If such a solution should be put into practice, it would involve extensive re-arrangements akin to filling a warehouse from scratch. While this is a suitable approach for filling an initially empty warehouse or to determine the potential for improvement, it can not be implemented easily in an already operating warehouse. In a real-world scenario the potential efficiency gains should instead be weighed against the re-arrangement efforts. In this paper we will extend the classic SLAP formulation for multi-period scenarios, thus allowing an evaluation of storage assignment strategies under changing conditions, and use an extended objective function that considers re-arrangement, putaway and order picking efforts.

The chapter is structured as follows: Section 7.2 gives an overview about previous research on slotting, with a focus on random (7.2.1), turn-over based (7.2.2) and affinity based (7.2.3) slotting. Since affinity based slotting strategies are a fairly recent scientific development, a detailed mathematical formulation of the Pick Frequency / Part Affinity (PF/PA) score as one representative of the group is provided in Section 7.2.4. In Section 7.3 we introduce the multi-period storage location problem (M-SLAP) and distinguish between re-warehousing, which involves the re-arrangements of large parts of the warehouse, and healing, which moves only a small number of goods at a time. To test the effectiveness of various slotting strategies on the M-SLAP, a benchmark dataset has been released, which is described in Section 7.4. Test configurations and results are given in Section 7.5, followed by a brief discussion and outlook in Section 7.6.

7.2 Introduction to Slotting

The goal of slotting - also called storage assignment optimization, inventory slotting, or inventory profiling - is to determine the *best* place to store each stock keeping unit (SKU) in a warehouse. Two of the most common incentives for companies to slot their warehouse include the need to squeeze more SKUs into an already overflowing warehouse and the desire to reduce overall handling costs and efforts [2]. In many order picking environments the travel time to retrieve an order has been found to be the largest component of labor, amounting to 50% or more of total order picking time [21]. Therefore this study will focus on the reduction of travel efforts, or more specifically aisle changes, as an approximation of travel time. It should be noted that other factors, such as load balancing accross warehouse zones, work ergonomics (e.g. to reduce bending and reaching activity [19]) or pre-consolidation (to reduce downstream sorting), can also be of importance for a particular scenario.

Generally speaking, slotting is a two-stage process that first assigns a SKU to a product class and afterwards assigns the class to storage locations within the warehouse. Within a class the SKUs are usually arranged via a simple policy such as random or closest location. If there is only a single class, the approach is called random storage, which is discussed in more detail in section 7.2.1. Conversely, if the number of classes equals the number of SKUs, the policy is called dedicated storage. Class-based storage is situated somewhere in between random and dedicated storage. Choosing the right number of classes, product-to-class assignment strategy and storage locations for each class is dependent on the particular warehousing scenario, in particular the layout, material handling equipment, routing strategy and order profile. Detailed reviews of the various storage assignment strategies can be found in [7] and [3]. The primary literature considered in this paper is also summarized in Table 7.1.

Table 7.1. Overview of slotting strategies with an emphasis on the Cube per Order Index (COI) and affinity-based methods such as Order Oriented Slotting (OOS) and Pick Frequency / Part Affinity (PF/PA) Slotting

Type	Year Citation	Method
Random	1996 C. Malmborg [16]	Random vs. dedicated storage
Turnover-based	1963 Heskett [9]	COI
	1976 Kallina and Lynn [10]	COI
Affinity-based	1989 Frazelle and Sharp [5]	Correlated storage
	2005 Garfinkel [6]	Correlated storage
	2007 Mantel, Schuur, and Heragu [17]	OOS
	2008 Kim and Smith [11]	Improving search
	2009 de Ruijter et al. [20]	OOS, Parameter tuning
	2010 Kofler et al. [13]	PF/PA Slotting
	2010 Wutthisirisart [24]	Minimum Delay Algorithm
	2011 Kofler et al. [14]	PF/PA Slotting, Healing
Review Articles	2007 Gu, Goetschalckx, and McGinnis [7]	Various Approaches
	2007 deKoster, Le-Duc, and Roodbergen [3]	Various Approaches

Most of the existing scientific studies present results obtained by completely re-slotting a warehouse according to some assignment strategy, a process commonly referred to as re-warehousing. Migrating from a randomly slotted warehouse to class-based storage, changing the number of classes or the class characteristics in a live operating warehouse is more problematic. Warehouse managers might wish to implement a target assignment gradually during normal operations, rather than interrupt the normal workflow to move hundreds of items. In addition, an optimal storage assignment created by re-warehousing might become outdated before long due to fluctuations in product demand caused by seasonal variations and product life cycle characteristics. In this study, we therefore also distinguish between construction and improvement approaches. Construction algorithms build a feasible slotting from scratch, assuming that the warehouse is initially empty. On the other hand, improvement algorithms try to enhance an existing feasible solution. We will compare the results of both construction and various improvement strategies, distinguishing further between greedy and heuristic search methods. The approaches are summarized in Table 7.1 and described in more detail in subsection 7.2.1 to 7.2.3.

7.2.1 Random Slotting

In the random storage paradigm incoming SKUs are assigned randomly to suitable, available storage locations. The advantages of random storage are ease of implementation and balanced picker traffic across the warehouse. On the downside this may result in longer travel times. Random storage is frequently used in practice and as a performance baseline in the scientific literature.

7.2.2 Slotting by Turnover Based Metrics

Early attempts to optimize slotting in dedicated storage warehouses were based on the idea that fast-moving items should be located in easily accessible pick areas. Heskett extended this simple policy and proposed the cube per order index (COI) rule [9], which ensures that heavy or fast-moving SKUs are stored in more desirable locations close to ground level. Kallina and Lynn discussed the implementation of the COI policy in practice and proved that the COI rule is optimal under certain conditions [10]. One such condition is that there is no dependency between picked items in the same tour, which is unfortunately not the case for most order picking scenarios [7]. Modifications of the COI rule have since been published, which also consider inventory costs, zoning constraints or work ergonomics [16].

7.2.3 Slotting by Affinity

In order picking environments a picker usually retrieves multiple items per order, processing the individual order lines according to some routing strategy. Items that are frequently ordered together are said to be correlated or affine [6]. The idea behind slotting by affinity is that storing affine items close to each other will reduce the total travel time. Unfortunately this is not universally true for all warehouse scenarios and depends on the specific warehouse layout, material handling equipment, picker routing strategy and order profile. Even if two items are ordered together the picker will not necessarily pick them in sequence [22]. Moreover, narrow aisles that do not allow reverse back out or large orders might require a full traversal of the warehouse anyway. In such scenarios storing by affinity could even have a negative effect on picking efficiency by causing congestion in aisles were many fast-moving items are stored together.

Slotting by affinity was first introduced by Frazelle et al., who implemented a class-based storage strategy called *correlated storage* [5]. The algorithm starts with the most popular product and subsequently adds affine items to the class until a capacity constraint is reached. The generated classes are then placed within the warehouse according to their total popularity.

Garfinkel developed a local search improvement heuristic based on 2-exchange and cyclic exchange with the goal to minimize multi-zone orders. The moves are evaluated with different correlation measures and the algorithm is benchmarked against various construction approaches [6]. Kim and Smith introduced a similar two-phase dynamic slotting heuristic procedure that generates an initial assignment using a pick frequency measure such as the COI rule and afterwards uses pairwise interchanges to move affine parts closer to each other [11].

Mantel, Schuur and Heragu created the order oriented slotting (OOS) problem [17]. They present integer linear programming (ILP) models for two small warehouse scenarios but also two heuristics for the optimization of larger, real-world instances. Their so-called *interaction frequency based quadratic assignment heuristic* (IFH-QAP) forces items that frequently occur in the same order to be close together, while at the same time ensuring that fast movers are not allocated too far from the I/O-point. The objective function is based on two measures that are multiplied with the routing specific distances for each SKU: The *popularity* of an SKU denotes how often it was ordered. The *interaction frequency* of two SKUs equals the number of orders that contain them both. The impact of each component on the target function is tuned via a weight parameter, which can be adjusted empirically or automatically as shown in [20]. Kofler et al. devised the pick frequency / part affinity (PF/PA) objective function [13], which extends the IFH-QAP heuristic such that one SKU may be stored in multiple locations throughout the warehouse. Moreover relative values of interaction frequency and popularity are used to get comparable results, independent of the time window that is considered for a given order profile.

Another interesting approach that considers SKU affinity is the Minimum Delay Algorithm (MDA) [24], which was inspired by linear placement algorithms. In this construction heuristic approach SKUs are placed in a fashion that reduces the delay (= additional traveling distance) for the other orders.

Due to the similarities between [6], [11], [17] and [13] the PF/PA score is used exemplarily in this study as a representative of improvement heuristics for affinity-based slotting. A detailed algorithm description is given in the following section.

7.2.4 Pick Frequency / Part Affinity Score

The pick frequency / part affinity (PF/PA) score was first introduced in [13] and slightly revised in [14]. The approach combines affinity based storage and storage by pick frequency. To increase comprehensibility and reproducibility of the results presented in this study, we provide a full mathematical formulation of the score, as previously published in [14]. Let's assume:

$$S = \text{set of all storage locations } s_k;$$
$$0 < k <= m \text{ where } m = |S| \tag{7.1}$$

$$dist(s_k, s_l) = \text{routing specific distance between} \tag{7.2}$$
$$\text{storage location } s_k \text{ and } s_l$$

In addition to the storage locations that can hold products, at least one input/output location *origin* must be defined that denotes the shipping dock. The travel times / distances between locations are routing-specific.

Usually, the storage locations and the distance matrix need to be determined only once for a given warehouse and can later be re-used for different problem instances. Conversely, the following parameters are likely to change over time and need to be retrieved from the enterprise resource planning or warehouse management system. Most important, the set P lists all products that are present in a particular assignment.

$$P = \text{set of all products } p_i; \tag{7.3}$$
$$0 < i <= n \text{ where } n = |P|$$

For each product p_i we need to know the number of picking orders in which the product occurs. We use the relative number of orders $orderRatio(p_i)$ to get comparable scores as results, independent of the time window that is considered from a given order profile. Similarly, the affinity matrix stores the ratio of all orders in which a particular product pair occurs together. Finally, the current warehouse assignment defines how many products p_i are stored at location s_k. The set of locations $L(p_i)$ stores all locations of a particular product.

$$orderRatio(p_i) = \text{relative number of orders in} \tag{7.4}$$
$$\text{which } p_i \text{ occurs}$$

$$affinity(p_i, p_j) = \text{relative number of orders in which} \tag{7.5}$$
$$\text{products } p_i \text{ and } p_j \text{ occur together}$$

$$quantity(p_i, s_k) = \text{number of packing units of} \tag{7.6}$$
$$\text{product } p_i \text{ stored at location } s_k$$

$$L(p_i) = \text{set of all locations where } quantity(p_i, s) > 0 \tag{7.7}$$

The entities defined in 7.3-7.7 can be calculated from order picking histories or demand forecasts and the current warehouse assignment. We can now define the objective functions in Equation 7.8 and 7.9.

$$PF = \sum_{i=0}^{n} \frac{orderRatio(p_i)}{|L(p_i)|} \cdot \sum_{s \in L(p_i)} dist(s, origin) \qquad (7.8)$$

The *total pick frequency score* (PF), as defined in Equation 7.8, ensures that frequently picked products are placed in more favorable storage locations near the I/O point. For each product it detects all current storage locations $L(p_i)$, calculates their distance to the origin and weights each distance with the expected number of picks given the number of previous orders. The picks are uniformly distributed on all storage locations, independent of the actual stored quantities in the different locations.

$$PA = \sum_{i=0}^{n} \sum_{j=0}^{n} \frac{affinity(p_i, p_j)}{|L(p_i)| \cdot |L(p_j)|}$$
$$\cdot \sum_{s_k \in L(p_i)} \sum_{s_l \in L(p_j)} dist(s_k, s_l) \qquad (7.9)$$

The *total part affinity score* (PA) takes all pairs of products p_i and p_j, and retrieves all respective storage locations $L(p_i)$ and $L(p_j)$ from the current assignment. The distance between each resulting storage location pair is calculated and weighted with the part affinity divided by the number of location pairs $|L(p_i)| \cdot |L(p_j)|$. The term reduces to zero for products with no part affinity, therefore the calculation can be sped up by only looking at products p_j that have an affinity greater than zero with a given product p_i.

The resulting multi-objective evaluation function for assignments is computed as weighted sum of the two objective functions given in Equation 7.8 and 7.9 such that

$$\text{PF/PA score} = \alpha \cdot PF + \beta \cdot PA. \qquad (7.10)$$

PF and PA have different ranges. The parameters α and β can be adjusted automatically ensure that both factors contribute equally to the objective as proposed in [20]. We prefer tuning them manually, usually assigning an intentionally higher weight to α as discussed in [14].

7.3   Multi-period Warehouse Slotting

Both turn-over and affinity based slotting strategies rely on historical SKU order profiles and/or demand forecasts for decision-making. However, slotting is usually not a one-off event since SKU demands are subject to change over time. An optimal storage assignment might become outdated due to demand fluctuations, modifications to the picking line, infrastructure changes, or variations in the order mix etc. We therefore propose an extension of the classical SLAP, called the multi-period storage location assignment problem (M-SLAP).

The M-SLAP problem formulation was inspired by the field of facility layout problems (FLP), where researchers face a similar challenge: Changing material flows between departments during the planning horizon might make re-arrangements of the departments necessary, in order to keep material handling costs low. In the so-called dynamic FLP formulation the planning horizon is split into multiple consecutive periods. Within each period, the estimated flow data is assumed to be constant. The optimization generates a separate layout for each of these periods, which can be treated as classic, static FLPs. However, re-arrangement costs caused by layout adjustments between periods of different material flows are considered in the evaluation of the dynamic layout. An overview of dynamic layout algorithms can be found in [1].

Likewise, the M-SLAP problem consists of a sequence of warehouse assignments and their associated SKU demand profiles. To evaluate the quality of an M-SLAP solution the picking effort for a set of picking orders is simulated on each generated assignment. The objective is to minimize the total travel distance of all pickers, caused by order picking and (re-)slotting. The aim of this new formulation is to evaluate mid- and long-term stability of warehouse assignments generated by various slotting approaches. Most important, it should yield a better cost-benefit analysis of re-slotting effort (= cost) vs. order picking savings (= benefit).

7.3.1 Re-warehousing

Most previous studies focused on achieving optimality in the allocation of SKUs and often re-ordered the existing layout to a very large degree. This process is commonly referred to as re-warehousing. The effort for re-warehousing a real-world logistics center is considerable and requires the movement of hundreds or thousands of items, thus blocking personnel and material handling equipment for an extensive time period. This is why it is conducted infrequently (quarterly, biannually, during holidays) in practice.

The only study known to us that considers re-warehousing over time was conducted by Neuhäuser and Wehking [18]. In this publication they developed a metric to determine suitable re-warehousing intervals for the food retail sector, which is characterized by strong seasonal demand and stock fluctuations. The authors assume that the optimal target storage zone or class is known for each stock keeping unit at each point in time, for instance based on COI, and calculate the (weighted) sum of displaced SKUs over all SKUs as a metric of warehouse entropy. By comparing the cost of operation in warehouses with high and low entropy and also considering the re-warehousing costs, they were able to determine suitable re-warehousing intervals for their scenario in a simulation study. Unfortunately, data and simulation model are not publicly available, which makes a comparison with the other approaches or replication of the results difficult.

7.3.2   Healing

From a warehouse manager's perspective, achieving 'optimality' might not be as important as finding a 'good enough' slotting that is both robust and requires only few stock transfers to implement. For instance, experiments in [14] showed that moving only a limited number of SKUs can significantly reduce the total travel distances. In this previous study picker travel distances could be reduced by 23% by moving only 60 pallets. The total distance optimization potential for the problem instance was over 60%, however this required the movement of 1400 pallets. For many real-world scenarios it might be more prudent to move only a few SKUs per day, but iteratively over a longer time period. We refer to this process as *healing* [14].

7.3.3   M-SLAP: Optimization and Evaluation

The typical M-SLAP optimization steps are depicted in Figure 7.1 for a single period. The result of this chain is a new warehouse assignment, which is used as input for the next period, plus cost estimates for re-arrangements, order picking and putaway effort. The total costs of an M-SLAP is the sum over all costs over all periods.

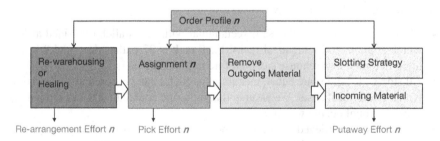

Fig. 7.1. The M-SLAP optimization process for a single period

We explain the individual steps in more detail, starting with the first period:

1. Re-warehousing or healing: This step is optional. In the first period, a warehouse assignment can either be retrieved from a real-world warehouse management system or generated via a heuristic (=re-warehousing). In each subsequent period, either re-warehousing, healing or no action can be conducted. If re-warehousing/healing is performed the re-arrangement effort counts towards the total costs.
2. Assignment 1: The assignment is then evaluated with the test order profile of period 1, resulting in a cost estimate for the pick process in this period.

3. Remove outgoing material: The test order profile should be representative for a period but does not have to include all pick operations that take place within the period. It is therefore assumed that all SKUs that are present in period 1 but not in stock in period 2 will be removed in one step. This is one of the abstractions of the M-SLAP. However, if the temporal spacing of the periods is very small, such as a day or an hour, this error becomes negligible. The removal has no associated costs since they are already considered in the picking process.
4. Likewise, all incoming SKUs that are newly available in period 2 will be added in one step. A slotting strategy is used to assign the new SKUs to locations, which can use the order profile as input to calculate intra-period turn-over rates or PF/PA scores. The result of this step is a new assignment, which is fed into period 2. The cycle starts over at step 1.

7.4   M-SLAP Benchmark Data

A benchmark instance for the M-SLAP was generated using anonymized data from the logistics center of an Austrian company in the automotive sector. The data was taken from the high-rack pallet warehouse, which is operated with man-to-goods order picking. In order to comply to confidentiality agreements and to make the dataset more suitable as a benchmark, we simplified the data and environmental constraints in the following way:

- We assume that the warehouse is rectangular with 12 parallel, two-sided aisles, which amounts to 24 identical racks. Each rack is 27 pallets deep and 9 levels high. In total the warehouse can therefore hold 5832 pallets.
- All storage locations are equal, meaning that SKUs can be stored in any location without additional costs. We assume that only one container size (a euro pallet) is used throughout the warehouse.
- The SKUs are retrieved from the warehouse with fork lifts. Aisle changes are the most time-consuming step for this particular scenario, therefore the target cost value is the number of aisle changes required to pick the orders, slot incoming material or perform re-arrangement operations. We assume that all products within one aisle are equally easy to reach.
- Each SKU is stored in only one storage location, independent of the stored amount. There is no mixed storage, meaning that SKUs cannot share a storage location.
- Storage locations can be empty. Due to inventory changes not all SKUs are in stock in all periods. However, it is assumed, that all SKUs that are requested in the test orders of are particular period are in stock in sufficient quantities.
- There is an interim storage area, where incoming/outgoing SKUs are placed, which is located at the head of the first aisle. The pickers always start from and return to the interim storage area.

- One order is picked by one picker (no zoning) in one go (no order splitting or batching).
- Pickers process the individual positions of an order ascending by aisle index.

The M-SLAP benchmark instance consists of five periods, each of which provides an initial (random) storage assignment and a 1-month demand forecast. The entire data set spans one year from December 2010 until December 2011. Over the year, almost 12,000 different SKUs are stored in the warehouse, however in any given period only 4,300 to 5,200 pallets are in stock. A detailed overview of the stock movements is given in Table 7.2 and Figure 7.2 also illustrates the fluctuating inventory levels.

Table 7.2. Inventory levels and incoming/outgoing SKUs between the periods

	Inventory	Outgoing	Incoming
DEC10	4708	-1316	+1696
MAR11	5088	-1199	+1417
JUN11	5306	-1171	+1080
AUG11	5215	-1061	+1199
DEC11	5353		

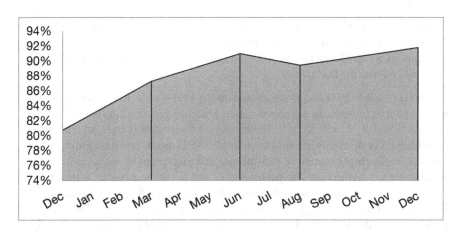

Fig. 7.2. Illustration of the inventory levels per period in the published M-SLAP benchmark instance

The M-SLAP benchmark instance and best known solutions will be published at http://dev.heuristiclab.com/trac/hl/core/wiki/AdditionalMaterial.

7.5 Experimental Setup and Results

7.5.1 Algorithms

All experiments were conducted using HeuristicLab [23], an open source framework for heuristic optimization. We implemented the random and cube per order index (COI) construction methods. In addition, we employed first improvement local search and simulated annealing [12] to optimize a randomly generated initial warehouse configuration subject to the PF/PA objective.

Simulated annealing (SA) is a metaheuristic that was modeled after the annealing process in metallurgy. One of the advantages of simulated annealing is that it offers a strategy to escape from local optima by employing a temperature parameter to guide the search. Contrary to greedy search techniques, which only accept moves that improve the fitness of a solution, SA accepts *uphill* moves with a certain probability. As the algorithm proceeds, a step-wise reduction of the temperature according to a pre-defined cooling scheme reduces the likelihood that bad moves get accepted. The simulated annealing algorithm runs were all configured in the following way: Exponential annealing scheme, start temperature 100, end temperature 1E-06, 4 million iterations.

The same random initial solution was used for all improvement algorithms to make the results comparable. The algorithms could be configured with two different move generators:

- Random Swap Move Generator: Generates one 2-swap move that switches the content of two random locations in the warehouse.
- Sampling Swap Move Generator: Generates n 2-swap moves, sorts them by quality and returns the best move.

We also varied the α and β parameters in the PF/PA objective function. Setting $\alpha = 1$ and $\beta = 0$ results in a turnover-based objective function similar to COI. Setting $\alpha = 0$ and $\beta = 1$ results in an optimization of part affinity only, which reduces the within-order distances but does not consider SKU turnover rates. Finding a good trade-off between placement by part affinity and placement by retrieval frequency is unfortunately not trivial. The parameters α and β can be adjusted automatically to ensure that both factors contribute equally to the objective as proposed in [20]. However, we found that empirically sampling the parameters produced better results as described in detail in [13]. We fixed $\beta = 1$ and conducted tests for $\alpha \in \{1, 2, 3, \ldots 20, 30, 40, 50, 60\ 70, 80, 90, 100\}$. For the investigated scenario a setting of $\alpha = 10$ and $\beta = 1$ was found to be most effective. All optimization runs were conducted in a high performance computing environment on an 8-core machine with 2x Intel Xeon CPU, 2.5 Ghz and 32GB memory.

7.5.2   Results

7.5.2.1   Re-warehousing

First we employed re-warehousing on each period separately to assess the maximum optimization potential. As shown in Table 7.3 and Figure 7.3, optimizing by part affinity alone (PA) did not produce very good results, only reducing the total travel distance by 11% to 21%, depending on the observed period, compared to an initial random slotting. Although part affinity slotting optimizes the number of aisle changes within an order, the placement relative to the shipping dock is not optimized. Conversely, the optimization potential for turn-over based or mixed approaches is quite large and reduced the number of aisle changes in picking by up to 97%.

Table 7.3. Re-warehousing with different slotting strategies can reduce the pick effort (given in aisle changes) significantly

	Random	PA	COI	PF	PF/PA
Dec 10	9,878	8,746	218	248	214
Mar 11	17,024	13,386	452	474	440
Jun 11	12,672	10,500	410	454	416
Aug 11	15,138	12,080	372	402	352
Dec 11	16,596	14,030	602	618	572

The COI construction heuristic and simulated annealing with pick frequency (PF) as target performed equally well, which is not surprising, since no weight information was available for the data set. In this case, slotting by COI and slotting by pick frequency reduce to the same objective function. A properly parameterized simulated annealing run with the pick frequency objective therefore converged towards the quality obtained with the COI construction heuristic. The combined PF/PA slotting performed marginally better than turnover based slotting (PF, COI) in all periods except one.

Although the improvements seem vast, all of the above approaches require a complete re-warehousing, meaning that no SKU is on its initial place in the target assignment. Table 7.4 lists the re-warehousing efforts to move all SKUs from their initial positions in the random slotting scenario to the respective target assignments, as generated by the different approaches. The re-warehousing effort assessment was conducted under idealized conditions, assuming that the two forklifts work in tandem and buffer the pallets in the intermediate storage area. Each forklift retrieves a pallet that needs to be moved to a different aisle, brings it to the intermediate storage area, picks up a waiting pallet and moves it to its target aisle, where - it is assumed - the next pallet is already waiting for transport to the intermediate storage area. Thus the fork lifts never run empty.

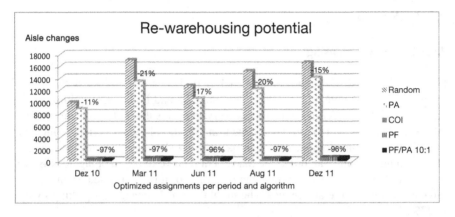

Fig. 7.3. Re-warehousing potential in each period: The number of picker aisle changes can be reduced by up to 97% compared to the initial random slotting, if a complete re-arrangement is conducted. (Data from Table 7.3)

Table 7.4. Number of aisle changes required to re-arrange the assignments from table 7.3

	PA	COI	PF	PF/PA
Dec 10	75,506	57,786	75,436	75,734
Mar 11	81,612	63,594	81,944	81,716
Jun 11	85,078	67,646	86,030	85,558
Aug 11	83,326	65,298	83,930	84,158
Dec 11	86,022	68,166	85,618	86,262

Table 7.4 shows that re-warehousing with PF, PA and PF/PA require approximately the same amount of effort in terms of aisle changes to realize the new assignment. Realizing a COI slotting is 'cheaper' but nevertheless the involved effort cannot be redeemed quickly. In each period we utilize an order picking preview of roughly one month. If we interpolate this data, it would take roughly five months for a complete COI re-warehousing to pay off - that is, if the order profiles do not change in the meantime. In the Section 7.5.2.2 we will therefore discuss a more efficient way to increase order picking efficiency by simply switching to a different putaway strategy for incoming SKUs.

7.5.2.2 Putaway

Since a complete re-warehousing is very time-consuming, the second set of experiments focused on the effect that different putaway strategies for incoming goods have on an existing layout. We used the assignments generated with random, COI or PF/PA slotting for the first period in Section 7.5.2.1 as starting point and observed how the warehouse performance develops when newly arriving SKUs are slotted randomly or according to COI. In these experiments we do not only consider the order picking efforts but also the putaway efforts for newly arriving SKUs between two periods. Re-warehousing efforts, on the other hand, are not considered, because they have already been investigated in the previous section.

Table 7.5. How the quality (= number of aisle changes) of an initial random, COI or PF/PA slotted warehouse develops over time when random or COI slotting strategies are used for incoming material

	Rand-Rand	Rand-COI	COI-Rand	COI-COI	PF/PA-Rand	PF/PA-COI
DEC 10	9,878	9,878	218	218	214	214
PUT1	18,716	12,214	24,766	18,612	20,920	14,890
MAR 11	17,868	13,642	15,058	6,578	12,782	6,996
PUT2	14,978	14,284	17,344	16,182	17,412	16,944
JUN 11	13,408	10,308	11,404	5,582	10,502	5,592
PUT3	11,860	8,998	12,420	9,690	13,166	10,820
AUG 11	14,868	10,654	12,076	5,696	11,422	5,802
PUT4	11,774	11,504	12,750	12,638	14,626	14,580
DEC 11	15,370	11,826	13,648	8,208	14,208	8,018
Total	128,720	103,308	119,684	83,404	115,256	83,860

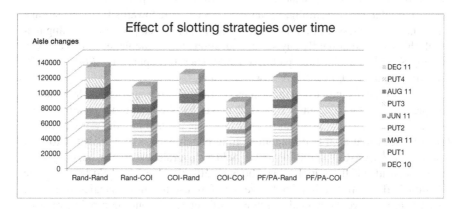

Fig. 7.4. Effects of putaway slotting strategies on an initial assignment (Illustration of Table 7.5)

Table 7.5 and Figure 7.4 summarize the results and illustrate how random slotting can defragment a warehouse after a while. For instance, in December 2010 almost 10,000 aisle changes are required to fulfill the demand forecast on a randomly slotted warehouse but less than 220 on a recently PF/PA or COI slotted warehouse. However after one year of random putaway for incoming SKUs the effect of re-warehousing has almost evaporated for the COI-Rand and PF/PA-Rand scenarios. The total annual number of aisle changes for these scenarios are 115,256 and 119,684, which amounts to a relative improvement of 10.5% and 7% compared to Rand-Rand. Most importantly, the initial re-warehousing efforts (cf. Section 7.5.2.1, Table 7.4) are not yet included in these numbers. If these are considered as well, COI-Rand and PF/PA-Rand actually perform worse than Rand-Rand. The conclusion is, that re-warehousing once per year does not pay off for this M-SLAP benchmark if random slotting is employed for putaway during the year.

Before the second set of results with COI as putaway stategy is discussed, we would like to point out a singularity of the test data set. We claim that order picking is the most effort-intensive process in the warehouse, yet the putaway effort in Table 7.5 is frequently larger than the picking effort. This can be easily explained: As discussed in Section 7.4 the available benchmark data only provides a 1-month demand forecast for each period and not all picking orders between the snapshots. The intervals between periods range between two and four months, therefore the pick efforts are roughly twice to four times as large in reality.

Figure 7.4 illustrates the same results as a graph, showing pick and putaway efforts in turns. One interesting result that the graph shows very well is that the putaway efforts between period 1 and period 2 (PUT1) are much larger for well slotted warehouses than for a randomly slotted warehouse. This can be explained in the following way: Both the COI construction heuristic and the PF/PA slotting algorithm will result in assignments where all the best locations are occupied with SKUs. An illustration of this effect can be seen in Figure 7.5 for a warehouse where the effort is calculated as distance (in meters) from the shipping dock.

We consider this overfitting, because all SKUs drift towards the good storage locations, and the less desirable storage locations are left empty. For incoming material, this means that hardly any storage locations are available, even if the newly arriving SKU has a very high pick frequency. Therefore, putaway efforts are very large directly after COI or PF/PA re-warehousing.

Figure 7.6 also summarizes the results of Table 7.5 but aggregates the putaway and pick efforts. Once again, it should be noted that the total order picking efforts over the entire year are larger since only forecasts for five months were available. It can be seen that an initial random slotting benefits most from a switch to COI slotting. Also, COI-COI and PF/PA-COI yield the overall best results, but one would have to consider if the initial re-warehousing efforts required would be worth the additional efforts. In Section 7.5.2.3 we therefore use healing strategies as another measure to improve the annual warehouse efficiency without the prohibitive costs of re-warehousing.

7.5.2.3 Healing

In the last set of experiments we investigated how conducting a small number of healing moves at the beginning of a period can impact the total warehouse performance. The goal of this additional step is to address changing demand patterns and re-slot SKUs already present in the warehouse according to the demand forecast

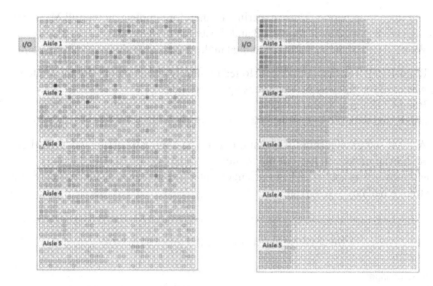

Fig. 7.5. In this illustration, the storage locations in a bird's eye view of a warehouse are colored according to the pick frequency of the assigned SKUs. The white locations indicate empty storage locations or locations with SKUs that are never ordered. An initial random slotting (left) is compared to an (turnover) optimized slotting (right). In the optimized assignment, all the closest locations to the shipping dock in the upper left are occupied.

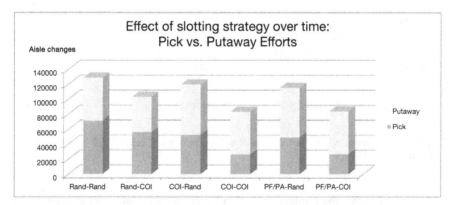

Fig. 7.6. Effects of putaway slotting strategies on an initial assignment: Putaway vs. pick efforts (Illustration of Table 7.5)

for the upcoming period. These experiments therefore implement the full M-SLAP process as described in Section 7.3.3 and illustrated in Figure 7.1.

We used two of the scenarios presented in the previous section as a starting point:

- Rand-Rand: The first period was slotted randomly and a random slotting strategy was used for putaway between subsequent periods.
- Rand-COI: Once again, the first period was slotted randomly, but putaway was conducted with the COI strategy between periods.

A local search algorithm was started at the beginning of period 2-5 to find 50 good 2-swap moves. In each of the 50 iterations the best of 10,000 randomly generated moves was applied. The re-arrangement effort was once again considered in the total efficiency assessment.

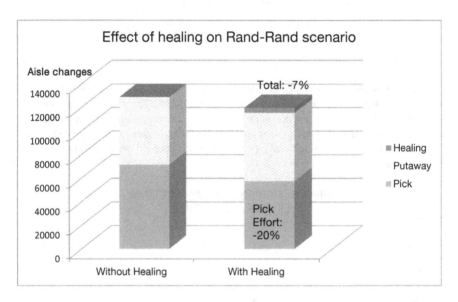

Fig. 7.7. Warehousing efforts for the Rand-Rand scenario from Section 7.5.2.2 without (left) and with healing (right)

Figure 7.7 shows the results for the Rand-Rand scenario with and without healing. It can be seen that healing reduced the number of aisle changes in order picking by 20%. However, the total improvement is only 7% compared to a scenario without healing, because the cleanup swaps require some effort themselves and the putaway effort rises slightly. This can be attributed to the overfitting phenomenon described in the previous section: Due to healing fewer 'good' storage locations will be available for putaway.

This effect is even more pronounced in the Rand-COI scenario as depicted in Figure 7.8. Once again, the order picking effort could be significantly reduced (-19%) compared to the same scenario without healing. However, simultaneously the

putaway effort increases by 13% and when the healing effort is considered the total
gain amounts to less than 1%.

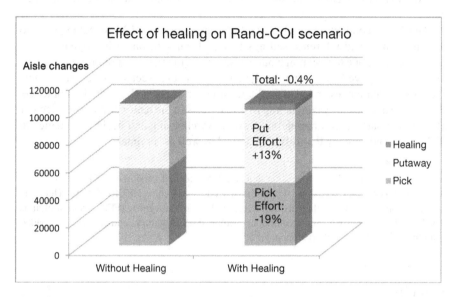

Fig. 7.8. Warehousing efforts for the Rand-COI scenario from Section 7.5.2.2 without (left)
and with healing (right)

The presented experiments are not extensive enough to offer guidance for the
'best' M-SLAP strategy, not even for the given scenario. However, they illustrate
clearly that focusing solely on pick efficiency improvements offers a very biased
warehouse efficiency assessment, especially if real-world warehouses should be optimized during on-going operations.

7.6   Conclusion and Outlook

In this paper we introduced a multi-period formulation of the classic storage location assignment, called M-SLAP, and released a first benchmark instance for the
problem. The M-SLAP consists of multiple planning periods with different inventory levels and SKU demand patterns. A new objective function does not only assess
order picking efforts but also considers re-arrangement and putaway efforts over all
periods. Some preliminary results for the benchmark instance were given, illustrat-
ing how different initial slottings (random, COI and PF/PA) develop over time given
a random or COI putaway strategy. Moreover, the impact of conducting a small
number of healing moves in each period was investigated. First results indicate that
considering only order picking efforts is a very limited view on the problem. Moreover, the dynamic change of demand patterns over time makes it even more crucial

to consider re-arrangement efforts and putaway efforts. In dynamic warehousing scenarios an 'optimal' slotting can degenerate very quickly, therefore the effort invested into re-arrangements must be weighed against both robustness and efficiency gains.

We plan to conduct comprehensive tests on the presented M-SLAP instance to find algorithms and reference settings that perform well on such a dynamic scenario. We also hope to encourage other researchers to model and optimize M-SLAP instances, since we believe that the dynamic multi-stage scenario offers many new algorithmic challenges and also facilitates transfer of research results into practice. Finally, we are currently conducting M-SLAP tests in the real-world logistics center of a project partner, combining putaway, re-warehousing and healing strategies to find a balanced strategy that is both practical as well as 'optimal'.

Acknowledgements. This paper is an updated and extended version of [15]. The work described in this chapter was done within the Josef Ressel-Centre HEUREKA! for Heuristic Optimization sponsored by the Austrian Research Promotion Agency (FFG). HeuristicLab is developed by the Heuristic and Evolutionary Algorithm Laboratory[1] and can be downloaded from the official HeuristicLab homepage[2].

References

1. Balakrishnan, J., Cheng, C.: Dynamic layout algorithms: a state-of-the-art survey. Omega 26(4), 507–521 (1998)
2. Bartholdi, J.J., Hackman, S.T.: Warehouse and distribution science (2010), Textbook available at http://www.warehouse-science.com (accessed September 22, 2012)
3. de Koster, R., Le-Duc, T., Roodbergen, K.J.: Design and control of warehouse order picking: A literature review. Eur. J. Oper. Res. 182, 481–501 (2007)
4. Frazelle, E.: Stock location assignment and order picking productivity. PhD thesis, Georgia Institue of Technology (1990)
5. Frazelle, E., Sharp, G.: Correlated assignment strategy can improve order-picking operation. Ind. Eng. 4, 33–37 (1989)
6. Garfinkel, M.: Minimizing multi-zone orders in the correlated storage assignment problem. PhD thesis, School of Industrial and Systems Engineering, Georgia Institute of Technology (2005)
7. Gu, J., Goetschalckx, M., McGinnis, L.F.: Research on warehouse operation: A comprehensive review. Eur. J. Oper. Res. 177(1), 1–21 (2007)
8. Hausman, W., Schwarz, L., Graves, S.: Optimal storage assignment in automatic warehousing systems. Manage Sci. 22(6), 629–638 (1976)
9. Heskett, J.: Cube-per-order index - a key to warehouse stock location. Transport Distrib. Manage 1963, 27–31 (1963)
10. Kallina, C., Lynn, J.: Application of the cube-per-order index rule for stock location in a distribution warehouse. Interfaces 7(1), 37–46 (1976)

[1] http://heal.heuristiclab.com/

[2] http://dev.heuristiclab.com/

11. Kim, B., Smith, J.: Dynamic slotting for zone-based distribution center picking operation. In: 10th International Material Handling Research Colloquium, Dortmund, Germany, pp. 577–599 (2008)
12. Kirkpatrick, S., Gelatt, C.D., Vecchi, M.P.: Optimization by simulated annealing. Science 220, 671–680 (1983)
13. Kofler, M., Beham, A., Wagner, S., Affenzeller, M., Reitinger, C.: Rassigning storage locations in a warehouse to optimize the order picking process. In: Proceedings of the 22th European Modeling and Simulation Symposium (EMSS 2010), Fez, Morocco (2010)
14. Kofler, M., Beham, A., Wagner, S., Affenzeller, M., Achleitner, W.: Re-warehousing vs. healing: Strategies for warehouse storage location assignment. In: Proceedings of the IEEE 3rd International Symposium on Logistics and Industrial Informatics (Lindi 2011), Budapest, Hungary, pp. 77–82 (2011)
15. Kofler, M., Beham, A., Wagner, S., Affenzeller, M., Achleitner, W.: The multi-period storage location assignment problem. In: Proceedings of IEEE APCAST 2012 Conference, Sydney, Australia, pp. 38–41 (2012)
16. Malmborg, C.: Storage assignment policy tradeoffs. Int. J. Prod. Res. 33, 989–1002 (1996)
17. Mantel, R., Schuur, P., Heragu, S.: Order oriented slotting: a new assignment strategy for warehouses. Eur. J. Ind. Eng. 1(3), 301–316 (2007)
18. Neuhäuser, D., Wehking, K.H.: Der Lagerorganisationsgrad als Steuerungsgröße für optimale Reorganisationszyklen in Kommissioniersystemen. Logist J. Proc. 7, 1–11 (2011), doi:10.2195/LJ_proc_neuhaeuser_de_201108_01
19. Petersen, C.G., Siu, C., Heiser, D.R.: Improving order picking performance utilizing slotting and golden zone storage. Int. J. of Oper. Prod. Manage 25(10), 997–1012 (2005)
20. de Ruijter, H., Schuur, P.C., Mantel, R.J., Heragu, S.S.: Order oriented slotting and the effect of order batching for the practical case of a book distribution center. In: Proceedings of the 2009 International Conference on Value Chain Sustainability, Louisville, Kentucky (2009)
21. Tompkins, J., White, J., Bozer, Y., Frazelle, E., Tanchoco, J., Trevino, J.: Facilities planning. Wiley, New York (1996)
22. Waescher, G.: Supply chain management and reverse logistics. In: Order Picking: A Survey of Planning Problems and Methods, pp. 323–347. Springer, Heidelberg (2004)
23. Wagner, S.: Heuristic optimization software systems - modeling of heuristic optimization algorithms in the heuristiclab software environment. PhD thesis, Institute for Formal Models and Verification, Johannes Kepler University, Linz, Austria (2009)
24. Wutthisirisart, P.: Relation-based item slotting. Master's thesis, University of Missouri (2010)

Chapter 8
Technological Infrastructure and Business Intelligence Strategies for the EDEVITALZH eHealth Delivery System

M.A. Pérez-del-Pino, P. García-Báez,
J.M. Martínez-García, and C.P. Suárez-Araujo

Abstract. In the recent years, the sociological importance of the elderly has grown significantly because of the increase of the prevalence of degenerative disorders, among which Mild Cognitive Impairment (MCI), Alzheimer's Disease (AD) and other cortical dementias should be highlighted. Using actual diagnostic criteria, by the time a patient is diagnosed with AD, the pathology has spread itself during several years. Besides, the unique certain AD diagnosis can only be performed *post-mortem*. Thus, it becomes absolutely necessary to accomplish early detection and MCI diagnosis. This chapter presents the advances achieved in the fields of Medical Informatics and Telemedicine by means of the EDEVITALZH eHealth Delivery System. On one hand, and of remarkable importance to succeed in a certain diagnosis, a standardization of the clinical protocol is proposed and digitally implemented to provide clinicians with the adequate procedures, methods and tools to perform more accurate early and differential diagnosis, and prognosis of patients affected by these neuropathologies. On the other hand, EDEVITALZH technological infrastructure is described, regarding computer architecture, databases and software engineering, with special focus on the embedded mechanisms that allow integration between EDEVITALZH Core components and the Intelligent Systems for Diagnosis (ISD) and the Intelligent Decision Support Tools (IDST), providing Computational Intelligence to the described virtual clinical environment. These integrations upgrade EDEVITALZH to an intelligent Clinical Workstation, being able to aid clinicians in decision making for early and differential diagnosis, prognosis and in performing evolution studies of patients and their pathologies, no matter which healthcare level patients are being assisted in, Primary or Specialized.

M.A. Pérez-del-Pino · J.M. Martínez-García · C.P. Suárez-Araujo
Institute for Cybernetics, University of Las Palmas de Gran Canaria,
Las Palmas de Gran Canaria, Canary Islands, Spain
e-mail: cpsuarez@dis.ulpgc.es, miguel.perez107@alu.ulpgc.es

P. García-Báez
Dept. of Statistics, Operative Research and Computation, University of La Laguna,
Santa Cruz de Tenerife, Canary Islands, Spain
e-mail: patricio@etsii.ull.es

R. Klempous et al. (eds.), *Advanced Methods and Applications in Computational Intelligence*, 145
Topics in Intelligent Engineering and Informatics 6,
DOI: 10.1007/978-3-319-01436-4_8, © Springer International Publishing Switzerland 2014

8.1   Introduction

The population of the developed occidental countries is suffering a remarkable aging process with a clear increasing tendency in the elderly census. Several research studies reveal an extraordinary increase of people's life expectancy and a significative decrease of the birth rate and of child mortality, [1]. In the recent years, the sociological importance of the elderly has grown significantly, mostly due to the increase of the prevalence of degenerative diseases and their secondary functional disorders, causing severe socio-sanitary consequences. Alzheimer's Disease (AD) must be pointed out among aging-associated pathologies; a slow evolution neurodegenerative disorder featured by the loss of memory, orientation and language skills. AD, Vascular Dementia (VD) as well as Mild Cognitive Impairment (MCI), both of high prevalence too, could be considered as cognitive disorders in the aging process of a healthy brain. These disorders make up the most interesting group, estimated to affect between 5%-10% of people aging more than 65, because of the damages they cause to the patient's family, and to his/her caregiver.

Currently-used diagnostic procedures are, by excellence, clinical, hispathological and imaging ones, becoming very complex to perform a correct differentiation between dementias. In the particular case of AD, the unique certain diagnosis, and even when knowledge about these disorders is rising in such a way never seen before, nowadays the *ante-mortem* diagnosis of probable AD is confirmed *post-mortem* in 80%-90% of cases only in specialized centers, estimating the existence of a high level of underdiagnosis, reaching 95% of the cases in normal conditions. The International Working Group for New Research Criteria for the Diagnosis of Alzheimer's Disease (IWGNRCDAD) has proposed a new terminology which considers symptomatology and the behavioral-cognitive evolution as well as the possibility of complementing diagnosis by employing protein biomarkers [2], being able to arise several diagnostic cathegories, even years before the disease has started to manifest symptomatically. On the other hand, several handicaps emerge in the way to accomplish a certain diagnosis of these disorders. In first place, the spread spectrum of usual diagnostic criteria (CAMDEX, DSM-IV, CIE-10) and the very limited coincidence between them [1], not higher than 5%, if taken as a set (not yet having specific clinical valid criteria for each dementia). Second, the so-cited clinical-pathological duality. It is estimated that, by the time a patient is diagnosed with AD, the disease has spread itself during several years.

Geriatric and neurological valorations should be diagnostic instruments to analyze biological, mental, functional and psychosocial disorders of the elderly, to achieve an adequate treatment plan and an optimal management of the sociosanitary resources. Primary Care (PC) centers do not have specialists in AD and other related pathologies at their disposal. Hence, used procedures, techniques and diagnostic tools are not especially conditioned for the care needs to perform an accurate diagnosis of patient's disease (type of dementia) neither to provide the disease stage. In the particular case of Specialized Care (SC) centers, diversity and variability of procedures and diagnostic methods employed by clinicians to perform patient-evaluation and valoration are of remarkable importance. Continuity in patient care, a detailed monitoring of any variable associated to the disorder, fomenting collaborative work between specialists and giving the patient's familiar environment the importance it deserves become a must to build a solid knowledge base about MCI, AD and other dementias. This knowledge base will be highly important in the path to achieve two of the main goals in the field. First one, to implement mechanisms to early detect the disease. Second one, to improve efficiency in decision making to accomplish a more accurate and trustable diagnosis by means of differential diagnosis of dementias. It becomes necessary to search for new methods and techniques to speed up and increase reliability of diagnosis of these pathological syndroms, gaining knowledge about MCI, AD and other dementias and, moreover, improving patient's quality of life. In the same line, these new tools should boost a better use of resources of the sanitary system, that is to say, new management models are needed which improve the relationship between healthcare quality, efficiency and budget, [16].

A great quantity of the data-harvesting work taken in consultation rooms is still performed in paper or small standalone media such as local databases stored in personal computers, [8]. Digital gathering, access and store of clinical data in a centralized and structured way decreases the time a clinical professional needs to manage patients' medical records, which instantly increases the time a clinician is able to dedicate to his/her patients. Grouping the information provided individually by all clinicians involved in a patient's diagnosis and treatment plan makes it able to find key relationships inside data, empowering knowledge about diseases. To achieve this, a standardization of the clinical protocol to succeed in a trustable and accurate diagnosis of MCI, AD and any other dementia is needed; a clinical protocol customized to be used in both PC and SC.

In this chapter, we present the technological advances performed over EDEVIT-ALZH, a Personalized, Predictive, Preventive, and Participatory Healthcare Delivery System (4P-HCDS) [11, 12, 15], in the fields of computer architecture, databases, software engineering and systems integration, modelling, managing and exploiting clinical information of patients potentially affected by MCI, AD and/or other dementias according to healthcare level and research needs. The first section presents a brief overview of the EDEVITALZH Environment, describing its structure and objectives. Next, EDEVITALZH Systems Tier is analyzed, detailing the implementation of a High Performance Computer, Grid type, especially focused to improve computing throughput in the execution of Intelligent Systems for Diagnosis (ISD) and Intelligent Decision Support Tools (IDST), which provide EDEVIT-ALZH with Computational Intelligence to aid in the tasks of diagnosis, prognosis, monitoring and patient/pathology evolution studies. Subsequently, EDEVITALZH Database Tier is presented, describing its data model, according to the Patient Management (PMS) and Clinical Workstation (CW) paradigms and being fundamented on the Global Clinical Protocol for Dementias (GCPD) [11, 12, 15], which provides a detailed set of clinical procedures, forms, tests and diagnostic criteria, validated by medical experts in the fields of Geriatrics and Neurology, reflecting in an schematic way specific data of interest focused on the diagnosis of MCI, AD and other dementias. Furthermore, it correlates clinical and therapeutical parameters simultaneously.

The following section describes the User Interface (UI), also according to GCPD, which makes EDEVITALZH a CW developed to guide and aid clinicians in their workflow. Besides, EDEVITALZH is considered an intelligent CW (iCW) because of the integration of the EDEVITALZH Core with ISDs-IDSTs to aid physicians in diagnosis, prognosis and studying how these disorders spread through time. Last section presents our conclusions and future work in this research field.

8.2 EDEVITALZH Clinical Environment

EDEVITALZH is a Personalized, Predictive, Preventive, and Participatory Healthcare Delivery System (4P-HCDS) following the phylosophy of CW [11, 12, 14, 15]. EDEVITALZH Environment provides an Electronic Medical Records Database (EMRDB) as well as the digital implementation of the Global Clinical Protocol for Dementias (GCPD), a detailed set of clinical procedures, forms, tests and diagnostic criteria, validated by medical experts in the fields of Geriatrics and Neurology, which correlate clinical and therapeutical parameters focused on the diagnosis of MCI, AD and other dementias [14, 15].

EDEVITALZH Environment is based on a robust, secure, scalable and fault-tolerant technological architecture which provides users around the globe the capability to connect from anywhere and at any time. Its implementation follows the Spanish LOPDCP15/1999 policy, one of the hardest information security policies in the world, providing cyphered transactions over the web, database and systems security policies, patient data anonimization on sharing data between systems, and users authentication, profiling and control. Moreover, the entire environment is constantly being monitored and controlled by several network security systems. The main set of systems, applications, tools and mechanisms for information management, exploitation and integration inside the EDEVITALZH Environment is known as EDEVITALZH Core, (Figs. 8.1, 8.3). Any system, application or tool that shares data from/to the EDEVITALZH-Core must follow the EDEVITALZH policies about abstraction, modularity and security. This way, the Production EDEVITALZH Environment has passed code, systems and security benchmarks to assure it is a stable and safe platform.

EDEVITALZH web applications have been developed using the Model-View-Controller (MVC) design pattern, (Fig. 8.2), offering capabilities to improve, update and scale the environment with new resources, wizards, assistants and/or tools. This software engineering model has reduced both development complexity and time, as well as the needs of hardware resources. EDEVITALZH Environment is considered to be an iCW thanks to the Intelligent Clinical Wizards and Assistants (ICWA), which consist of the integration of EDEVITALZH-Core UI applications (ICWA-App) with ISDs-IDSTs, [5, 6, 14], providing the environment of Computational Intelligence. In this line, EDEVITALZH-Core has embedded the needed business logic to handle several ISDs-IDSTs working simultaneously on different requests

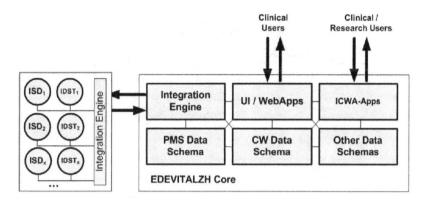

Fig. 8.1. High Level Architectural Diagram of the EDEVITALZH Environment

according to the clinical criteria selected, (Figs. 8.1, 8.3). Furthermore, EDEVIT-ALZH incorporates the Electronic Medical Interconsultation (EMI), (Fig. 8.8). EMI empowers collaboration between clinicians by means of an internal messaging system, providing mailboxes and message send/receive operations at users' disposal so they can ask for colleagues' opinions, participate in discussions and collaborate in the monitorization of colleagues' patients. EMI is a powerful and valuable tool for professionals at PC and other welfare centers where there are no specialist physicians in MCI, AD and other dementias, giving the PC clinicians the possibility to assist patients who suffer these neuropathologies with a greater level of quality, efficiency and accuracy in the diagnosis and prognosis processes.

In short, thanks to the provided clinical protocol guideline, wizards, assistants and ISDs-IDSTs, EDEVITALZH is able to aid clinicians and researchers in decision making, focusing on the diagnosis and prognosis tasks, improving accuracy in early and differential diagnosis, reducing the needed time to perform a certain diagnosis as well as in carrying out evolution studies of patients and their pathologies.

8.2.1   EDEVITALZH Systems Tier

EDEVITALZH Systems Tier (ST) is the basis over which the clinical environment is built. Emerging from basic project concept requirements, there exist great influences which have important effects on the way systems architecture have been designed and constructed. Hence, it was taken into consideration that the EDEVITALZH Environment must:

- Provide services of PMS to manage patients' medical records in a computerized way. This item is directly related to database systems and storage.

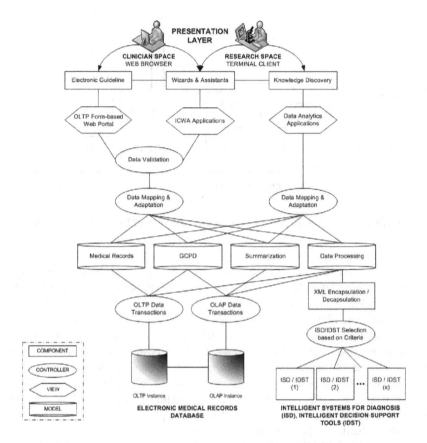

Fig. 8.2. Model-View-Controller Diagram of the EDEVITALZH Environment

- Provide services of CW by means of the set of clinical procedures, forms, tests and diagnostic criteria used by physicians to accomplish patient's diagnosis. This item has direct implications with UI Applications, database systems and storage.

- Provide the necessary technological mechanisms to integrate and communicate EDEVITALZH-Core with ISDs-IDSTs, enabling EDEVITALZH to become an iCW. This item is directly related to processing systems, database systems and communication procedures.

- Be a multi-user environment, allowing several clinicians to request, simultaneously, aid in the diagnosis process.

One of the basic concept requirements of EDEVITALZH forces to exploit the architectural model of the Internet network, that is to say, to take approach of influences, both technological and of resources, that global interconnection has provided through the last 40 years, [10]. In this line, ST design should take approach of the advantages the Distributed Computing Model (DCM) offers, [13], enhancing

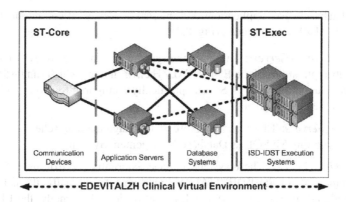

Fig. 8.3. EDEVITALZH Modular High-Level Organization regarding Systems Architecture: Core Module (ST-Core), ISDs and IDSTs Execution Module (ST-Exec)

Modularity. This means that every part of the environment has to be developed as self-reliant. To support this independence, two determined parameters become of maximum importance:

- *Fault Tolerance*: The clinical environment should be capable of minimizing damage upon failure of any of its modular parts. Furthermore, it should be provided of the needed redundancy on its critical components.

- *Scalability*: ST must be able to support the requirements of processing, storage and networking that the EDEVITALZH Environment may have through its lifetime. A scalable infrastructure, capable of being increased or decreased in physical resources according to its workload (processing cores, storage systems, communication and security devices) must be proposed.

Fig. 8.3 shows a high-level modular vision of the two main environment components, especifically:

- *EDEVITALZH ST Core Module (ST-Core)*: It consists of the main hardware systems that support Application Servers, Database Servers and Communication Devices, which allow users' requests to be served.

- *EDEVITALZH ST ISDs-IDSTs Execution Module (ST-Exec)*: It consists of the hardware systems that support processing and storage of ISDs-IDSTs and related components.

8.2.1.1   ST-Core: Components of the Systems Tier Core

The Systems Tier Core (ST-Core) is the hardware basis to support the EDEVITALZH-Core, (Figs. 8.1, 8.3). It consists of the group of systems that

manage databases, user applications and internal and external communications of
the EDEVITALZH environment, (Fig. 8.3):

- *Web Application Servers*: These are the systems serving the pages of the different web applications of the EDEVITALZH environment, which run instances of
  Apache Foundation's HTTP Server specifically configured for every web application of the environment.

- *Database Servers*: These systems store and manage the data schemas shown in
  Fig. 8.1, running MySQL as Database Management System.

- *Networking Devices*: These systems connect internally and externally the several components of the EDEVITALZH environment. Internally, the network is
  switched and monitored by 1Gbps Ethernet devices. Externally, the EDEVIT-
  ALZH Environment is protected by a firewalling device and connected to the
  mainstream network of the University of Las Palmas de Gran Canaria, (Fig. 8.5).

8.2.1.2 ST-Exec: Components of the Systems Tier ISDs and IDSTs Execution Module

The Decision Support Tools Execution Module (ST-Exec) consists of the group of
systems which handle processing load and storage by means of execution policies
regarding ISDs-IDSTs workload. As every part of the environment has to be developed as self-reliant, each ISD-IDST component must be developed as an autonomous entity which receives input data under a user request and returns results
according to that data. This way, every time a user demands a *Decision Supported
Operation* (DSO), one or several ISDs-IDSTs are first selected upon a selection procedure and then asked to generate certain results according to the provided inputs.
This description defines perfectly the concept of *execution job*. Hence, DSOs will be
considered as execution jobs, demanding a certain amount of computing resources
to accomplish their task.

Taking into account that EDEVITALZH is considered a multi-user platform, the
environment must be able to handle several simultaneous DSOs. This condition demands EDEVITALZH to be capable of planning and managing its global processing
resources to handle all user DSOs in an efficient mode, hence requiring execution
policies to queue, prioritize and execute DSOs.

To satisfy the described requirements, a High Performance Computing architecture is proposed, (Fig. 8.5). It is needed to implement this architecture in the
Grid Computing (GC) scope to be capable of having different physical computing resources placed and operating in different geographical locations. The term
Grid refers to an infrastructure that allows integration and collective use of high
performance systems, communication networks and databases, shared by several
institutions. Because collaboration between institutions means data exchange and
computing time, GC proposal is to easily integrate computational resources by
means of any type of software which implements this concept. GC allows to use,
in a coordinated way, all types of computing resources (processors, storage systems

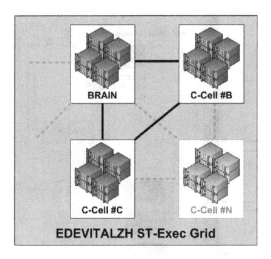

Fig. 8.4. Grid Computing Topology used in the implementation of ST-Exec. Several C-Cells can cooperate, being able to disconnect already-linked C-Cells or connect new ones to increase computing power. The environment offers scalability and fault tolerance.

and specific applications, among others) without being subject to a centralized control. In this line, it can be considered as a new way of distributed computing in which resources can be heterogeneous (different computing architectures) and be connected by means of Wide Area Networks as the Internet. Computing power offered by a crowd of GC interconnected systems is practically ilimited. Some of the main advantages of GC are listed in the following lines:

- *Load Balancing*: There is no need to calculate systems' capabilities in function of workload peaks. Resources can be reassigned according to needs and priorities.

- *High Availability and Fault Tolerance*: If a system fails, assigned tasks to that system will be reassigned to the rest of Grid nodes according to workload.

- *High Performance and Low Implementation Costs*: Thanks to distribution and heterogeneity properties, a performance similar to big mainframes can be gathered by using low-cost resources.

It is a high scalable solution, potent and flexible, that will avoid lack-of-resources problems and will never be obsolete due to the possibility of modifying the quantity and features of its components. On the other hand, communications and networking between systems become highly critical. Therefore, GC is a versatile, scalable, fault tolerant, distributed computing paradigm by means of the combination of many local systems to achieve a great global performance.

ST-Exec has been implemented using Oracle Grid Engine (OGE), an open-source distributed process management system capable of creating and managing GC architectures, [18]. Architecturally speaking, ST-Exec consists of Computing Cells

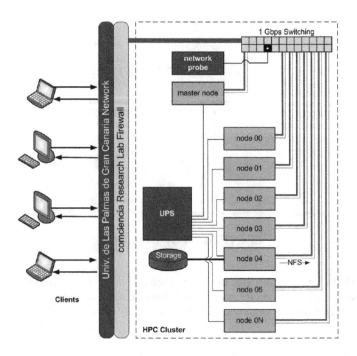

Fig. 8.5. Topology of 'BRAIN', the main EDEVITALZH Grid Computer C-Cell

(C-Cell), (Fig. 8.4). Every C-Cell is made up of a farm of servers running Linux operating system and connected by a high-speed communication network, linked to the Internet. Every C-Cell, identified by a Cell-ID parameter, is made up of the following structural components and parameters:

- *Master Host*: It is the responsible for managing, scheduling and assigning execution jobs. In practice, C-Cells can have one active Master Node and several shadowed Master Nodes for Fault Tolerance.

- *Execution Hosts*: They are responsible for execution of tasks assigned by the Master Host. They receive the needed binary executable image and inputs to run the process and gather results.

The main EDEVITALZH C-Cell has been identified as 'BRAIN' (Fig. 8.4) and it is located at COMCIENCIA Lab. It consists of two different computing architecture servers: ix86 and x86_64. 'BRAIN' is set up by 1 Master Host (no shadowed Master Hosts were provided at this stage of the project) and 5 Execution Hosts, (Fig. 8.5). A distributed filesystem has been configured by using Network File System (NFS). It is a stable and widely used filesystem. Following the same phylosophy, Network Information Service (NIS) has been configured as user credential management service.

It can be said that the max C-Cell performance will be the one of the node with highest performance. The performance of each node will integrally depend on its job-type orientation and configuration. To gather the performances of the different nodes integrated into the C-Cell, Intel Math Kernel Library - LINPACK has been used, [17]. LINPACK describes performance to solve a 3-sizes dense general matrix problem $Ax = b$. Table 8.1 shows the minimum and maximum execution results of LINPACK, individually on every node that make up the implemented C-Cell. Results show that, aparently, same hardware configuration nodes offer different performance ratios, although reality is that systems have been configured to increase performance in determined scopes; that is to say, determined nodes were enhanced for I/O performance, while other ones were enhanced for execution performance.

Table 8.1. LINPACK Execution Statistics on 'BRAIN' C-Cell Nodes

Node	Arch	CPUs	Cores	CLK (Ghz)	LINPACK min	LINPACK max
axon	x86	2	2	2.791	0.2535	4.1106
hipocampo	x86_64	2	8	2.660	16.4393	20.2175
nucleo	x86_64	2	8	2.494	31.2284	70.8108
sinapsis	x86_64	2	4	1.994	18.7492	27.8162
soma	x86	2	2	2.791	0.2624	4.3343

8.2.2 EDEVITALZH Databases Tier

The Database Tier (DBT) consists of the Electronic Medical Records Database Infrastructure (EMRDB) that stores, manages and relates the social, administrative and clinical data of EDEVITALZH patients. DBT is able to provide both online transactional (OLTP) and analytical (OLAP) processing mechanisms by means of different database instances, (Fig. 8.6). OLTP is focused onto single-patient data entry and retrieval, while OLAP is focused on knowledge discovery, performing tasks over summarized and aggregated data, data processing, visualization and mining.

Thus, EMRDB is the resulting database structure of representing the EDEVITALZH Data Model (DM), which models the patients' electronic medical records and their detailed set of clinical procedures, forms, tests and diagnostic criteria defined by GCPD [11, 12, 15]. GCPD, which has been validated by medical experts in the fields of Geriatrics and Neurology, reflects in an schematic way specific data of interest focused on the diagnosis of AD and other dementias, correlating clinical and therapeutical parameters simultaneously. DM is based on 2 principal entities, formally named Axis Entities: the 'Patient' entity and the 'Clinical Episode' entity. Around them, another set of 13 entities coexist (alphabetically listed): 'Analytics', 'Caregiver ', 'Complementary Test', 'Diagnosis', 'Disease', 'Exploration', 'Family', 'Habit', 'Interconsultation', 'Physician', 'Prescription', 'Symptom' and 'Test'. The SQL implementation of DM produces two types of tables:

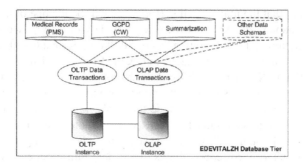

Fig. 8.6. EDEVITALZH Database Tier High-Level Diagram. On the upper part, Data Schemas. On the middle part, Transactional Components. On the down part, Databases.

- *Master Tables*, containing default values for GCPD parameters and other ones relating EDEVITALZH configuration needs. The number of Master Tables in EDEVITALZH DM is 39.
- *Activity Tables*, the ones where inserted data will be stored. They will contain data about patients and their medical records. The number of Activity Tables in EDEVITALZH DM is 18.

To provide acceleration in data access to EMRDB, two sets of customized data views have been developed: 'Summarization' and 'Data Processing', (Figs. 8.2, 8.6). 'Summarization' views speed up access to global clinical-parameter statistics. 'Data Processing' views help in the data extraction stage, previous to data encapsulation and transfer to ISDs-IDSTs.

The EDEVITALZH DBT provides mechanisms to manage and store relevant data for patients- and pathologies monitoring as well as to speed-up access to data of interest during the execution of ICWAs.

8.2.3 Presentation Tier: User Interfaces (PT-UI)

User Interfaces should be considered as the set of controls and tools which allow representing information that both, human and machine, will share and exchange in a work process, the work-flow, during which the machine will be the human's guide to accomplish a certain goal. Douglas Engelbart mentioned *Augmenting Human Intellect* [3, 4] in the beginning of the 1960s, where he proposed *'increasing the capability of a man to approach a complex problem situation, to gain comprehension to suit his particular needs, and to derive solutions to problems'*, [3]. Engelbart's essay set a start point on research about interfaces and how to improve human results by using computers. In this line, EDEVITALZH virtual environment can help clinicians to improve their capabilities to perform evaluation of patients, by supporting them in decision making for a more accurate diagnosis and prognosis

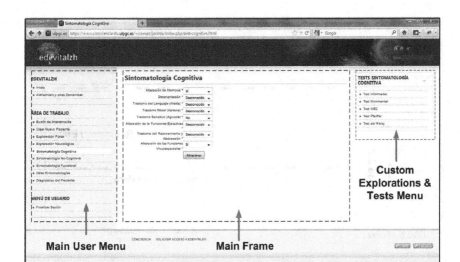

Fig. 8.7. EDEVITALZH User Interface Screenshot representing the *Cognitive Symptomath-ology* iCW Section. Left frame shows the Main User Menu, which allows users to browse through the different iCW GCPD-defined sections. Main frame depicts the clinical forms relating the requested section. Right frame offers explorations and test to perform custom patient evolution.

of their patients; furthermore, to provide computational intelligent tools for them to carry out evolution studies about patients and their pathologies.

Following the technological evolution of other disciplines, in PC medicine as well as in SC one, information systems have evolved by using new paradigms to develop them. Despite during the 1990s the client-server architecture was the primary tendency, since the last half decade of 2000s until today, web application modelling has burst into the information systems development business sector. EDEVITALZH applications have been developed using this paradigm by means of the MVC design pattern. This development model provides several advantages:

- Costs of installation and support are extremely reduced. Any computer platform or operating system is valid. Just a web browser is needed. No additional software or special package download is required.

- Any authorized user, no matter where he/she is geographically located, is able to access and use the provided applications and tools.

- Software updates, changes or even any problem arised during any of the software production stages, can be managed by developers and published for every user instantly. Furthermore, all these operations are performed in just one place, where the web applications and their generated pages are served, thus minimizing the required updating time.

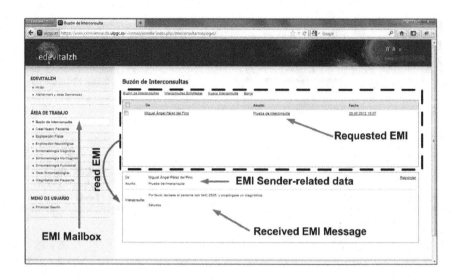

Fig. 8.8. EDEVITALZH User Interface Screenshot: Electronic Medical Interconsultation Mailbox

- Needed resources at clients is minimum; most of the business logic is performed by servers. The storage requirements for client devices is highly minimized because applications and pages are stored in servers, not client devices.

- Collaboration and information sharing between users of the service is empowered.

Hence, EDEVITALZH PT-UI consists of the web applications that make up the iCW UI. PT-UI represents a set of web forms and screen controls that guide clinical users in their workflow, improving patient dedication and consultation time. Two main web applications have been implemented, based on the GCPD. The PC web application (PCapp), focused on PC clinicians, manages a reduced set of the GCPD which groups the needed medical criteria to perform a basic diagnosis at any not-specialized center. The SC web application (SCapp), focused on SC clinicians, handles the complete version of GCPD to perform a more accurate differential diagnosis, prognosis and to carry out evolution studies of patients and their pathologies. Both applications follow a 3-frame screen scheme, thought to facilitate user experience on EDEVITALZH, (Fig. 8.7):

- The Main User Menu Frame, containing labels linked to all GCPD sections.

- The Main Frame, which represents each GCPD section upon request. This frame groups the set of related parameters and criteria. Furthermore, it can also handle customized tools as EMI, (Fig. 8.8).

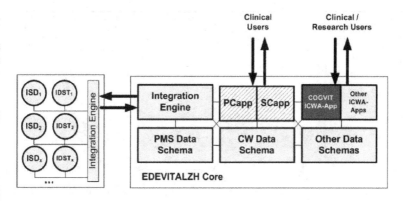

Fig. 8.9. High Level Architectural Diagram of the EDEVITALZH Environment with PCapp, SCapp and COGVIT ICWA embedded. Filled with light pattern, PCapp and SCapp web applications. In dark grey, COGVIT Sanger artificial neural network based IDST and COGVIT ICWA-App are highlighted.

- The Custom Explorations and Tests Menu, which can be shown or not, lists all the provided procedures and methods to gather information and perform certain evaluations of patients.

EDEVITALZH has electronic medical interconsultation (EMI) capabilities, (Fig. 8.8). It can be performed shared consultation by several physicians, at PC or SC, or a colleague can be asked for opinion or specialized valoration. Moreover, since both applications, PCapp and SCapp, share the database environment, physicians can refer their patients to any other specialist physician in a simple and easy way.

EDEVITALZH PT-UI supports ICWA-Apps, which make use of the EDEVIT-ALZH Integration Engine, to perform decision supported analytical- and research operations, aiding clinicians in carrying out customized studies about patients evolution and/or pathology spreading (Fig. 8.9), such as COGVIT ICWA-App, (Fig. 8.10). COGVIT extracts knowledge about the patient's symptomathology, visualizes the relationship between the symptomathology and the used clinical criteria by means of a representation of them in a transformed space. In concrete, in the Principal Component Analysis space provided by the analysis performed by the Sanger neural network.

8.2.4   Integration Mechanisms

EDEVITALZH integration mechanisms are grouped in the logical component of Integration Engine, (Fig. 8.1). This logical component specifies processes relating

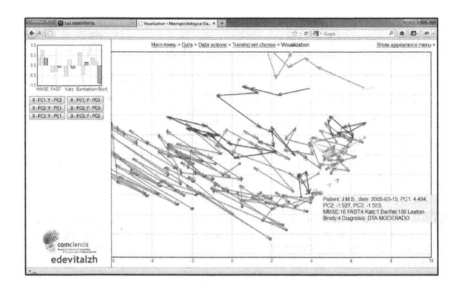

Fig. 8.10. COGVIT ICWA-App Screenshot. It allows to evaluate and visualize the evolution of patients according to the monitoring of their clinical episodes.

extraction, deidentification (if needed), encapsulation and transfer of information between ICWA-Apps, hence the EDEVITALZH-Core, and ISDs-IDSTs, (Fig. 8.11).

EDEVITALZH processes for data extraction are implemented as SQL views inside the database and located in the logical *Data Processing* data schema. Each one of these views is custom developed to extract a specific data set, according to the input needs of the ISD-IDST or ISDs-IDSTs that are going to be requested. Thus, there exists a logical mapping between SQL views and ISDs-IDSTs. The exploited data from the EMRDB will be deidentified if necessary, to preserve patient's anonimity. Subsequently, the data set will be encapsulated as an XML structure and saved into a physical file in the EDEVITALZH-ST storage system. In case there exists any restriction (i.e. special parametrization of any ISD-IDST), all XML files contain an *ISD-IDST ID* parameter section, which will allow ISDs-IDSTs to identify when they should operate or not upon a certain request.

Once the file is generated, the corresponding ICWA-App will send an execution job order to the Cluster Submit Node, asking for a determined process or set of processes (matching each requested ISD-IDST) to be executed with the dumped XML file as input. This way, the needed ISD-IDST or ISDs-IDSTs will be addressed. The Cluster Submit Node will handle the requested job and assign computing resources for its execution when available. Once results are ready, the requested ISD-IDST or ISDs-IDSTs will call the Write-Later Database Driver, writing the generated results to the corresponding patient's medical record in the EMRDB, (Fig. 8.11).

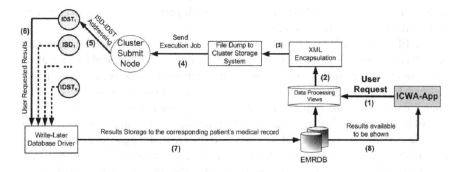

Fig. 8.11. Mechanism to integrate Intelligent Systems for Diagnosis (ISD) and Intelligent Decision Support Tools (IDST) with the corresponding ICWA Application to provide aid in decision making tasks. The flow diagram shows the operative since the clinical user requests aid in decision making until results are available to be used.

This integration protocol has several benefits to the way EDEVITALZH functions, among which the following ones can be highlighted: First, it offers the possibility to request a certain support for decision making to accomplish, while the user is performing any other task without interfering with his/her workflow. Second, it allows the ST-Exec to handle requests in an asynchronous way; that is to say, there is no need for the requester application to wait until there are free computing resources to attend the requested jobs. Results will be notified to EMRDB when available.

8.3   Conclusion

In this chapter, the technological fundamentals of the EDEVITALZH environment, a Personalized, Predictive, Preventive, and Participatory eHealth Delivery System have been presented. EDEVITALZH is considered an intelligent Clinical Workstation able to aid clinicians in early and differential diagnosis and prognosis of MCI, AD and other dementias. The proposed environment presents advances in the field of Medical Informatics, focused on the Geriatrics and Neurology scopes.

The EDEVITALZH Systems Tier consists of the Core, made up by the database and application servers as well as by all the networking devices involved in communications, and the Execution infrastructure, a High Performance Cluster, based on the Grid Computing paradigm, to manage, plan and execute Decision Supported Operations by means of Intelligent Systems for Diagnosis (ISD) and/or Intelligent Decision Support Tools (IDST). Its distributed architecture allows to scale the entire system by adding computing resources, even located in different geographical places, by means of the Internet or any wide area network. Furthermore, the architectural model of EDEVITALZH offers high performance, high availability, high scalability and fault tolerance.

The EDEVITALZH Database Tier stands the Electronic Medical Records Database. Its data model is based on GCPD, the clinical protocol validated by specialists in the fields of Geriatrics and Neurology that standardizes the set of clinical procedures, forms, tests and diagnostic criteria needed by Primary or Specialized Care clinicians to gather the required clinical data to perform a more accurate evaluation of patients regarding MCI, AD and other dementias. Its implementation relates specific data of interest focused on early and differential diagnosis and prognosis of patients affected by these neuropathologies. Moreover, several customized SQL views have been constructed, to speed up access to data of interest during the execution of Decision Supported Operations by means of Web Applications or Intelligent Clinical Wizards and Assistants (ICWA) by user request.

EDEVITALZH Presentation Tier is based on the web applications paradigm, which means the entire environment is accessible just by using a web browser on any device or operating system connected to the Internet. Therefore, this tier consists of web applications that make the environment a Clinical Workstation. In a first stage, the main web applications provide web forms and screen controls to guide clinicians in their workflow. Two main web applications are considered: PCapp, which manages a reduced set, and SCapp, which manages the entire parameter set, both representations of the GCPD in terms of user interface. In a second stage, and improving user experience, EDEVITALZH is able to support ICWAs based on two well-differentiated but integrated components: ICWA-Apps, representing the required user interface, and ISDs and/or IDSTs, providers of intelligence to our environment and responsible for EDEVITALZH to be an intelligent Clinical Workstation.

Last, EDEVITALZH presents a potent and flexible Integration Engine, consisting of the embedded mechanisms that allow ICWA-Apps and ISDs-IDSTs to communicate with each other to provide aid in clinicians' decision making tasks.

The result is a light-weighted distributed-computing web clinical environment, which implements a standardization of the needed diagnostic criteria to perform a more accurate evaluation of patients, no matter what healthcare level the patient is being assisted in, Primary or Specialized Care. EDEVITALZH aids clinical professionals in a more accurate decision making, focusing on early and differential diagnosis and prognosis of MCI, AD and other dementias. In addition, it provides intelligent wizards and assistants to carry out complex tasks such as patient- and pathology spreading analytical studies by means of ISDs and IDSTs. The most relevant effect of EDEVITALZH is that it ensures that no patient with MCI, AD or other dementias, or with the potential to suffer it, will go without the adequate medical attention, without a proper diagnosis and finally, without the appropriate treatment to alleviate their condition, due to lack of human and/or necessary clinical resources.

Acknowledgements. We would like to thank Canary Islands Government and EU Funds (FEDER) for their support under Research Project SolSubC200801000347.

Acronym Glossary

- *AD*: Alzheimer's Disease.
- *App*: Application.
- *C-Cell*: Computing Cell.
- *CW*: Clinical Workstation.
- *DBT*: Database Tier.
- *DCM*: Distributed Computing Model.
- *DM*: Data Model.
- *DSO*: Decision Supported Operation.
- *EDV*: EDEVITALZH.
- *EMI*: Electronic Medical Interconsultation.
- *EMRDB*: Electronic Medical Records Datababase.
- *GCPD*: Global Clinical Protocol for Dementias.
- *iCW*: Intelligent Clinical Workstation.
- *ICWA*: Intelligent Clinical Wizards and Assistants.
- *IDST*: Intelligent Decision Support Tools.
- *ISD*: Intelligent Systems for Diagnosis.
- *IWGNRCDAD*: International Working Group for New Research Criteria for the Diagnosis of Alzheimer's Disease.
- *MCI*: Mild Cognitive Impairment.
- *MVC*: Model-View-Controller Model.
- *NFS*: Network File System.
- *NIS*: Network Information Service.
- *OGE*: Oracle Grid Engine.
- *OLAP*: OnLine Analytical Processing.
- *OLTP*: OnLine Transaction Processing.
- *PC*: Primary Care.
- *PMS*: Patient Management System.
- *SC*: Specialized Care.
- *ST*: Systems Tier.
- *UI*: User Interface.
- *VD*: Vascular Dementia.

References

1. Campillo-Páez, M., San Laureano-Palomero, T., Sánchez de la Nieta-Martín, J.: ¿Hacemos correctamente la valoración geriátrica en Atención Primaria? Revista Centro de Salud, Marzo, 157–162 (2001)
2. Dubois, B., Feldman, H.H., et al.: Revising the definition of Alzheimer's disease: a new lexicon. The Lancet Neurology 9(11), 1118–1127 (2010)
3. Engelbart, D.C.: Augmenting Human Intellect: A Conceptual Framework (1962), http://www.dougengelbart.org/pubs/papers/scanned/ Doug_Engelbart_AugmentingHumanIntellect.pdf (cited May 27, 2012)

4. Engelbart, D., Landau, V., Clegg, E.: The Engelbart Hypothesis: Dialogs with Douglas Engelbart. NextPress, Berkeley (2009)
5. García Báez, P., Fernández Viadero, C., del Pino, M.A.P., Prochazka, A., Suárez Araujo, C.P.: HUMANN-based systems for differential diagnosis of dementia using neuro-psychological tests. In: 14th International Conference on Intelligent Engineering Systems (INES), pp. 67–72 (2010) ISBN: 978-1-4244-7650-3
6. García Báez, P., del Pino, M.A.P., Viadero, C.F., Araujo, C.P.S.: Artificial Intelligent Systems Based on Supervised HUMANN for Differential Diagnosis of Cognitive Impairment: Towards a 4P-HCDS. In: Cabestany, J., Sandoval, F., Prieto, A., Corchado, J.M. (eds.) IWANN 2009, Part I. LNCS, vol. 5517, pp. 981–988. Springer, Heidelberg (2009)
7. Herrera-Rivero, M., Hernández-Aguilar, M.E., Manzo, J., Aranda-Abreu, G.: Enfermedad de Alzheimer: inmunidad y diagnóstico. Neurología 51(3), 153–164 (2010)
8. Hernandez Salvador, C.: Modelo de historia clínica electrónica para teleconsulta médica. Doctoral Thesis. Polytechnical University of Madrid (2004)
9. Papez, V., Jaros, P., Suárez Araujo, C.P., del Pino, M.A.P., García Báez, P.: A new cognitive component of visualization tool for diagnosis and monitoring Alzheimer's disease and other dementias: COGVIT. Front. Neuroinform. Conference Abstract: 4th INCF Congress of Neuroinformatics, doi:10.3389/conf.fninf.2011.08.00012
10. del Pino, M.A.P., García Báez, P., Fernández Viadero, C., Suárez Araujo, C.P.: Self-Organizing Maps for Early Detection of Denial of Service Attacks. In: Fodor, J., Klempous, R., Araujo, C.P.S. (eds.) Recent Advances in Intelligent Engineering Systems. SCI, vol. 378, pp. 195–219. Springer, Heidelberg (2012)
11. del Pino, M.A.P., Suárez Araujo, C.P., García Báez, P., Fernández López, P.: EDEVIT-ALZH: an e-Health Solution for Application in the Medical Fields of Geriatrics and Neurology. In: 13th International Conference on Computer Aided Systems Theory (EUROCAST 2011), pp. 382–383 (2011) ISBN: 978-84-693-9560-8
12. del Pino, M.A.P., García Báez, P., Fernández López, P., Suárez Araujo, C.P.: Towards a Virtual Clinical Environment to Assist the Diagnosis of Alzheimer's Disease and other Cortical Dementias for use in Telemedicine. In: The 3rd Students' Scientific Conference Man-Civilization-Future (KNS 2005), vol. 1, pp. 177–183. Politechnika Wroclawska (2005) ISSN: 1732-0240
13. Stalling, W.: Comunicaciones y Redes de Computadores, 6th edn. Prentice Hall (2000) ISBN: 84-205-2986-9
14. Suárez Araujo, C.P., del Pino, M.Á.P., García Báez, P., Fernández López, P.: EDEVIT-ALZH: Predictive, Preventive, Participatory and Personalized e-Health Platform to Assist in the Geriatrics and Neurology Clinical Scopes. In: Moreno-Díaz, R., Pichler, F., Quesada-Arencibia, A. (eds.) EUROCAST 2011, Part II. LNCS, vol. 6928, pp. 264–271. Springer, Heidelberg (2012)
15. Suárez Araujo, C.P., del Pino, M.A.P., García Báez, P., Fernández López, P.: Clinical Web Environment to Assist the diagnosis of Alzheimer's Disease and other Dementias. WSEAS Transactions on Computers 6(3), 2083–2088 (2004) ISSN: 1109-2750
16. Weir, D.R., Wallace, R.B., Langa, K.M., Plassman, B.L., Wilson, R.S., Bennett, D.A., Duara, R., Loewenstein, D., Ganguli, M., Sano, M.: Reducing case ascertainment in U.S. population studies of Alzheimer's disease, dementia, and cognitive impairment Part 1. Alzheimer's & Dementia 7, 94–109 (2011)
17. Intel Math Kernel Library (LINPACK) for Linux, http://software.intel.com (cited May 27, 2012)
18. ORACLE Grid Engine, http://www.oracle.com/us/products/tools/oracle-grid-engine-ds-075946.pdf (cited May 27, 2012)

Chapter 9
Correlation of Problem Hardness and Fitness Landscapes in the Quadratic Assignment Problem

Erik Pitzer, Andreas Beham, and Michael Affenzeller

Abstract. Estimating hardness based on intrinsic characteristics of problem instances plays an important role in algorithm selection and parameter tuning. We have compiled an extensive study of different fitness landscape and problem specific measures for the quadratic assignment problem to predict and correlate problem instance hardness for several different algorithms. We combine different fitness landscape measures that provide a generally applicably problem characterization independent of problem class with problem specific measures that have been defined for the quadratic assignment problem specifically. These measures are used to create separate and combined regression models for several different hardness measures of several different metaheuristic algorithms and are compared against each other.

9.1 Introduction

The task of problem hardness calculation is as old as the use of optimization algorithms and since the use of stochastic algorithms also hardness estimation has been used. Many attempts have been made to find a good descriptive value that is easily derived from a problem instance to describe its difficulty. There have been, essentially, two different schools: One is the scattered area of problem-specific hardness prediction where for every problem class a different collection of methods has been proposed to estimate the problem instance's difficulty. On the other hand, we have the broad and generally applicable area of fitness landscape research which is currently undergoing a renascence as sufficient computing power as well as powerful

School of Informatics, Communication and Media,
University of Applied Sciences Upper Austria, Research Center Hagenberg
Softwarepark 11, 4232 Hagenberg, Austria
e-mail: {erik.pitzer,andreas.beham,michael.affenzeller}
@fh-hagenberg.at

R. Klempous et al. (eds.), *Advanced Methods and Applications in Computational Intelligence*,   165
Topics in Intelligent Engineering and Informatics 6,
DOI: 10.1007/978-3-319-01436-4_9, © Springer International Publishing Switzerland 2014

data base systems are now available, which make many of the past approaches much more feasible and interesting.

On top of the hardness estimation problem are even more interesting applications like the algorithm selection problem or the parameter selection problem. For this, however, it has to be ascertained that problem hardness should not be seen as a property of the particular problem instance, but in combination with a certain algorithm. Some problems might be easy for one but difficult for another algorithm as already detailed in the No Free Lunch Theorem [53]. However, by proper and efficient examination, using some of the methods described in this article, it might be possible to score a free appetizer [11] by choosing a relatively good algorithm or a good parameter set to start with.

9.2   Previous Approaches

The quadratic assignment problem (QAP) is a very interesting case as it is relatively simple to formulate but hard to solve in general. There are degenerate cases which are easily solvable and there are efficient branch and bound algorithms for obtaining exact solutions for smaller problem sizes [14, 30, 13, 29, 34]. For larger problem instances, metaheuristics are usually employed. In [14] a construction heuristic is used, moreover popular metaheuristic algorithms have also been successfully employed to solve large QAP instances e.g tabu search [16], robust taboo search [36], as well as simulated annealing [52, 9], genetic algorithms [45] and memetic algorithms [32] have all been successfully used to solve large QAP instances.

For problem hardness estimation one can use general fitness landscape analysis (FLA) methods [43, 35]. Interesting approaches are the theoretical decomposition into elementary landscapes, introduced in [51, 42], which have already been applied to the quadratic assignment problem [8] to allow efficient exact calculation of the ruggedness. In [32] a comprehensive analysis of the quadratic assignment problem (QAP) using general fitness landscape analysis (FLA) combined with problem specific (PS) measures has already been carried out. However, they rely on some FLA methods that are hard to obtain for new problem instances such as the fitness distance correlation coefficient [24]. In a previous study [37], we have already compared the correspondence between fitness landscape measures and problem specific (PS) measures, while here, we will perform an extended modeling approach comparing fitness landscape analysis (FLA) and problem specific (PS) measures for predicting the hardness of different instances of the quadratic assignment problem for different metaheuristics.

For the quadratic assignment problem, hardness prediction has often been attempted with problem specific measures [48, 27, 1]. Moreover, general fitness landscape analysis measures are also applicable and will be examined in detail in the following sections.

9.3   Fitness Landscape Analysis

The story of fitness landscapes begins with the first mention in [54] where natural populations and their fitness evolution are characterized. Since then, the term has been used informally and formally to describe the inherent structure of the solution space when subjected to optimization.

Formally, a fitness landscape $\mathscr{F}$ is composed of a solution space S and a fitness function $f : S \to \mathbb{R}$ that assigns a fitness to every solution candidate $x \in S$. To complete the formal description of the fitness landscape a notion of connectivity $\mathscr{X}$ between different solution candidates is required. This connectivity is often implicitly assumed in the form of a neighborhood relation $N : S \to \mathscr{P}(S)$, where $\mathscr{P}(S)$ is the power set of S, or a distances function $d : S \times S \to \mathbb{R}$ that is used to compare solution candidates and gives rise to notions of locality inside the solution space and, hence, inside the fitness landscape.

In [41, 4] and [8], theoretical approaches to fitness landscape analysis have been conducted. In this article, however, we want to contrast the problem specific measures that have to be defined for every problem anew with the generally available and applicable fitness landscape analysis methods. While, obviously, problem-specific measures have the advantage of great insight into the concrete problem class and knowledge of specific characteristics, this is at the same time their main weakness. For every new problem class, these measures have to be found and their usefulness has to be estimated. Fitness landscape analysis measures, on the other hand, are general and readily applicable to any problem class. Therefore, no theoretical up-front work has to be conducted to apply them to a new problem class. Moreover, results from one problem class can be compared with the results of another problem class as comparable quantities are measured. The only difficulty here are different encodings and neighborhoods for the same problem instance which complicate comparability.

9.3.1   Trajectories

A random sample of solution candidates and their fitness values can provide a preliminary first view of a problem instance. However, to examine the *landscape* we also need to incorporate a notion of coherence, which is done by incorporating the neighborhood structure. Therefore, instead of taking a random sample for further analysis, we create a trajectory of connected samples by stepping from one solution candidate to a neighboring one, based on the distance or neighborhood structure we used in the definition of the fitness landscape $\mathscr{F}$.

Different trajectory-based exploration strategies are introduced in the next sections. Each of which can present a new perspective of the fitness landscape and can greatly enhance the resulting insight.

9.3.1.1   Random Walks

One very frequently employed trajectory is the random walk. Very often, this is actually the only trajectory that is examined. It is very simple: Based on the neighborhood structure a neighboring solution candidate is used at random e.g. by applying a mutation operation or choosing a new solution candidate with a small distance to the current one. This provides a very simple and effective method of obtaining a connected but diverse and relatively unbiased sample of the whole landscape. Care has to be take in case the mutation operators include a certain heuristic or direction that cannot be reversed by re-applying it. In such cases a different mutation should be chosen or the operator can be combined with another one that exhibits a complementary heuristic or direction.

9.3.1.2   Adaptive and Up-Down Walks

To obtain samples that resemble the view of the fitness landscape obtained by an optimization algorithm we can use an adaptive sampling strategy. In this case, similar to a simple hill climber, the next solution candidate is chosen as the best of a few neighbors. To increase the exploratory nature of the adaptive walk, instead of using the best of many or even all neighbors, which is often the case in optimization algorithms, we can limit the number of neighbors to just a handful. This trajectory faces similar problems as the optimization algorithms themselves. After a few iterations it is likely that the trajectory becomes stuck in a local optimum. Optimization algorithms include countless methods to escape form this situation, however, our purpose is not to find an optimal solution but an optimal insight into the landscape. Therefore, a very radical *escape* strategy has been proposed in [23] where instead of trying to circumvent a local optimum and continuing to converge to another one, it is proposed to simply change the direction of optimization. This yields a very insightful *up-down* walk, where the search for a minimum is followed by a search for a maximum, hence, yielding a trajectory that exhibits features seen during optimization without the danger of becoming trapped in a local optimum. Instead of being stuck in such a local optimum, the optimization objective is simply negated.

9.3.1.3   Neutral Walks

Before we introduce neutral walks we first have do define the notion of neutrality. This seemingly trivial property can play an important role in the success or failure of a heuristic optimization algorithm. Neutrality describes the areas in a fitness landscape where neighboring solution candidates have equal fitness [38, 2, 25].

Superficially, neutrality seems to be undesirable because it deprives the optimization scheme of a direction towards the next local optimum. And, indeed, for simple algorithms, neutrality can cause stagnation and premature convergence. On the other hand, an intelligent population-based algorithm might be able to use these neutral areas to its advantage a explore it by spreading across this area.

The analysis of neutral areas can be performed by specially crafted neutral walks that try to remain inside a neutral area as long as possible without getting stuck in place by continuously moving away from the entry point into the neutral area. This scheme crosses several neutral areas diametrically and can collect statistical information about these areas which can facilitate further understanding and provides yet another perspective of the fitness landscape of a particular problem instance.

9.3.2 Measures

Many different measures have been defined to characterize fitness landscapes. Many have been introduced along a random walk trajectory but can be reused for other trajectories as well. In the following sections we will describe the measures we have used for the analysis of the quadratic assignment problem.

9.3.2.1   Autocorrelation and Correlation Length

When thinking of traversal of a landscape, the term *ruggedness* was one of the first to describe the encountered difficulties to home in on the global optimum. While in a simple definition, ruggedness can be be seen, simply as the degree of multimodality or the number of local optima, it has later been formalized into the measures of autocorrelation [51] and correlation length [51, 23]. The autocorrelation function describes the self similarity of the fitness values along the trajectory when shifted against itself a certain number of steps as shown in Eq.(9.1). Of course, when not shifted, the function fully resembles itself, so the autocorrelation at step zero is always one. However, the further the function is shifted, or the more steps are taken, the more the resulting fitness differs until the difference is not statistically significant any more. This point also defines the correlation length [20]. Typically, the autocorrelation at one step is examined which corresponds to the average fitness difference between neighboring solution candidates.

$$\rho(\varepsilon) := \frac{E(f_i \cdot f_{i+\varepsilon}) - E(f_i) \cdot E(f_{i+\varepsilon})}{\mathrm{Var}(f_i)} \tag{9.1}$$

where f_i is the fitness trajectory, $f_{i+\varepsilon}$ is the series of fitness values shifted by ε, $E(x)$ is the expected value of x and $\mathrm{Var}(x)$ is the variance of x.

9.3.2.2   Information Analysis

A slightly different view is provided by the information analysis of fitness trajectories. Here, instead of a detailed examination of relative difference values of neighboring fitness candidates, the slope changes are examined [47]. For this, the series of fitness values, $\{f_i\}_{i=0}^n$, is first transformed into a series of fitness changes $\{d_i\}_{i=1}^n = \{f_i - f_{i-1}\}_{i=1}^n$. This series is then further discretized into a series of slope change symbols with the help of the relaxed sign function shown in Eq.(9.2). This sequence of slopes depends on the choice of ε which can, hence, be used to zoom in and out to view slopes of different extends in the landscape.

$$\text{sign}_\varepsilon(x) := \begin{cases} 1 & \text{if } x > \varepsilon \\ -1 & \text{if } x < \varepsilon \\ 0 & \text{otherwise} \end{cases} \qquad (9.2)$$

The first measure that can derived from this sequence of slopes is the number of slope changes which is called the *partial information content*. Moreover, we can look at the smallest ε for which no slope changes remain, or, in other words, the largest fitness difference along the entire trajectory, which is called the *information stability*.

Moreover, the following entropy measures have been defined in [47] which characterize the compressibility of the different shapes. The function $H(\varepsilon)$, shown in Eq. (9.3), is called the *information content* and measures the entropy of different consecutive slopes or the prevalence of *rugged* steps while $h(\varepsilon)$, shown in Eq. (9.4) measures the entropy of equal consecutive slopes and is called the *density basin information* as it captures the information of smooth points.

$$H(\varepsilon) := -\sum_{p \neq q} P_{[pq]} \log_6 P_{[pq]} \qquad (9.3)$$

$$h(\varepsilon) := -\sum_{p = q} P_{[pq]} \log_3 P_{[pq]} \qquad (9.4)$$

Another measure described in [47] is the *regularity* which is the number of unique fitness values contained in the trajectory. This value, while seemingly trivial, can have a large influence on the performance of population-based optimization methods. A larger number of different fitness values make different, distant solution candidates better comparable as it gives a finer-grained view of the different fitness values and, hence, facilitates the choice of which features to select.

9.3.2.3 Intermediate Walk Lengths and Distances

While for the random walks, the previously described measures are the only pieces of information that can be harvested, for the more advanced walk types, like up-down or neutral walks, additional values can be obtained. In the case of the up-down walks we can count the number of steps taken to reach the next minimum or maximum as well as analyze the achieved fitness levels. For the neutral walk we can also count the steps that were made inside the neutral area. Moreover, we can look at the distances between entry and exit points of the neutral areas. As shown previously [3], the distances must not necessarily correspond to the number of steps as the neutral areas can be convoluted and contain a much higher number of steps yet have smaller distances. For these step counts, fitness values, and distances we can then include the averages as well as their variances as additional measures to describe the fitness landscape of a particular problem instance.

9.3.3   Landscape Variants

As defined in the previous section, the fitness landscape of a problem instance is not only defined by the fitness function itself but also by the neighborhood structure. To obtain a fuller view of a particular problem instance, one can, therefore, not only examine the concrete landscape which will be used for optimization, which is implicitly defined by the move or mutation operator, but also different landscape variants which are obtained by using different operators and, hence, by examining different neighborhoods. Figure 9.1 shows the different distributions of auto correlation values obtained by using different landscape variants. This is actually one of the issues which makes it harder to use fitness landscape analysis methods across different problem classes, because the choice of landscape variant or neighborhood structure can have a very large impact on the obtained measures. The figure shows a violin plot [18] which is a combination of a box plot and a density distribution.

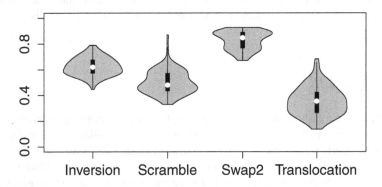

Fig. 9.1. Distributions of auto correlations for different landscape variants induced by using different move operators when applied to the same collection of problems

9.4   Quadratic Assignment Problem

The quadratic assignment problem (QAP) is quite an old problem, first described in [26]. It has been applied to model many different actual problems from facility layout planning to keyboard design [12, 18, 7, 6]. Because of its NP-hardness [39], exact algorithms are hardly used for practically interesting problem of sizes 20 or larger. Therefore, real-world quadratic assignment problems (QAPs) are most often optimized using metaheuristics such as Simulated Annealing [26], Tabu Search [15, 36, 21] or Genetic Algorithms [19, 10].

Simply put, the quadratic assignment problem is to find an optimal assignment of a set of facilities to an equal number of locations. The difficulty of the problem lies in the fact that there are two associated networks of flows and distances. There are different flows between every pair of facilities and different distances between every pair of locations. Each assignment incurs the flow between all facilities over the distances between their respective locations. The objective is to find an assignment that minimizes the flow times the distances, or in other words, assigns larger flows to smaller distances.

The QAP can be formalized as a double sum over the product of assigned flows times distances as shown in Eq.(9.5).

$$f(\pi) := \sum_{i=1}^{N} \sum_{j=1}^{N} w_{ij} \cdot d_{\pi(i)\pi(j)} \tag{9.5}$$

Here, the permutation π is the assignment of facilities to locations. The element j at position i in the permutation represents the facility j that will be placed at location i. The flows are encoded as elements of the matrix w_{ij} and the distances in the distance matrix d_{ij}. It is possible to define a quadratic assignment problem where the number of facilities does not equal the number of locations. In this case additional dummy facilities with zero flows or additional dummy locations with very high distance can be added.

9.4.1   QAPLIB

As the quadratic assignment problem is one of the hardest optimization problems, a large and diverse collection of problem instances has been collected as a testbed for many metaheuristic algorithms. This collection, the Quadratic Assignment Problem Library (QAPLIB) has been described in [5] and is now hosted at the University of Pennsylvania[1]. It contains a total of 137 instances from different applications and detailed reference of the problem instances' origins. In our experiments, we have used a representative subset of 102 instance all of which have a size between 12 and 50.

9.4.2   Problem Specific Measures

While fitness landscape analysis provides a very general tool set for exploring the characteristics of different problem instances, as long as only a single problem class

---

[1] http://www.seas.upenn.edu/qaplib

is examined, we can also use problem-specific (PS) characteristics. In the past, problem-specific (PS) characteristics have been the most straightforward method for predicting problem instance hardness or measuring similarities between different problem instances of the same class. Therefore, we are also looking at measures which pertain particularity to the QAP.

In the past, several PS measures have been proposed already. Among them, the most widely known is the *flow dominance* which uses the coefficient of variation ($c_v := \sigma/\mu$) of the flow matrix to derive a single characteristic value [48]. A similar measure can be derived from the distance matrix [27]. Moreover, in a more theoretic analysis, the cost's mean and variance have been derived which analyze the distribution of the resulting fitness function by combining statistics of the distance and weight matrices [1]. We have further simplified this calculation to make it more practical and, hence, only very roughly estimate the mean and variance of the fitness function by using the means and variances of the given matrices as shown in Eqs. (9.6) and (9.7). While this value is not actually the variance of the cost function, it was used as an additional, very easily obtainable measure of a QAP instance. The actual variance of the cost function has been described in [1] for symmetric QAPs and is given in Eq. (9.8). One of the drawbacks of this method is that it has only been defined for symmetric instances and that it still takes $O(n^4)$ steps instead of only $O(n^2)$ steps to calculate. In practice, however, compared to other analysis methods and even in comparison with fitness landscape methods, the additional effort is negligible.

$$E[c] := E[w_{ij}] \cdot E[d_{ij}] \tag{9.6}$$

$$\widehat{Var}[c] := E[w_{ij}]^2 \cdot Var[d_{ij}] + E[d_{ij}]^2 \cdot Var[w_{ij}] + Var[w_{ij}] \cdot Var[d_{ij}] \tag{9.7}$$

$$
\begin{aligned}
Var[c] &:= [S_0 + S_1 + S_2]/n! - \mu^2 \\
S_0 &= 4(n-4)! \sum_{\cap_0} f_{ij} f_{rs} \sum_{\cap_0} d_{ij} d_{rs} \\
S_1 &= (n-3)! \sum_{\cap_1} f_{ij} f_{rs} \sum_{\cap_1} d_{ij} d_{rs} \\
S_2 &= 2(n-2)! \sum_{\cap_2} f_{ij} f_{rs} \sum_{\cap_2} d_{ij} d_{rs} \\
\cap_k &= \left\{ ((i,j),(r,s)) \ \middle| \ 1 \leq i,j,r,s \leq n, |(i,j) \cap (r,s)| = k \right\}
\end{aligned}
\tag{9.8}
$$

In addition to these documented problem specific values, it is straightforward to add other distribution characteristics of the matrices such as skewness and kurtosis as well as the *p*-Value of a simple normality test [22] which compares the skewness and kurtosis with a standard normal distribution as shown in Eq. (9.9) where S is the sample skewness and K is the sample kurtosis also shown in Eq. (9.9).

$$
\begin{aligned}
JB &= \tfrac{n}{6} \left(S^2 + \tfrac{1}{4}(K-3)^2 \right) \\
S &= \frac{\frac{1}{n} \sum_{i=1}^{n} (x_i - \bar{x})^3}{\left(\frac{1}{n} \sum_{i=1}^{n} (x-\bar{x})^2 \right)^{3/2}} \\
K &= \frac{\frac{1}{n} \sum_{i=1}^{n} (x_i - \bar{x})^4}{\left(\frac{1}{n} \sum_{i=1}^{n} (x-\bar{x})^2 \right)^2}
\end{aligned}
\tag{9.9}
$$

174   E. Pitzer, A. Beham, and M. Affenzeller

Moreover, when considering the structure and use of the matrices other typical
matrix measures can be added to the QAP matrices as well as additional throughput
measures which are described in the following:

- Dominance: In addition to the popular flow dominance, we have also included
  distance dominance which is defined as the coefficient of variation over either
  the flow matrix (w_{ij}) or the distance matrix (d_{ij}) as shown in Eq. (9.10).

$$\text{dominance}(x_{ij}) := \frac{Var[x_{ij}]}{E[x_{ij}]} \tag{9.10}$$

- Distributions: For each matrix we have calculated the mean, variance, skewness,
  kurtosis and normality p-value as described above.
- Sparsity: From each matrix a measure of how sparse it is can be derived. This
  is simply the ratio of zero entries over the whole number of values. For the flow
  matrix w_{ij}, for example, this is defined as follows: $|\{x \mid x \in w_{ij}, x = 0\}|/n^2$.
- Symmetry: For each of the two matrices we can define the *total asymmetry* as the
  sum of differences between opposing flows or distances as shown in Eq. (9.11).
  Similarly, we can define the *relative asymmetry count*, shown in Eq. (9.12) that
  does not measure the total magnitude but only the relative occurrence.

$$\frac{2}{n \cdot (n-1)} \sum_{1 \leq i < j \leq n} |w_{ij} - w_{ji}| \tag{9.11}$$

$$\frac{2}{n \cdot (n-1)} \sum_{1 \leq i < j \leq n} \text{sgn}(|w_{ij} - w_{ji}|) \tag{9.12}$$

- Transitivity: Another interesting property, when thinking of flow networks in the
  quadratic assignment problem is the transitivity or, when thinking in terms of
  difficulty measures, the *intransitivity* shown in Eq. (9.13) and the *intransitivity
  count* in Eq. (9.14).

$$\sum_{1 \leq i,j,k \leq n} \max(0, w_{ik} - (w_{ij} + w_{jk})) \tag{9.13}$$

$$\sum_{1 \leq i,j,k \leq n} \text{sgn}(|w_{ik} - (w_{ij} + w_{jk})|) \tag{9.14}$$

- Throughput: Finally, we can define a few node-specific throughput measures that
  can be included in the problem specific analysis. For each node (both in the
  distance and in the weight matrix) we can define the total *in flow* (Eq. (9.15)),
  out flow (Eq. (9.16)), and based on these values, the *flow sum* (Eq. (9.17)) and
  the *flow surplus* (Eq. (9.18)) which are simple sums over rows and columns of
  the matrices.

$$w_{i\_} := \sum_j w_{ij} \tag{9.15}$$

$$w_{\_j} := \sum_i w_{ij} \tag{9.16}$$

$$w_k := w_{k_-} + w_{\_k} \qquad (9.17)$$

$$w_{k+} := w_{k_-} - w_{\_k} \qquad (9.18)$$

Values extracted from the matrices will usually have a high variability due to only the magnitudes and distribution of the number in the matrix. Therefore, we have first standardized the values in the matrix before extracting most of the measures by subtracting the average and dividing by the standard deviations. This reduces the influence of arbitrary scaling factors that could otherwise hide important statistical properties.

Finally, it has to be emphasized that these measures have not been devised to provide specific insights into particular properties of the problem instances themselves but rather to extract characteristic properties regardless of their particular meaning. These values, similarly to some fitness landscape analysis values, can be seen more like fingerprints. They do not convey any particular information when looked at in isolation, however, when compared with other values from other problem instances, they can help in obtaining a more complete picture and, hence, facilitate comparison.

9.5   Algorithms

9.5.1   Robust Taboo Search

The standard tabu search is a very popular metaheuristic. It is a clever extension of a local search [15]. To prevent becoming stuck in a local optimum, tabu search forbids a certain number of previous moves. This enables the tabu search to escape from local optima instead of moving back and forth between a local optimum and its best neighbor. Therefore, the tabu search algorithm has one additional parameter, the tabu list length and needs the implementation of a tabu criterion which determines whether a move is tabu or not. A very simple tabu criterion is to forbid revisiting previously seen solution candidates. In case of the quadratic assignment problem and its permutation encoding a different strategy is usually employed. Here it is not the particular solution candidate that is tabu but the assignment of a certain facility to a certain location.

Later, in 1991, tabu search itself has been extended to Robust Taboo Search [sic] in [36]. The first addition was to make the tabu list length a random variable instead of a fixed length. Additionally, a new diversification mechanism has been added. While the tabu list length tries to prevent recent moves, after a number of iterations, a new aspiration mechanism forces a move that has not been performed for a long time to create additional diversity.

In summary the robust taboo search contains two parameters, the *maximum tabu tenure* which determines the distribution parameter for the tabu tenure random variable and the aspiration tenure which determines the number of steps after which a diversification step with a move the has not been seen for a long time is enforced.

9.5.2 Simulated Annealing

Another popular metaheuristic algorithm is simulated annealing. The basic idea is to model the physical properties of a hot material that is cooling down. While still hot, any "shape" is acceptable as the material is still "fluid", with decreasing temperature, however, the material, lacking sufficient intrinsic energy, can only settle for more stable and, hence, less energetic shapes. It is a simple extensions of local search in which the temperature models the acceptance probability of an inferior step. While the search is still hot, many inferior solutions will be accepted. As the search cools down, however, this probability is reduced, until the search resembles a local search that converges to the next local optimum.

This algorithm, first proposed in [26] as a variation of the Metropolis-Hastings algorithm [33], is modeling a thermodynamic system with interchangeable fitness function. Several extensions of this algorithm are possible. One very popular extension, similar to local search is the addition of multiple restarts. Another, slightly more advanced version includes several, so-called, re-heating phases, where the temperature is raised according to different conditions during the optimization.

9.5.3 Genetic Algorithms

We have also used different parameterizations of genetic algorithms [19]. In contrast to the other optimization schemes, instead of relying on a single trajectory of optimization, this class of algorithms uses a whole population of concurrently evolving individuals. These individuals are selected in pairs according to their fitness for reproduction. Using a certain cross-over operator, each pair is then used to produce new solution candidates inheriting and combining properties from both parents. After this recombination step, with a certain probability, some of these offspring solution candidates are further changed using a mutation operator. Afterwards, some or all of the previous generation of individuals are replaced by the newly generated solution candidates.

9.6 Experiments

To determine the correlation between simple fitness landscape analysis and their correspondence to optimization algorithms we have conducted a large number of experiments. Typically, only a fraction of these experiments would have been carried out by an experienced researcher. He or she might have been able to intelligently tune the parameters to achieve better and faster convergence. However, in this series of experiments we are chiefly interested in the hardness of the problem itself. Therefore, we have tried to establish a frame of reference as the average performance of a certain algorithm irrespective of parameter settings. The choice of parameter is very important and even the range of parameters we have used in these experiments plays an important role in the perception of problem difficulty.

We have performed a preliminary manual study to determine initial parameter ranges for each of the algorithms. Moreover, we have used these initial tests to determine critical parameter settings for the quadratic assignment problem, while other parameters have been left fixed at reasonable default values.

Three different algorithms were used to optimize the instances of the QAPLIB. Each with several different parameter configurations. Moreover, each configuration has been tried 20 times to account for stochastic aspects of the metaheuristics. For the robust taboo search we used different values for maximum tabu tenure and alternative aspiration tenure, giving a total of 90 different parameter settings. The first variant of simulated annealing had its start and end temperature varied yielding a total of 25 different configuration, in a second attempt we also varied the annealing curve using three different slopes totaling 36 different configurations. For the genetic algorithm we varied the cross-over operator, the mutation probability and the mutation operator.

Table 9.1. Algorithms and Parameter Configurations

Algorithm & Parameter		Value(s)
RTS	Max Tabu Tenure	25, 50, 100, 150, 200, 300, 400, 600, 800
RTS	Alternative Aspiration	100, 500, 750, 1000, 1500, 2000, 3500, 5000, 7500, 10000
RTS	Maximum Iterations	10000
SA-1	Start Temperature	1, 2, 4, 8, 16 [%]
SA-1	End Temperature	0.03125, 0.06250, 0.12500, 0.25000, 0.50000 [%]
SA-2	Start Temperature	1, 4, 16, 32 [%]
SA-2	End Temperature	0, 0.03125, 0.125 [%]
SA-2	Base (Curve Shape)	1, 0.01, 0.00001
GA	Population Size	500
GA	Selection	Proportional
GA	Crossover	Cyclic, Partially-Matched, Uniform-Like
GA	Mutation	Swap2, Scramble, Inversion, Insertion
GA	Mutation Rate	1, 5, 10, 15 [%]

For the robust taboo search we have tried several different parameter configurations. This algorithm has been designed specifically for solving instances of the quadratic assignment problem and can, therefore, and because of its very good performance, be considered as a sort of gold standard in terms of performance on most of the problem instances. The robust taboo search has essentially two parameters, the maximum tabu tenure and the alternative aspiration tenure. We have used nine different values for the tabu tenure and ten different settings of the maximum aspiration tenure. Moreover, we have limited the maximum number of steps to 10,000 and have stopped algorithm execution once the best known solution or global optimum had been found.

We have employed two variants of a Simulated Annealing algorithm. One was parameterized with an implicit cooling schedule derived by fitting an exponential function between variable start and end temperatures and another variant that also varied the shape of the temperature curve between those start and end temperatures. The start and end temperatures were selected relative to the problem instance instead of a fixed set of temperatures as this value corresponds directly to the obtained fitness values which can vary strongly between different problem instances and are defined as the probability of accepting an inferior successor based on the average fitness. We used five different start temperatures and five different end temperatures in the first trial, and three start and four end temperatures in the second trial. Moreover, we used three different exponent bases in the second trial.

We also used several parameterizations of a Genetic Algorithm that exposes the highest number of parameters. Therefore, we have restricted the variation to the most influential parameters. We have kept the population size constant at 500, and always used proportional selection. We have, however, varied the cross-over operator, the mutation rate and the mutator to explore some parts of the parameter landscape trying a total of 48 different parameter configurations. In Table 9.1, a detailed summary of all algorithms and parameter configurations is shown.

In each of these tests, we recorded several performance indicators. The first indicator was the number of iterations or generations until convergence. Moreover, in case the algorithm configuration did not converge to the global optimum, we also recorded the scaled difference to the global optimum or in some cases to the best known solution. Figure 9.2 shows most of the algorithm and parameter configurations for one of the problem instances.

To put the performance results obtained from the optimization algorithms into perspective we have conducted several runs of fitness landscape analysis. A very important concern was the run-time of fitness landscape analysis in comparison to the optimization process. As our long-term goal is to facilitate algorithm and parameter selection we wanted fast results from the fitness landscape analysis. Otherwise, if we had the time, we could just run repeated optimization experiments instead. Therefore, we have only used sampling based analysis that can be automatically and quickly performed.

We have used all of the analysis methods described in Section 9.3, random walks, up-down walks, and neutral walks and collected all mentioned FLA values for each walk type. For the analysis, we have also included the autocorrelation coefficient c which is a transformation of the auto correlation at offset one, $\rho(1)$, as $c = 1/(1 - \rho(1))$. In preliminary analyzes we have also seen a strong correlation between these values and the problem size so we have included a normalized auto correlation coefficient and correlation length, each divided by the problem size to magnify the remaining effective values. Table 9.2 summarizes the run-times of the different analysis algorithms and configurations.

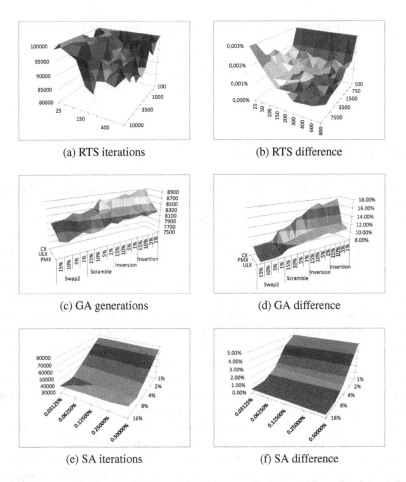

(a) RTS iterations   (b) RTS difference

(c) GA generations   (d) GA difference

(e) SA iterations   (f) SA difference

Fig. 9.2. Parameter landscapes for the problem instance `chr25a` which can be characterized as an *average* problem instance from our experience. The horizontal axes contain the various parameters while the vertical axis shows the number of steps (either iterations or generations) needed to find the global optimum or best known solution, while the graphics on the right side show the average difference to the global optimum for a particular configuration. As can be seen, robust taboo search performs much better than the other algorithms.

While the problem specific measures were calculated on a single machine within seconds for the whole QAPLIB, total fitness landscape analysis took 7 minutes and the effort prediction took 11 hours for the robust taboo search, the first simulated annealing batch 90 hours, the second one around 8.7 hours and the genetic algorithm runs took 114 hours, each on a computer cluster with 112 cores at 2.5GHz running HeuristicLab Hive [41, 50]. The resulting experiment configurations can be downloaded from our homepage[2].

---

[2] http://dev.heuristiclab.com/AdditionalMaterial

Table 9.2. Trajectory configurations and average run-times per problem instance

trajectory	steps	neighbors	average run-time
random	100,000	1	59 [s]
up-down	100,000	10	190 [s]
neutral	10,000	100	140 [s]

9.6.1 Hardness Measurement

The first important question, before we can start to estimate hardness, is the definition of a quantifiable amount that represents problem instance hardness. For our experiments we have defined several hardness measures. One additional aspect that has to be considered is the non-comparability of fitness values of different QAP instances. As the fitness of a QAP instance is the sum over a matrix product, the values depend strictly on the distributions of the values inside those matrices. Therefore, all values of the fitness function have to be put into perspective with respect to this distribution. Therefore, we have, of course, not used the final fitness values as an estimate of problem hardness. The relative difference $(f(s) - f^*)/f^*$ of the final solution candidate s is the difference to the fitness of the best known or optimal solution f^*, divided by this best known or optimal solution. This measure, enables us to compare different algorithms on the same problem and allows as to follow convergence towards zero instead of an arbitrary problem dependent number. However, its magnitude is still very dependent on the problem instance. To overcome this issue, we have used another measure, the scaled relative difference $(f(s) - f^*)/(\hat{f} - f^*)$, which additionally considers the average fitness of a random solution candidate $\hat{f}$ as the base line for convergence. If, on average, the fitness values of random samples are already quite close to f^* this scaled difference provides a better means of discerning convergence quality than the relative difference alone, when comparing results across different problem instances.

The base for our investigations of problem hardness are the performance indicators directly obtained from the optimization runs. These are the numbers of iterations or generations until the best known or globally optimal solution have been found or otherwise the scaled difference to the global optimum in case the algorithm did not find it on time. In the end, however, we want to obtain a single number that characterizes the hardness of a particular problem instance with respect to a particular algorithm or parameter setting.

For this purpose, we can use the averages over the repetitions of iterations and scaled differences. The achieved total performance of an algorithm configuration ranges from iterations from 0 to 10000 followed by a scaled difference going from 0 to 1. Figure 9.3 shows the choice we are faced with in case the algorithm does not consistently converge after a certain number of iterations. In this case, we have to find a combination of the two performance factors, before and after algorithm termination. In our case, much greater emphasis was put on the quality of the result than on the speed. Therefore, we have chosen this scaling point much in favor of the scaled difference. Moreover, the outer ends of both scales are hardly ever reached. It is equally unlikely for an algorithm to terminate after just a few iterations as it is for an algorithm to return a final quality comparable to the quality of a random point in the fitness landscape.

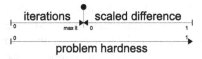

Fig. 9.3. Scaling Point Between Iterations and Scaled Difference. The problem hardness should be a single value but in case the global optimum cannot be found the two parts "before" and "after" algorithm termination have to be combined.

After combination of pre and post termination measures we have one number that describes the problem's hardness with respect to a certain algorithm configuration. From these configuration hardnesses we have further derived the *average hardness* over all configurations which can be seen as the general hardness of a problem for a particular algorithm or solution approach regardless of the particular parameter settings. Moreover, we can look at the best achievable performance or the *best hardness* which reflects the performance for very good parameter settings. Moreover, another interesting measure can be derived as the difference between these two values. If a problem is very hard on average but has one specific parameter setting that is able to achieve good results it means that a considerable effort for parameter tuning is necessary for this problem instance. Therefore, we have also included this *specificity* of the best parameter setting in comparison to the average performance over all configurations.

For exact algorithms and simple heuristics it is often sufficient to measure only the performance of this algorithm or heuristic, again, by counting the number of steps or evaluations until a certain quality is reached or to measure the quality reached after a certain number of steps [31, 55, 40]. However, typical metaheuristics have many parameters that need to be tuned for a particular problem class or even for a particular problem instance. Therefore, we have chosen to evaluate the performance on a whole parameter grid as visualized in Figure 9.2.

9.7 Results

We have performed two sets of final analyzes. One was the direct correlation of various measurements and properties which is described in the following section, followed by a more advanced regression model containing a sophisticated multi-stage variable selection in Section 9.7.2.

9.7.1 Simple Correlations

Probably the first interesting aspect is the correlation of problem hardnesses across different algorithms. In Tables 9.3a and 9.3b, the correlation of problem hardness between different algorithms is shown. Unsurprisingly, the hardness between the two simulated annealing variants correlates quite well. Because of the relatively poor performance of the standard genetic algorithm, the hardness measures do not correlate well with the exceptionally performing robust taboo search. Interestingly, simulated annealing correlates on average with the genetic algorithm but can come closer to the performance of the robust taboo search when comparing only the best parameter set. When considering that robust taboo search performs quite well on average and is not very sensitive to parameter settings, this makes sense.

Table 9.3. Correlation of hardnesses for different algorithms

(a) average hardnesses

	RTS	SA	SA2
SA	.39		
SA2	.59	.94	
GA	.55	.61	.70

(b) minimum/best hardness

	RTS	SA	SA2
SA	.79		
SA2	.79	.93	
GA	.36	.38	.49

Another interesting analysis is the discrepancy of hardnesses between different algorithms, as shown in Figure 9.4. The bur group of problems is relatively easy for any algorithm with little variance. The chr groups shows a little more variability, els19 (left of rou15, without a label in the chart) is indeed very special, relatively hard for the simulated annlealing while easier for both genetic algorithm and robust taboo search. The esc instances are rather easy, but especially so for the genetic algorithm. The lipa instances are on average quite hard, but provide some advantage when solved with a robust taboo search.

In general, there is a relatively clear tendency between the different hardness measures, so one could even define some sort of general average hardness. In view of the No Free Lunch Theorem, however, this will probably reduce, more or less, to the problem size.

Table 9.4 shows the list of the highest correlations between different measures. There, we take not only Pearson's r to correlate the different observations but also Spearman's rank correlation ρ. The minimum correlation $\min(|r|, |\rho|)$ of both Pearson's r and Spearman's ρ was used for the final ordering to enhance the confidence of correlation. Please note, that, in contrast to later comparisons, we use the

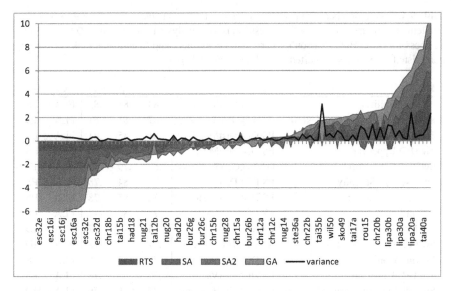

Fig. 9.4. Overview of average problem hardnesses for different algorithms. The values are normalized per algorithm and hence are not directly comparable with each other. Rather they show the difference in relative hardness for different algorithms.

correlation coefficent r instead of the more stringent r^2 as in the comparisons of the regression models. Among all of the highest correlations we only find FLA values with either problem specific measures or problem hardness measures.

Figure 9.5 exemplifies one of the issues of the simple correlation coefficients. Even though, the figures are relatively high, the correlation coefficient alone does not guarantee a nice correspondence. Instead, some extreme outliers correspond quite well, which reduces the penalty on mismatches for most of the remaining instances.

In total, we have analyzed 156 fitness landscape values by combining four mutation operators which each examine a slightly different landscape along three different exploration trajectories analyzing four ruggedness values, five information analysis values, and a total of 12 walk dependent analysis values.

Additionally, we used 64 problem specific measures, categorized as two dominance values, two sparsity values, ten distribution measures, four symmetry estimates, four measures of transitivity, 40 values summarizing node centered throughput, and two values combining the distinct matrices' values into cost related values. These values have then been compared to a total of seven different hardness measures for four different algorithms. From these different types of measures we have examined interrelation or correlations between these types. Figure 9.6 gives an overview of the correlations between different types of measures compared to their minimum correlation coefficient.

Table 9.4. Correlation between pairs of values from different "classes", where PS stands for problem specific measure, FLA for fitness landscape analysis measures and HD for problem hardness measure. Both Pearson's r correlation coefficient and Spearman's rank correlation ρ are listed. The values are sorted according to $\min(|r|, |\rho|)$.

Measure 1	Measure 2	r	ρ	min	Classes	
* * Inf Stability	PS Dist DistMean	.83	.85	.83	FLA	PS
Scr Neut AutoCorr1 Coeff Norm	GA avgGenHardness	-.87	-.81	.81	FLA	HD
Trn U/D UpperVar	PS Dist DistMean	.81	.85	.81	FLA	PS
Inv U/D UpperVar	PS Dist DistMean	.80	.85	.80	FLA	PS
Sw2 U/D AutoCorr1	GA avgGenHardness	.82	.79	.79	FLA	HD
Sw2 Neut InfStability	PS Dist DistMean	.78	.84	.78	FLA	PS
Sw2 U/D UpperVar	PS Dist DistMean	.78	.85	.78	FLA	PS
* * Inf Stability	PS Dist DistVar	.77	.87	.77	FLA	PS
Scr Neut AutoCorr1 Coeff Norm	GA Hardness	-.85	-.77	.77	FLA	HD
Sw2 U/D AutoCorr1 Coeff	RTS avgHardness	.77	.84	.77	FLA	HD
* * Inf Stability	PS Dist DistVar	.76	.87	.76	FLA	PS
Trn U/D UpperVar	PS Dist DistVar	.76	.87	.76	FLA	PS
Sw2 Rnd DensityBasinInf	GA avgGenHardness	.78	.76	.76	FLA	HD
Inv U/D UpperVar	PS Dist DistVar	.76	.87	.76	FLA	PS
Trn Neut AutoCorr1 Coeff Norm	GA avgGenHardness	-.76	-.77	.76	FLA	HD
Scr U/D InfStability	PS Dist DistVar	.75	.87	.75	FLA	PS
Sw2 U/D PartialInfContent	GA avgGenHardness	-.80	-.75	.75	FLA	HD
Trn Neut DensityBasinInf	SA2 avgItHardness	.80	.75	.75	FLA	HD
Sw2 U/D InfContent	GA avgGenHardness	-.75	-.75	.75	FLA	HD
Scr Neut AutoCorr1 Coeff Norm	SA2 avgItHardness	-.79	-.75	.75	FLA	HD
Trn Neut AutoCorr1 Coeff Norm	GA Hardness	-.75	-.76	.75	FLA	HD
Sw2 Neut InfStability	PS Dist DistVar	.74	.86	.74	FLA	PS

From the correlations between pairs in Figure 9.6 we have analyzed the frequency of these pairs regarding their minimum correlation coefficient in Table 9.5. We found, that true to their original intend, many fitness landscape analysis values correlate indeed with problem hardness estimates, especially when compared to problem specific measures. This is not surprising since problem specific measures are not intendet to directly measure problem hardness but describe particular intrinsic properties which can be used to compare and group problem instances with each other.

As a side note, we found that several FLA results correlate with each other, which is not surprising as most of them analyze exactly the same data, namely the sequence of fitness values of the various walk types. In particular we found that (partial) information content correlates with auto correlation, and auto correlation correlates with problem size and other correlation lengths. For the subsequent model building using linear regression we have, therefore, also tried to exclude highly correlated parameters to avoid complicating the parameter selection process.

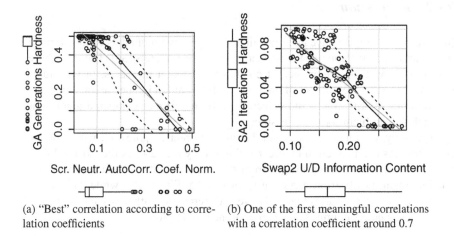

(a) "Best" correlation according to correlation coefficients

(b) One of the first meaningful correlations with a correlation coefficient around 0.7

Fig. 9.5. Correlation vs. Meaningful Correspondence. Even though correlation coefficients for single values are high, they are often due to outliers.

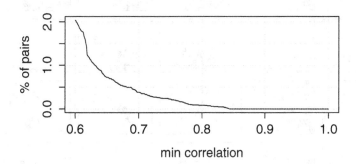

Fig. 9.6. Percentage of correlated pairs out of 16144 possibilities between value classes (problem-specific, FLA, and hardness) above a certain minimum correlation threshold.

Table 9.5. Number of correlated pairs from different classes. The number in parenthesis is the percentage of these counts from the number of possible pairs for these classes. Most direct correlations were observed between FLA values and hardness estimators, also FLA values correlate well with problem specific values. Interestingly, only few problem specific values directly correlate with problem hardness.

Pairs	$r \geq 0.5$		$r \geq 0.6$		$r \geq 0.7$		$r \geq 0.8$		total
hardness vs. problem specific	13	(0.72%)	0	(0%)	0	(0%)	0	(0%)	1792
FLA vs. problem-specific	354	(3.54%)	162	(1.62%)	25	(0.25%)	1	(0.01%)	9984
hardness vs. FLA	479	(10.97%)	173	(3.96%)	39	(0.89%)	13	(0.30%)	4368

9.7.2 Regression

In a previous study we have already built simple regression models using a limited set of variables only from the fitness landscape analysis results [36]. There, we have already observed the importance of combining different measures. In this study, we have extended the number of analyses to different landscape variants by using different neighborhood structures, increased the number of compared algorithms and included problem specific measures.

Previously, we found that problem hardness correlates quite well with problem size alone. As can be seen in Figure 9.7, this is only the case for the Robust Taboo Search. Another interesting insight from this chart is that problem hardness correlations stem mostly from particularity easy or hard instances, while the average instances' hardnesses varies widely between different algorithms which can be observed as unshapely clumps in the center of the scatter plots.

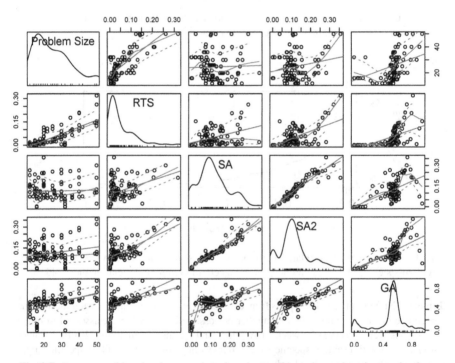

Fig. 9.7. Average problem hardnesses plotted against problem size and against each other.

One of the most difficult problems when using linear regression for prediction is variable selection. For this reason we have used several methods to limit the number of variables in the final models. The first measure to limit the number of variables was to build models for either FLA values or problem-specific values only. These models were further separated. Fitness landscape models were partitioned into mutation operators and trajectories and different combinations first. For problem specific models, the values were also split and recombined according to measure groups such as transitivity or throughput. Additionally, highly correlated input values were eliminated before automated variable selection. Different values for these input correlations were used. Finally, for a total of 4260 variable pre-selections, least angle regression using the Lasso method [46] was conducted.

Table 9.6. Classes of Variables: "×' denotes full combination, "1" denotes inclusion as-is, and "$\mathscr{P}$" denotes powerset expansion (except $\emptyset$) and full combination with all other variables

Class	Collection	Comb	Values
Hardness	Algorithm	×	RTS, SA, SA2, GA
	Measure	×	avg hardness, best hardness, specificity
Fitness Landscape	Variant	×	Swap2, Translocation, Inversion, Scramble
	Walk Type	$\mathscr{P}$	Random, Up-Down, Neutral
	Measures	×	Ruggedness, Information, Walk Lengths, . . .
Problem-Specific	PS1	1	Dominance, Flow
	PS2	1	PS1, Sparsity, Distribution
	PS3	1	PS2, Symmetry, Transitivity, Throughput

FLA values where partitioned into classes according to different neighborhoods and all permutations of different walk type. Problem-specific measures where directly assigned to different groups (PS1, PS2, and PS3) as shown in Table 9.6. Additionally, the problem size was allowed in any of the classes. The final variable selection was then filtered using the correlations between the variables using the thresholds 1, 0.99, and 0.9.

Table 9.7a shows the results when taking the best model after cross validation of least angle regression for each of the before-mentioned model classes and an unrestricted model that allows use of any combination of variables. The simple models using only previous problem specific values are not very apt in modeling any of the hardness values, even including simple distribution measures does not help a lot. Only the full array of problem specific measures can consistently create good models for different hardness values. Using the fitness landscape analysis models, we are able to build similarly apt models that are on average even slightly better. In view of their general applicability this is good news. While for problem specific values we have to think up new descriptors for every problem class, we can just re-use existing fitness landscape measures.

Another very promising aspect is that the combination of both FLA and PS measures yields significantly improved predictions as compared to the separate views. This supports the assumption that FLA and PS provide quite different perspectives of the problem instances.

Table 9.7b shows the qualities of only the best models and the averages over different algorithms. This allows to look at the difficulty of predicting certain hardness measures. In general the average hardness and average scaled difference can be

Table 9.7. Coefficients of determinations (R^2) of the best models

(a) Restrictions with different sets of variables

Algorithm	Measure	PS1	PS2	PS3	FLA	Full
RTS	avgH	.53	.66	**.81**	.76	.89
	avgSD	.37	.60	**.77**	.69	.82
	avgIt	.37	.40	**.64**	.61	.74
	H	.22	.22	.44	**.54**	.60
	spec	.39	.50	**.78**	.56	.78
SA	avgH	.04	.31	.47	**.59**	.60
	avgSD	.04	.35	**.46**	**.46**	.60
	avgIt	.12	.21	.64	**.72**	.79
	H	.22	.24	.48	**.69**	.77
	spec	.13	.41	**.58**	.55	.71
SA2	avgH	.07	.36	.50	**.58**	.69
	avgSd	.06	.44	**.50**	.45	.69
	avgIt	.19	.43	.68	**.74**	.84
	H	.25	.22	.58	**.65**	.79
	spec	.09	.46	**.52**	.45	.67
GA	avgH	.29	.48	.69	**.82**	.88
	avgSd	.41	.55	**.89**	.65	.92
	avgGen	.16	.30	.59	**.82**	.87
	H	.27	.43	.60	**.82**	.85
	spec	.00	.17	**.43**	.19	.46
Average		.21	.39	.60	**.62**	.75
>0.5		1	3	14	16	19

(b) Averages over algorithms for different
hardness measures.

Measure	RTS	SA	SA2	GA	Avg
avgH	.89	.60	.69	.88	.76
avgSD	.82	.60	.69	.92	.76
avgIt/Gen	.74	.79	.84	.87	.81
H	.60	.77	.79	.85	.75
spec	.78	.71	.67	.46	.65

satisfactorily be predicted. Average iterations, or average generations for the genetic
algorithm, seem to be easy to predict too. However, the main reason for their easier
prediction for SA and GA are mostly due to the fact that these algorithms often run
until the maximum number of iterations, which is, therefore, quite easy to predict.
On average, parameter specificity is the most difficult value to predict. This value
supposedly requires not only knowledge of the examined problem instance but also

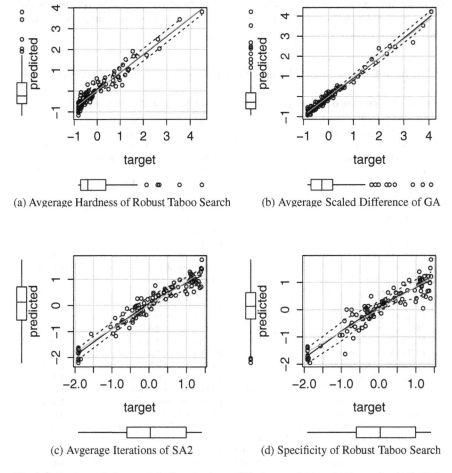

(a) Avgerage Hardness of Robust Taboo Search   (b) Avgerage Scaled Difference of GA

(c) Avgerage Iterations of SA2   (d) Specificity of Robust Taboo Search

Fig. 9.8. Best correlation models found using multi-stage variable selection using both FLA and problem-specific measurements

the specific difficulties for a certain algorithm and its robustness concerning parameter changes.

Some of the best regression models shown in Table 9.7b are shown in Figure 9.8 for visual inspection. As we have seen with the correlations with single values, a high correlation does not necessarily result in a good prediction due to various reasons explained earlier. However, the best models for each of the hardness measures using the full power of both FLA and problem-specific measures show acceptable performances and good visual correlation.

For reference, two of these models are provided in detail in Table 9.8. It is very instructive to look at the coefficients for these models with normalized variables as they can give a hint on the performance of the overall predictor and, hence, on the importance of the variable itself in predicting problem hardness. Apparently, the throughput and cost measures, as well as information and auto correlation values have a very large impact.

Table 9.8. Best models allowing all variables for different hardness estimates

(a) Average Hardness of Robust Taboo Search		(b) Avg Scaled Difference of GA	
(Intercept)	-2.2e+13	(Intercept)	-5.9e+11
Sw2.Rnd.InformationContent	9.5e+00	Inv.Neut.AutoCorrelation1	-6.8e-03
Sw2.Rnd.Regularity	4.3e-06	Inv.Neut.CorrelationLength	-3.3e-04
Sw2.U/D.DownWalkLenVar	-3.4e-02	Inv.Neut.DensityBasinInformation	1.1e+00
Problem Size	8.6e-02	Inv.Neut.Regularity	2.1e-05
Sw2.Rnd.AutoCorrelation1.Coeff.Norm	-2.6e+00	Inv.U/D.DownWalkLenVar	-4.2e-02
Sw2.U/D.AutoCorrelation1.Coeff	2.6e-02	Problem Size	4.8e-02
PS.Sparcity.Dist	-6.1e-01	Inv.U/D.AutoCorrelation1.Coeff.Norm	-6.2e-01
PS.FlowMean	-4.4e-05	PS.Sparcity.Dist	-2.8e-01
PS.FlowVar	-3.8e-10	PS.FlowSkew	-8.2e-02
PS.FlowKurt	-8.9e-03	PS.DistSkew	-6.7e-03
PS.FlowNormality	7.5e-01	PS.DistKurt	-1.9e-04
PS.DistKurt	2.1e-03	PS.DistNormality	-1.1e-01
PS.DistNormality	-5.7e-01	PS.Asymmetry.Flow	2.6e+00
PS.Asymmetry.FlowCount	-8.6e+00	PS.Asymmetry.FlowCount	9.8e+00
PS.Intransitivity.Flow	-1.2e+00	PS.Intransitivity.FlowCount	-3.4e-01
PS.Intransitivity.DistCount	-5.2e-02	PS.Intransitivity.DistCount	-5.9e-01
PS.Thru.FlowInFlowMean	9.2e+11	PS.Thru.FlowInFlowVar	-3.9e-03
PS.Thru.FlowOutFlowVar	-5.1e-03	PS.Thru.FlowInFlowSkew	-3.3e-02
PS.Thru.FlowOutFlowSkew	1.1e-01	PS.Thru.FlowInFlowNormality	1.1e-01
PS.Thru.FlowOutFlowKurt	3.2e-02	PS.Thru.FlowFlowSurplusSkew	6.1e-01
PS.Thru.FlowFlowSumNormality	-7.2e-02	PS.Thru.FlowFlowSurplusKurt	1.1e-02
PS.Thru.FlowFlowSurplusMean	2.1e+14	PS.Thru.DistInFlowMean	6.4e+12
PS.Thru.FlowFlowSurplusVar	3.6e-03	PS.Thru.DistInFlowVar	-1.4e-03
PS.Thru.DistInFlowMean	-3.7e+13	PS.Thru.DistFlowSurplusSkew	-3.8e-01
PS.Thru.DistInFlowSkew	-7.0e-02	PS.Thru.DistFlowSurplusKurt	-5.5e-02
PS.Thru.DistInFlowKurt	4.3e-02	PS.Thru.DistFlowSurplusNormality	1.8e-01
PS.Thru.DistOutFlowSkew	-1.5e-04	PS.Cost.Avg	1.2e+29
PS.Thru.DistFlowSurplusVar	-2.8e-03	PS.Cost.Var	5.9e+11
PS.Cost.Avg	-2.2e+29		
PS.Cost.Var	2.2e+13		

Examining all best models and the frequency of various analysis methods (data not shown), neighborhoods, trajectories and value classes shows that problem specific values are very frequently used, especially throughput, distribution, transitivity and asymetry, so essentially everything except the popular flow dominance. FLA trajectories as well as different neighborhoods are used approximately equally likely. It seems that PS values can only be used in combination, either with each other or with FLA values, as their correspondence with hardness is less direct than for FLA methods.

An important aspect to consider when building large regression models is the chance of building good models by chance. Figure 9.9a shows the performance of 200 completely random models with random targets (all assumed normally distributed) and the distribution of the resulting training qualities for different numbers of input variables. To ensure that we have not fallen prey to this issue we have verified that we are not using too many variables at once. Figure 9.9b shows the best models' coefficients of determination plotted against the number of variables. The line shows the lower bounds of the 99% confidence interval on the minimum number of variables necessary to build a random model with a certain quality. Most models can, therefore, be considered to carry significantly more information than a random model, and the identified correlations are likely to exhibit true links between the contained variables.

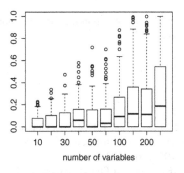

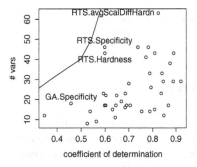

(a) Performances of random models. As the number of variables increases, even the quality of a random distribution can be estimated more and more easily.

(b) Coefficient of Determination (r^2) vs. problem size for the best models obtained. The line is the minimum number of variables in a random model (99% confidence interval)

Fig. 9.9. Performance comparison with random models

9.8   Conclusion

We have performed an extensive analysis of different easily obtainable generally applicable fitness landscape analysis values as well as problem specific measures for a collection of instances of the quadratic assignment problem and compared and correlated them with hardness measures of several algorithms. To a certain extent, fitness landscape analysis methods are somewhat better suited to predict problem hardness when used in isolation as compared to problem specific values. However, when used in conjunction the predictive performance is increased significantly which hints at the complementarity of the different measurement classes. While the prediction of problem hardness for different algorithms and different problem classes is still a difficult problem, we have seen, that by combing many different measures a decent prediction is possible. Even in isolation, fitness landscape analysis measures show promising results and set the stage for more advanced forecasting such as across different problem classes.

Important problems for future work are the determination of suitable parameter spaces and the large effort required to build solid data bases of previous solution attempts that are necessary for predicting unknown problems' hardnesses. Moreover, we have seen that different algorithms can have quite different "opinions" of whether a particular instance is difficult. In the future, the application of a similar methodology cannot only be used to predict problem instances' hardnesses but, based on this prediction, predict suitable algorithms and parameters. This might be what was previously called the "free appetizer" by Droste and colleagues which might be had even though there is no free lunch.

Acknowledgements. This work was in part supported by the Austrian Research Promotion Agency (FFG) within the Josef Ressel Center "Heureka!".

References

1. de Abreu, N.M.M., Netto, P.O.B., Querido, T.M., Gouvea, E.F.: Classes of quadratic assignment problem instances: isomorphism and difficulty measure using a statistical approach. Discrete Applied Mathematics 124, 103–116 (2002)
2. Barnett, L.: Ruggedness and neutrality—the NKp family of fitness landscapes. In: AL-IFE: Proceedings of the Sixth International Conference on Artificial Life, pp. 18–27. MIT Press, Cambridge (1998)
3. Beham, A., Pitzer, E., Affenzeller, M.: A new metric to measure distances between solutions to the quadratic assignment problem. In: Proceedings of the IEEE 3rd International Symposium on Logistics and Industrial Informatics, LINDI 2011 (2011)
4. Borenstein, Y., Poli, R.: Decomposition of fitness functions in random heuristic search. In: Stephens, C.R., Toussaint, M., Whitley, L.D., Stadler, P.F. (eds.) FOGA 2007. LNCS, vol. 4436, pp. 123–137. Springer, Heidelberg (2007)
5. Burkard, R.E., Karisch, S.E., Rendl, F.: QAPLIB - A quadratic assignment problem library. Journal of Global Optimization 10(4), 391–403 (1997)
6. Burkard, R.E., Offermann, J.: Entwurf von Schreibmaschinentastaturen mittels quadratischer Zuordnungsprobleme. Zeitschrift fr Operations Research 21, B121–B132 (1977)
7. de Carvalho Jr., S.A., Rahmann, S.: Microarray layout as quadratic assignment problem. In: Proceedings of the German Conference on Bioinformatics (GCB). Lecture Notes in Informatics, vol. P-83 (2006)
8. Chicano, F., Luque, G., Alba, E.: Elementary landscape decomposition of the quadratic assignment problem. In: Proceedings of the 12th Annual Conference on Genetic and Evolutionary Computation, GECCO 2010, New York, NY, USA, pp. 1425–1432 (2010)
9. Connolly, D.T.: An improved annealing scheme for the QAP. European Journal of Operational Research 46, 93–100 (1990)
10. Drezner, Z.: Extensive experiments with hybrid genetic algorithms for the solution of the quadratic assignment problem. Computers & Operations Research 35(3), 717–736 (2008); Part Special Issue: New Trends in Locational Analysis

11. Droste, S., Jansen, T., Wegener, I.: Perhaps not a free lunch but at least a free appetizer. In: Banzhaf, W., Daida, J., Eiben, A.E., Garzon, M.H., Honavar, V., Jakiela, M., Smith, R.E. (eds.) Proceedings of the Genetic and Evolutionary Computation Conference, vol. 1, pp. 833–839. Morgan Kaufmann (1999)
12. Elshafei, A.N.: Hospital layout as a quadratic assignment problem. Operational Research Quarterly 28(1), 167–179 (1977)
13. Gavett, J.W., Plyter, N.V.: The optimal assignment of facilities to locations by branch and bound. Operations Research 14, 210–232 (1966)
14. Gilmore, P.C.: Optimal and suboptimal algorithms for the quadratic assignment problem. SIAM Journal on Applied Mathematics 10, 305–313 (1962)
15. Glover, F.: Tabu search – part I. ORSA Journal on Computing 1(3), 190–206 (1989)
16. Glover, F., Laguna, M.: Tabu Search. Kluwer (1997)
17. Hahn, P.M., Krarup, J.: A hospital facility layout problem finally solved. Journal of Intelligent Manufacturing 12, 487–496 (2001)
18. Hintze, J.L., Nelson, R.D.: Violin plots: a box plot-density trace synergism. The American Statistician 52(2), 181–184 (1998)
19. Holland, J.H.: Adaptation in Natural and Artificial Systems. University of Michigan Press (1975)
20. Hordijk, W.: A measure of landscapes. Evol. Comput. 4(4), 335–360 (1996)
21. James, T., Rego, C., Glover, F.: Multistart tabu search and diversification strategies for the quadratic assignment problem. Systems, Man and Cybernetics, Part A: Systems and Humans, IEEE Transactions on 39(3), 579–596 (2009)
22. Jarque, C.M., Bera, A.K.: Efficient tests for normality, homoscedasticity and serial independence of regression residuals. Economics Letters 6(3), 255–259 (1980)
23. Jones, T.: Evolutionary algorithms, fitness landscapes and search. Ph.D. thesis, University of New Mexico, Albuquerque, New Mexico (1995)
24. Jones, T., Forrest, S.: Fitness distance correlation as a measure of problem difficulty for genetic algorithms. In: Proceedings of the 6th International Conference on Genetic Algorithms, pp. 184–192. Morgan Kaufmann (1995)
25. Katada, Y.: Estimating the degree of neutrality in fitness landscapes by the Nei's standard genetic distance – an application to evolutionary robotics. In: 2006 IEEE Congress on Evolutionary Computation, pp. 483–490 (2006)
26. Kirkpatrick, S., Gelatt, C.D., Vecchi, M.P.: Optimization by simulated annealing. Science 220, 671–680 (1983)
27. Knowles, J.D., Corne, D.W.: Towards Landscape Analyses to Inform the Design of a Hybrid Local Search for the Multiobjective Quadratic Assignment Problem. In: Soft Computing Systems: Design, Management and Application, pp. 271–279. IOS Press (2002)
28. Koopmans, T.C., Beckmann, M.: Assignment problems and the location of economic activities. Econometrica, Journal of the Econometric Society 25(1), 53–76 (1957)
29. Land, A.M.: A problem of assignment with interrelated costs. Operations Research Quarterly 14, 185–198 (1963)
30. Lawler, E.L.: The quadratic assignment problem. Management Science 9, 586–599 (1963)
31. Macready, W.G., Wolpert, D.H.: What makes an optimization problem hard? Complexity 5, 40–46 (1996)
32. Merz, P., Freisleben, B.: Fitness landscape analysis and memetic algorithms for the quadratic assignment problem. Evolutionary Computation, IEEE Transactions on 4(4), 337–352 (2000)

33. Metropolis, N., Rosenbluth, A.W., Rosenbluth, M.N., Teller, A.H., Teller, E.: Equation of state calculations by fast computing machines. The Journal of Chemical Physics 21, 1087–1092 (1953)

34. Nugent, C.E., Vollmann, T.E., Ruml, J.: An experimental comparison of techniques for the assignment of facilities to locations. Journal of Operations Research 16, 150–173 (1969)

35. Pitzer, E., Affenzeller, M.: A comprehensive survey on fitness landscape analysis. In: Fodor, J., Klempous, R., Suárez Araujo, C.P. (eds.) Recent Advances in Intelligent Engineering Systems. SCI, vol. 378, pp. 161–191. Springer, Heidelberg (2012)

36. Pitzer, E., Beham, A., Affenzeller, M.: Generic hardness estimation using fitness and parameter landscapes applied to the robust taboo search and the quadratic assignment problem. In: Proceedings of the Genetic and Evolutionary Computation Conference (GECCO 2012). ACM (2012) (in press)

37. Pitzer, E., Vonolfen, S., Beham, A., Affenzeller, M., Bolshakov, V., Merkurieva, G.: Structural analysis of vehicle routing problems using general fitness landscape analysis and problem specific measures. In: Proceedings of the 14th International Asia Pacific Conference on Computer Aided System Theory (2012) (accepted)

38. Reidys, C.M., Stadler, P.F.: Neutrality in fitness landscapes. Applied Mathematics and Computation 117(2-3), 321–350 (1998)

39. Sahni, S., Gonzalez, T.: P-complete approximation problems. Journal of the Association of Computing Machinery 23, 555–565 (1976)

40. Smith-Miles, K., Lopes, L.: Review: Measuring instance difficulty for combinatorial optimization problems. Comput. Oper. Res. 39(5), 875–889 (2012)

41. Stadler, P., Wagner, G.: The algebraic theory of recombination spaces. Evol. Comp. 5, 241–275 (1998)

42. Stadler, P.F.: Linear operators on correlated landscapes. J. Phys. I France 4, 681–696 (1994)

43. Stadler, P.F.: Fitness Landscapes. In: Biological Evolution and Statistical Physics, pp. 187–207. Springer (2002)

44. Taillard, E.D.: Robust taboo search for the quadratic assignment problem. Parallel Computing 17, 443–455 (1991)

45. Tate, D.M., Smith, A.E.: A genetic approach to the quadratic assignment problem. Computers and Operations Research 22, 73–83 (1995)

46. Tibshirani, R.: Regression shrinkage and selection via the lasso. Journal of the Royal Statistical Society B 58(1), 267–288 (1996)

47. Vassilev, V.K., Fogarty, T.C., Miller, J.F.: Information characteristics and the structure of landscapes. Evol. Comput. 8(1), 31–60 (2000)

48. Vollmann, T.E., Buffa, E.S.: The facilities layout problem in perspective. Management Science 12(10), 450–468 (1699)

49. Wagner, S.: Heuristic optimization software systems - Modeling of heuristic optimization algorithms in the HeuristicLab software environment. Ph.D. thesis, Johannes Kepler University, Linz, Austria (2009)

50. Wagner, S., Beham, A., Kronberger, G.K., Kommenda, M., Pitzer, E., Kofler, M., Vonolfen, S., Winkler, S.M., Dorfer, V., Affenzeller, M.: Heuristiclab 3.3: A unified approach to metaheuristic optimization. In: Actas del sptimo congreso espaol sobre Metaheursticas, Algoritmos Evolutivos y Bioinspirados (MAEB 2010), p. 8 (2010)

51. Weinberger, E.: Correlated and uncorrelated fitness landscapes and how to tell the difference. Biological Cybernetics 63(5), 325–336 (1990)

52. Wilhelm, M.R., Ward, T.L.: Solving quadratic assignment problems by simulated annealing. IEEE Transactions 19, 107–119 (1987)
53. Wolpert, D.H., Macready, W.G.: No free lunch theorems for optimization. IEEE Transactions on Evolutionary Computation 1(1), 67–82 (1997)
54. Wright, S.: The roles of mutation, inbreeding, crossbreeding and selection in evolution. In: Proceedings of the Sixth International Congress of genetics, vol. 1, pp. 356–366 (1932)
55. Xin, B., Chen, J., Pan, F.: Problem difficulty analysis for particle swarm optimization: deception and modality. In: Proceedings of the First ACM/SIGEVO Summit on Genetic and Evolutionary Computation, GEC 2009, pp. 623–630 (2009)

Chapter 10
Architecture and Design of the HeuristicLab Optimization Environment

S. Wagner, G. Kronberger, A. Beham, M. Kommenda, A. Scheibenpflug, E. Pitzer, S. Vonolfen, M. Kofler, S. Winkler, V. Dorfer, and M. Affenzeller

Abstract. Many optimization problems cannot be solved by classical mathematical optimization techniques due to their complexity and the size of the solution space. In order to achieve solutions of high quality though, heuristic optimization algorithms are frequently used. These algorithms do not claim to find global optimal solutions, but offer a reasonable tradeoff between runtime and solution quality and are therefore especially suitable for practical applications. In the last decades the success of heuristic optimization techniques in many different problem domains encouraged the development of a broad variety of optimization paradigms which often use natural processes as a source of inspiration (as for example evolutionary algorithms, simulated annealing, or ant colony optimization). For the development and application of heuristic optimization algorithms in science and industry, mature, flexible and usable software systems are required. These systems have to support scientists in the development of new algorithms and should also enable users to apply different optimization methods on specific problems easily. The architecture and design of such heuristic optimization software systems impose many challenges on developers due to the diversity of algorithms and problems as well as the heterogeneous requirements of the different user groups. In this chapter the authors describe the architecture and design of their optimization environment HeuristicLab which aims to provide a comprehensive system for algorithm development, testing, analysis and generally the application of heuristic optimization methods on complex problems.

S. Wagner · G. Kronberger · A. Beham · M. Kommenda · A. Scheibenpflug · E. Pitzer ·
S. Vonolfen · M. Kofler · S. Winkler · V. Dorfer · M. Affenzeller
Heuristic and Evolutionary Algorithms Laboratory, School of Informatics,
Communication and Media, University of Applied Sciences Upper Austria, Softwarepark 11,
4232 Hagenberg, Austria
e-mail: heal@heuristiclab.com

R. Klempous et al. (eds.), *Advanced Methods and Applications in Computational Intelligence*,   197
Topics in Intelligent Engineering and Informatics 6,
DOI: 10.1007/978-3-319-01436-4_10, © Springer International Publishing Switzerland 2014

10.1 Introduction

In the last decades a steady increase of computational resources and concurrently an impressive drop of hardware prices could be observed. Nowadays, very powerful computer systems are found in almost every company or research institution, providing huge processing power on a broad basis which was unthinkable some years ago. This trend opens the door for tackling complex optimization problems of various domains that were not solvable in the past. Concerning problem solving methodologies, especially heuristic algorithms are very successful in that sense, as they provide a reasonable tradeoff between solution quality and required runtime.

In the research area of heuristic algorithms a broad spectrum of optimization techniques has been developed. In addition to problem-specific heuristics, particularly the development of metaheuristics is a very active field of research, as these algorithms represent generic methods that can be used for solving many different optimization problems. A variety of often nature inspired archetypes has been used as a basis for new optimization paradigms such as evolutionary algorithms, ant systems, particle swarm optimization, tabu search, or simulated annealing. Several publications show successful applications of such metaheuristics on benchmark and real-world optimization problems.

However, JÃŒrg Nievergelt stated in his article in 1994 *"No systems, no impact!"* [30] and pointed out that well-engineered software systems are a fundamental basis to transfer research results from science to industry. Of course, this statement is also true in the area of heuristic optimization. In order to apply effective and enhanced metaheuristics on real-world optimization problems, mature software systems are required which meet demands of software quality criteria such as reliability, efficiency, usability, maintainability, modularity, portability, or security. Although these requirements are well-known and considered in enterprise software systems, they are not yet satisfactorily respected in the heuristic optimization community. Most heuristic optimization frameworks are research prototypes and are funded by national or international research programs. In such scenarios it is very hard to establish a continuous development process, to take into account many different users, to provide comprehensive support for users, or to reach the maturity of a software product. Therefore, these systems cannot be easily integrated into an enterprise environment.

Another major difficulty regarding the design of general purpose heuristic optimization software systems is that there is no common model for metaheuristics. Due to the heterogeneous nature of heuristic optimization paradigms, it is hard to identify and generalize common concepts which imposes a challenging problem on software developers. Many existing software frameworks focus on one or a few particular optimization paradigms and miss the goal of providing an infrastructure which is generic enough to represent all different kinds of algorithms. The variety of existing frameworks makes it very difficult for researchers to develop and compare their algorithms to show advantageous properties of a new approach. A unified software platform for heuristic optimization would improve this situation, as it would enable algorithm developers to assemble algorithms from a set of ready-to-use components and to analyze and compare results in a common framework.

Therefore, the development of high quality and mature heuristic optimization software systems would lead to a win-win situation for industry and for science. In this chapter, the authors describe their efforts towards this goal in the development of the flexible architecture and generic design of the HeuristicLab optimization environment. Instead of trying to incorporate different heuristic optimization algorithms into a common model, HeuristicLab contains a generic algorithm (meta-)model that is capable of representing not only heuristic optimization but arbitrary algorithms. By this means HeuristicLab can be used to develop custom algorithm models for various optimization paradigms. Furthermore, state-of-the-art software engineering methodologies are used to satisfy additional requirements such as parallelism, user interaction on different layers of abstraction, flexible deployment, or integration into existing applications.

10.1.1   Related Work

Modern concepts of software engineering such as object-oriented or component-oriented programming represent the state of the art for creating complex software systems by providing a high level of code reuse, good maintainability and a high degree of flexibility and extensibility (see for example [29, 20, 8, 17]). However, such approaches are not yet established on a broad basis in the area of heuristic optimization, as this field is much younger than classical domains of software systems (e.g., word processing, calculation, image processing, or integrated development environments). Most systems for heuristic optimization are one man projects and are developed by researchers or students to realize one or a few algorithms for solving a specific problem. Naturally, when a software system is developed mainly for personal use or a very small, well-known and personally connected user group, software quality aspects such as reusability, flexibility, genericity, documentation and a clean design are not the prime concern of developers. As a consequence, seen from a software engineering point of view, in most cases these applications still suffer from a quite low level of maturity.

In the last years and with the ongoing success of heuristic algorithms in scientific as well as commercial areas, the heuristic optimization community started to be aware of this situation. Advantages of well designed, powerful, flexible and ready-to-use heuristic optimization frameworks were discussed in several publications [34, 28, 37, 22, 44, 13, 31], identifying similar requirements as described in this chapter. Furthermore, some research groups started to head for these goals and began redesigning existing or developing new heuristic optimization software systems which were promoted as flexible and powerful white or even black box frameworks, available and useable for a broad group of users in the scientific as well as in the commercial domain. In comparison to the systems available before, main advantages of these frameworks are a wide range of ready-to-use classical algorithms, solution representations, manipulation operators, and benchmark problems which make it easy to start experimenting and comparing various concepts. Additionally, a high degree of flexibility due to a clean object-oriented design makes it easy for users to implement custom extensions such as specific optimization problems or algorithmic ideas.

One of the most challenging tasks in the development of such a general purpose heuristic optimization framework is the definition of an object model representing arbitrary heuristic optimization paradigms. This model has to be flexible and extensible to a very high degree so that users can integrate non-standard algorithms that often do not fit into existing paradigms exactly. Furthermore the model should be very fine-grained so that a broad spectrum of existing classical algorithms can be represented as algorithm modules. Then, these modules can serve as building blocks to realize different algorithm variations or completely new algorithms with a high amount of reusable code.

Consequently, the question is on which level of abstraction such a model should be defined. A high level of abstraction leads to large building blocks and a very flexible system. A lower level of abstraction supports reusability by providing many small building blocks, but the structure of algorithms has to be predefined more strictly in that case which reduces flexibility. As a consequence, these two requirements are contradictory to some degree.

Taking a look at several existing frameworks for heuristic optimization, it can be seen that this question has been answered in quite different ways. Several publications have been dedicated to the comparison and evaluation of common frameworks for heuristic optimization and show that each of the existing systems is focused on some specific aspects in the broad spectrum of identified requirements [34, 38, 49, 13, 31]. None of the frameworks is able to dominate the others in all or at least most of the considered evaluation criteria. This indicates that time is ripe for consolidation. The lessons learned in the development of the different frameworks as well as their beneficial features should be shared and incorporated into the ongoing development. As a basis for this process detailed documentation and comprehensive publications are required which motivate and describe the architecture and design as well as the specific features of the systems. Therefore this chapter represents a step towards this direction and describes the HeuristicLab optimization environment in detail and highlights the features that set HeuristicLab apart from other existing systems.

10.1.2 Feature Overview

In contrast to most other heuristic optimization systems, the development of HeuristicLab targets two major aspects: On the one hand HeuristicLab contains a very generic algorithm model. Therefore HeuristicLab is not only dedicated to some specific optimization paradigm (such as evolutionary algorithms or neighborhood-based heuristics) but is capable of representing arbitrary heuristic optimization algorithms. On the other hand HeuristicLab considers the fact that users of heuristic optimization software are in many cases experts in the corresponding problem domain but not in computer science or software development. Therefore HeuristicLab provides a rich graphical user interface which can be used not only to parameterize and execute algorithms but also to manipulate existing or define new algorithms in a graphical algorithm designer. Furthermore, the HeuristicLab user interface also provides powerful features to define and execute large-scale optimization experiments and to analyse the results with interactive charts. In this way also users who do not have a profound background in programming can work with HeuristicLab easily.

In general the most relevant features of HeuristicLab can be roughly summarized as follows:

- **Rich Graphical User Interface**
  A comfortable and feature rich graphical user interface reduces the learning effort and enables users without programming skills to use and apply HeuristicLab.
- **Many Algorithms and Problems**
  Several well-known heuristic algorithms and optimization problems are already implemented in HeuristicLab and can be used right away.
- **Extensibility**
  HeuristicLab consists of many different plugins. Users can create and reuse plugins to integrate new features and extend the functionality of HeuristicLab.
- **Visual Algorithm Designer**
  Optimization algorithms can be modeled and extended with the graphical algorithm designer.
- **Experiment Designer**
  Users can define and execute large experiments by selecting algorithms, parameters and problems in the experiment designer.
- **Results Analysis**
  HeuristicLab provides interactive charts for a comfortable analysis of results.
- **Parallel and Distributed Computing**
  HeuristicLab supports parallel execution of algorithms on multi-core or cluster systems.

10.1.3 Structure and Content

The remaining parts of this chapter are structured as follows: In Section 10.2 the main user groups of HeuristicLab are identified and their requirements are analyzed. The architecture and design of HeuristicLab is discussed in Section 10.3. Thereby, a main focus is put on the presentation of the flexible and generic algorithm model. Section 10.4 outlines how different metaheuristics can be modeled and Section 10.5 exemplarily describes some optimization problems which are included in HeuristicLab. Finally, Section 10.6 summarizes the main characteristics of HeuristicLab and concludes the chapter by describing several application areas of HeuristicLab and future work.

10.2   User Groups and Requirements

Developing a generic software system for heuristic optimization such as Heuristic-Lab is a challenging task. The variety of heuristic optimization paradigms and the multitude of application domains make it difficult for software engineers to build a system that is flexible enough and also provides a large amount of reusable components. Additionally, the users of heuristic optimization software systems are also very heterogeneous concerning their individual skills and demands. As a consequence, it is essential to study the different user groups in detail in order to get a clear picture of all requirements.

When considering literature on optimization software systems for different heuristic algorithms, some requirements are repeatedly stated by researchers. For example in [13], Christian Gagné and Marc Parizeau define the following six genericity criteria to qualify evolutionary computation frameworks: generic representation, generic fitness, generic operations, generic evolutionary model, parameters management, and configurable output. Quite similar ideas can also be found in [21, 37, 23, 44, 31].

Although most of these aspects reflect important user demands, none of these publications sketch a clear picture of the system's target users groups. As a consequence, without a precise picture of the users it is hard to determine whether the list of requirements is complete or some relevant aspects have been forgotten. Thus, before thinking about and defining requirements, it is necessary to identify all users.

10.2.1   User Groups

In general, users of a heuristic optimization system can be categorized into three often overlapping groups: practitioners, trying to solve real-world optimization problems with classical or advanced heuristics; heuristic optimization experts, analyzing, hybridizing and developing advanced algorithms; and students, trying to learn about and work with heuristic optimization algorithms and problems. Therefore, these three groups of users called *practitioners*, *experts* and *students* and their views on the area of heuristic optimization as well as their individual needs are described in detail in the following.

10.2.1.1   Practitioners

Practitioners are people who have encountered some difficult (often NP-hard) optimization problem and who want to get a solution for that problem. Hard optimization problems can be found in almost every domain (for example in engineering, medicine, economics, computer science, production, or even in arts), so this group is huge and very heterogeneous. Due to that heterogeneity, it is not possible to list all the domains where heuristic optimization algorithms have already been successfully applied or even to think of all possible domains in which they might be applied successfully in the future. Therefore, further refinement and categorization of the members of this user group is omitted.

Seen from an abstract point of view, practitioners work in a domain usually not related to heuristic optimization or software engineering. They normally have very little knowledge of heuristic algorithms but a profound and deep knowledge of the problem itself, its boundary conditions and its domain. This results in a highly problem-oriented way of thinking; in this context heuristic optimization algorithms are merely used as black box solvers to get a solution.

Usually, practitioners want to get a satisfactory solution to their problem as quickly as possible. Each second spent on computation means that some real-world system is running in a probably sub-optimal state. To them, time is money, and thus their number one concern is performance. For example, the operators of a production plant are interested in an optimal schedule of operations in order to minimize tardiness of orders and to deliver on time. Any time a machine breaks down, a new order is accepted, or the available capacity changes, a new schedule is needed at once. Each minute production is continued without following an optimized schedule may lead to the wrong operations being chosen for production. This may finally result in a higher tardiness on some high priority orders causing penalties and a severe loss of money.

Parallelism and scalability are of almost equal importance. In a production company the heuristic optimization software system used to compute optimized schedules should be able to provide equally good results, even if business is going well and there are twice as many production orders to be scheduled. A simple equation should hold: More computing power should either lead to better results or to the possibility to solve larger problems. Therefore, it has to be possible to enhance the optimization system with some additional hardware to obtain better performance.

Next, practitioners require a high level of genericity. Due to the heterogeneous nature of domains in which optimization problems might arise, a software system has to support easy integration of new problems. Usually this integration is done by implementing problem-specific objective functions, custom solution representations, and a generic way to introduce new operations on these solutions. For example, new solution manipulation operators might have to be developed, respecting some constraints a feasible solution has to satisfy.

Another important aspect is the integration of a heuristic optimization software system. Usually an optimization system is not a stand-alone application. Data defining a problem instance is provided by other existing software systems and solutions have to be passed on to other applications for further processing. For this reason, in most real-world scenarios heuristic optimization software systems have to be integrated into a complex network of existing IT infrastructure. Well-defined interfaces and technology for inter-system communication and coupling are necessary.

Finally, due to the highly problem-oriented focus of practitioners, they should not have to deal with the internals of algorithms. After a problem has been defined and represented in a heuristic optimization software system, the system should provide a comprehensive set of classical and advanced optimization algorithms. These algorithms can be evaluated on the concrete problem at hand and a best performing one can be chosen as a black box solver for live operation on real-world data.

10.2.1.2 Experts

Experts are researchers focusing on heuristic optimization algorithm engineering. Their aim is to enhance existing algorithms or develop new ones for various kinds of problems. Following the concept of metaheuristics, especially problem-independent modifications of algorithms are of major interest. By this means different kinds of optimization problems can benefit from such improvements. As a result and in contrast to practitioners, experts consider algorithms as white box solvers. For them, concrete optimization problems are less important and are used as case studies to show the advantageous properties of a new algorithmic concept, such as robustness, scalability, and performance in terms of solution quality and runtime. In many cases, problem instances of well-known benchmark optimization problems, as for example the traveling salesman problem or n-dimensional real-valued test functions, are used. Ideally, a comprehensive set of benchmark problems is provided out of the box by a heuristic optimization software system.

Due to the focus on algorithms, one main concern of experts is genericity. A heuristic optimization software system should offer abilities to integrate new algorithmic concepts easily. There should be as few restrictions in the underlying algorithm model as possible, enabling the incorporation of techniques stemming from different areas such as evolutionary algorithms, neighborhood-based search or swarm systems (hybridization), the flexible modification of existing algorithms, or the development of new ones. The sequence of operations applied to one or more solutions during algorithm execution, in other words the algorithm model, should be freely configurable. Furthermore, experts, similarly to practitioners, demand the integration of new operations, solution representations and objective functions.

One main task of experts is testing algorithms on different kinds of problems, as empirical evaluation is necessary to analyze properties of a new algorithm. Thus, automation of test case execution and statistical analysis play an important role. Thereby, performance in terms of execution time is usually just of secondary importance, as time constraints are not that financially critical as they are for practitioners.

In order to get some hints for algorithm improvements, experts have to use various tools to obtain a thorough understanding of the internal mechanisms and functionality of an algorithm. Since many heuristic optimization algorithms are very sensitive to parameters values, a generic way of parameter management is of great value, owing to the fact that various parameters have to be adjusted from test run to test run. Furthermore, stepwise algorithm execution and customizable output are the basis for any in-depth algorithm analysis, in order to get a clearer picture of how an algorithm is performing and what effect was caused by some parameter adjustment or structural modification.

Finally, replicability and persistence have to be mentioned: Each algorithm run has to be reproducible for the sake of later reference and analysis. A persistence mechanism is thus an essential means by which a run can be saved at any time during its execution and can be restored later on.

10.2.1.3   Students

Students entering the area of heuristic optimization are users that are located between the two user groups described above. In the beginning, they experiment with various heuristic optimization techniques and try to solve well-known benchmark problems. Therefore, a comprehensive set of classical heuristic optimization algorithms and benchmark problems should be provided by a heuristic optimization software system.

During several experiments and algorithm runs, students gain more and more insight into the internal functionality of algorithms and the interdependency between diversification and intensification of the search process. As the mystery of heuristic optimization is slowly unraveled, their view of algorithms changes from black box to white box. Hence, requirements relevant to experts - such as genericity, parameter management, automation, or customizable output - become more and more important to these users as well.

Additionally, when using heuristic optimization techniques and a heuristic optimization software system for the first time, an easy-to-use and intuitive application programming interface (API) is helpful to reduce the necessary learning effort. Even more, a graphical user interface (GUI) is extremely advantageous, so that students do not have to worry about peculiarities of programming languages and frameworks. Instead, with a GUI they are able to concentrate on the behavior of algorithms on an abstract level. Especially, studying charts and logs of the internal state of an algorithm during its execution is very effective to gain a deeper understanding of algorithm dynamics.

10.2.2   Requirements

Based on the analysis of the three user groups, the following requirements can be defined for heuristic optimization software systems (some of these requirements can be found in similar form in [22, 37, 23, 44, 31]). The requirements are listed alphabetically and the order does not reflect the importance of each requirement.

- **Automation**
  As heuristic algorithms are per se non-deterministic, comparison and evaluation of different algorithms requires extensive empirical tests. Therefore, a heuristic optimization software system should provide functionality for experiment planning, automated algorithm execution, and statistical analysis.
- **Customizable Output**
  In a real-world scenario heuristic optimization never is an end in itself. To enable further processing of results with other applications, the user has to customize the

output format of a heuristic algorithm. Furthermore, user defined output is also required by experts to visualize the internal mechanisms of algorithms by logging internal states such as distribution of solutions in the solution space, similarity of solutions, or stagnation of the search.

- **Generic Algorithm Model**
  In order to represent different kinds of heuristic optimization algorithms, the main algorithm model has to be flexible and customizable. It should not be dedicated or limited to any specific heuristic optimization paradigm. Especially, this aspect should also be kept in mind in terms of naming of classes and methods, so that users are not irritated or misled by the API.

- **Generic Operators**
  Related to the demand for a generic algorithm model, the operations applied by a heuristic algorithm should be generic as well. The user has to be able to implement either problem-specific or generic operations in an easy and intuitive way. There has to be a standardized interface for all operations and a uniform way how data is represented, accessed, and manipulated.

- **Generic Objective Functions**
  To enable integration of custom optimization problems, a generic concept of objective functions has to be available. Users should be able to add custom methods for quality evaluation easily. The quality evaluation mechanism has to be based on a clearly defined interface, so that all other operations depending on quality values - such as selection or heuristic manipulation operations - do not have to take care of how the quality of a solution is calculated in detail. In that context, working with the quality of a solution has to be abstracted from the concrete quality representation. For example, there should be no difference for selection operations whether the optimization problem is single-objective or multi-objective or a minimization or maximization problem. Therefore, a generic way of comparing two solutions is necessary.

- **Generic Solution Representations**
  As users need to integrate custom optimization problems, not only generic objective functions but also a generic way of solution representation is required. Ideally, the user should be able to assemble a custom solution representation by combining different standard data representations such as single values, arrays, matrices or enumerations of different data types. As an alternative, solution representations using complex custom data structures independent of any predefined ones should also be supported. This requirement has a strong impact on the requirement of generic operators, as crossover or manipulation operators have to work on solutions directly.

- **Graphical User Interface**
  To pave the way to heuristic optimization for users not so familiar with software development or specific programming languages, a heuristic optimization software system needs to be equipped with a graphical user interface (GUI). Users should be able to modify or develop algorithms without depending on any

specific development environment. Furthermore, a GUI is also very helpful for experimenting and rapid prototyping, as algorithms can be modeled and visualized seamlessly directly within the system.

- **Integration**
  After their development, heuristic optimization algorithms for real-world optimization problems have to be integrated into some existing information technology landscape. Hence, a heuristic optimization software system should be modular to be able to integrate just the parts required in a custom scenario. Generic communication protocols for the system and its environment are necessary for passing new optimization tasks into and getting results out of the system.

- **Learning Effort**
  Users of a heuristic optimization software system should be able to start to work with the system quickly. Only little knowledge of programming languages and just basic skills in programming and software development should be necessary. The API of a heuristic optimization software system should therefore be intuitive, easy to understand, and should follow common design practices. Additionally, a high level user interface should be provided to decouple algorithm and problem engineering from software development.

- **Parallelism**
  A heuristic optimization software system should be scalable in terms of computing power. Using additional computing resources should either enable the user to solve larger problems, or to achieve better solution quality. Consequently, exploitation of parallelism is an important success factor. A heuristic optimization software system has to offer a seamless integration of parallelism for development and execution of parallel algorithms. Ideally, the user just has to define which parts of an algorithm should be executed in parallel, without having to think about how parallelization is finally done. Due to the duality of high-performance computing systems (shared memory multi-core CPUs versus distributed memory cluster or grid systems) parallelization concepts for both architectures should be provided.

- **Parameter Management**
  Many heuristic optimization algorithms offer several parameters to influence their behavior. As performance of most algorithms is very sensitive in terms of parameter values, users need to run an algorithm many times to find an optimal configuration. Consequently, a generic parameter management facility is required to change parameters without needing to modify any program code or to recompile operators or, even worse, the whole system.

- **Performance**
  Heuristic optimization applications are usually time-critical. Many objective functions of real-world optimization problems as well as heuristic optimization algorithms are very expensive in terms of execution time. Thus a heuristic optimization software system should support runtime efficient implementation and execution of algorithms.

- **Predefined Algorithms and Problems**
  To enable solving of optimization problems or comparison of algorithms out of the box, a heuristic optimization software system has to provide a broad spectrum of predefined algorithms and problems. Especially, it should be possible to use parts of existing algorithms and problems as a basis for further development or hybridization.
- **Replicability and Persistence**
  As experimental evaluation is a substantial task in heuristic algorithm development, test runs of algorithms have to be reproducible. Users should be able to save an algorithm and to restore it later on. The software system should therefore also enable stopping and saving algorithms during execution at any time. In that context, random number generators have to be handled with care, depending on whether an algorithm should be replayed with the same or a different random number sequence.

10.3   Architecture and Design

In order to fulfill the requirements identified in Section 10.2, the authors work on the development of an advanced generic and flexible environment for heuristic optimization called HeuristicLab. HeuristicLab has continuously evolved in the last ten years and three major versions have been developed until now which are referred to as HeuristicLab 1.x, HeuristicLab 2.x, and HeuristicLab 3.x. In the following, the previous versions are briefly covered and especially the newest version, HeuristicLab 3.x, is presented in detail.

10.3.1   HeuristicLab 1.x

The development of HeuristicLab 1.x [40, 44] started in 2002 as a programming project at the Johannes Kepler University Linz, Austria. The main goal of the project was to develop a generic, extensible and paradigm-independent environment for heuristic optimization that can be used by researchers in scientific and industrial projects to develop and compare new optimization algorithms and by students in lectures and exercises.

Microsoft® .NET and the C# programming language were chosen as the development platform for HeuristicLab 1.x. Reasons for this decision were that HeuristicLab 1.x had a strong focus on a graphical user interface (GUI) right from the start to enhance usability and to provide a shallow learning curve especially for students. Consequently, a powerful GUI framework was required that is well integrated into the runtime environment and provides an authentic look and feel of applications. Concerning these aspects, back in 2002 the Microsoft® .NET platform provided a more promising approach than other alternatives such as Java^{TM} or C++[1]. Other aspects as for example platform independence were of minor importance as the developers were always focused on the Windows$^{®}$ operating system.

Similarly to some other heuristic optimization frameworks such as the Templar framework described in [21, 22], the main idea of the HeurisiticLab 1.x architecture is to provide a clear separation of problem-specific and problem-independent parts. A user should be able to develop a new heuristic optimization algorithm and to test and compare it with several existing optimization (benchmark) problems. Furthermore, a new problem should be easy to integrate and to solve with existing algorithms. By this means, this concept leads to a significant level of code reuse, as heuristic optimization algorithms can be used without any modification to solve new optimization problem and vice versa.

In order to realize this concept, HeuristicLab 1.x offers two abstract base classes called *Algorithm* and *Problem* from which every new extension, either optimization algorithm or problem, has to be inherited. Furthermore, another base class *Solution* represents data entities that are created and evaluated by problems and manipulated by algorithms. Any specific solution encoding has to be inherited from that class. On top of these basic framework classes, the HeuristicLab 1.x GUI layer is located. It provides two more base classes for visualizing algorithms (*AlgorithmForm*) and problems (*ProblemForm*). These classes represent forms in terms of the Microsoft$^{®}$ .NET WinForms framework and are presented in the GUI. As each algorithm is executed in its own thread, transport objects called *Results* are used to inform an AlgorithmForm about the progress of its algorithm by propagating values such as the actual number of evaluated solutions or the current solution quality. To support easy integration of new algorithms and problems, HeuristicLab 1.x additionally is equipped with a plugin mechanism enabling users to add custom extensions without knowing or even having access to the whole source code. As a summary of the HeuristicLab 1.x core architecture, Figure 10.1 schematically shows all these basic classes and their interactions.

As algorithms and problems are loosely coupled to be able to exchange both parts at will, communication between algorithms and problems is realized by delegates (i.e., types representing methods). An algorithm defines method interfaces it expects in order to be able to do its work (e.g., evaluation operators, manipulation operators, or neighborhood operators). On the other side, a problem provides implementations of these interfaces. If implementations of all required delegates are available, a problem can be solved by an algorithm. Linking between delegates (interfaces) and delegate implementations (operators) is done dynamically at runtime using code attributes and reflection. The whole application does not have to be compiled again when integrating new algorithms or problems.

---

[1] It has to be mentioned that today a very powerful and flexible GUI framework and a comfortable development environment is also available on the Java^{TM} side in form of the Eclipse^{TM} IDE and the Eclipse^{TM} Rich Client Platform. So in fact concerning GUI support the choice of an appropriate runtime environment would not be that easy today, as both solutions, Java^{TM} and Microsoft$^{®}$ .NET, are well developed and reached a high degree of maturity.

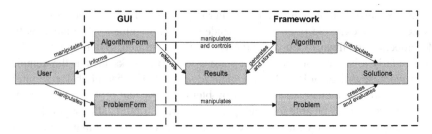

Fig. 10.1. Interaction of HeuristicLab 1.x classes

Beside this basic object model for representing arbitrary heuristic optimization algorithms and problems, HeuristicLab 1.x also includes another front-end for batch execution of algorithms called *TestBench*. In the HeuristicLab 1.x GUI multiple algorithm and problem configurations can be saved in a comma-separated text format (CSV) and can be executed in batch mode using the TestBench. This feature is especially useful for large scale experiments to compare different heuristic optimization algorithms. Furthermore, another sub-project called HeuristicLab Grid [42, 43] offers distributed and parallel batch execution of algorithm runs in a client-server architecture.

In the past, HeuristicLab 1.x has been intensively and successfully used by the research group of Michael Affenzeller in many research projects as well as in several heuristic optimization lectures. A broad spectrum of more than 40 plugins providing different heuristic optimization algorithms and problems has been developed. For example, various genetic algorithm variants, genetic programming, evolution strategies, simulated annealing, particle swarm optimization, tabu search, and scatter search are available. Furthermore, several heuristic optimization (benchmark) problems - for example, the traveling salesman problem, vehicle routing, n-dimensional real-valued test functions, the Boolean satisfiability problem, scheduling problems, or symbolic regression - are provided as plugins. An exemplary screenshot of the HeuristicLab 1.x GUI is shown in Figure 10.2.

10.3.2 HeuristicLab 2.x

Although HeuristicLab 1.x was extensively used in several research projects and was continuously extended with new algorithm and problem plugins, the authors identified a few drawbacks of HeuristicLab 1.x during its development and productive use. The most important ones of these issues are listed and discussed in the following:

- **Monolithic Plugins**
  As HeuristicLab 1.x provides a very high level of abstraction by reducing heuristic optimization to two main base classes (Algorithm and Problem), no specific

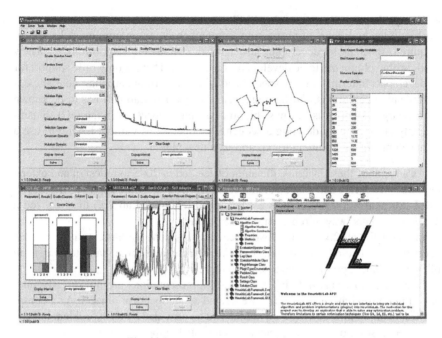

Fig. 10.2. Screenshot of HeuristicLab 1.x

APIs for particular optimization paradigms are available. For example, if a new variant of genetic algorithms is developed that differs from an existing algorithm in just some small aspects (for example using two different selection schemes for selecting solutions for reproduction [45]), the whole algorithm has to be provided in a separate and independent plugin. As more and more algorithms were added to HeuristicLab 1.x, this situation led to a severe amount of code duplication and to a significant reduction of maintainability.

- **Strict Separation of Algorithms and Problems**
  HeuristicLab 1.x requires strict separation of algorithms and problems and loose coupling between these two parts based on delegates. This approach makes it rather complicated to integrate heuristic optimization algorithms such as tabu search or hybrid algorithms that contain problem-specific concepts. A tighter interaction between algorithms and problems should be possible on demand.
- **Interruption, Saving and Restoring of Algorithms**
  The internal state of an algorithm during its execution cannot be persisted in HeuristicLab 1.x. As a consequence, it is not possible to interrupt, save and restore an algorithm during its execution. For example, if an algorithm has to be stopped because computation resources are temporarily required for some other task, the whole run has to be aborted and cannot be continued later on. As algorithm runs might take quite a long time in several application scenarios of heuristic optimization, this behavior turned out to be quite uncomfortable for users.

- **Comprehensive Programming Skills**
  Due to the high level of abstraction, comprehensive programming skills are required especially for developing new heuristic optimization algorithms; each new algorithm plugin has to be developed from scratch. There is only little support for developers in terms of more specialized APIs supporting particular heuristic optimization paradigms. Furthermore, also the HeuristicLab API has to be known to a large extent. It is not possible to assemble algorithms in the GUI dynamically at runtime by defining a sequence of operations without having to use a development environment for compiling a new plugin.

As a result of these insights, the authors decided in 2005 to redesign and extend the core architecture and the object model of HeuristicLab. Based on the Microsoft® .NET 2.0 platform a prototype was developed (HeuristicLab 2.x) that realized a more fine-grained way of representing algorithms.

In HeuristicLab 2.x so-called workbenches represent the basic entities of each heuristic optimization experiment. A workbench contains three main items in form of an algorithm, a problem and a solution representation.

In contrast to HeuristicLab 1.x, solution representations are not directly integrated into problems anymore, but are treated as independent objects. As a consequence, manipulation concepts do not have to be provided by each problem but can be shared, if solutions to a problem are represented in the same way.

Algorithms are no longer represented as single classes inherited from an abstract base class. Instead, each algorithm consists of several operators. Each operator works on either one or more solutions and represents a basic operation (e.g., manipulating or evaluating a solution, selecting solutions out of a solution set, or reuniting different solution sets). Furthermore, an operator may contain and execute other operators, leading to a hierarchical tree structure. By this means, the development of high level operators is possible which represent more complex operations and can be specialized with more specific operators on demand. For example, a solution processing operator can be defined that iterates over a solution set and executes an operator on all contained solutions (for example an evaluation operator). Another example is a mutation operator that executes a manipulation operator on a solution with some predefined probability. This concept leads to a fine-grained representation of algorithms and to better code reuse. Additionally, the strict separation between problem-specific and problem-independent parts is softened, as operators are able to access problem-specific information via the workbench.

The main idea of this enhanced algorithm model is to shift algorithm engineering from the developer to the user level. Complex heuristic optimization algorithms can be built be combining different operators in the GUI of HeuristicLab 2.x (see for example Figure 10.3). This aspect is especially important to support users who are not so well familiar with programming and software development (as for example many practitioners who are experts in some problem domain but not in software engineering). Furthermore, it also enables rapid prototyping and evaluation of new

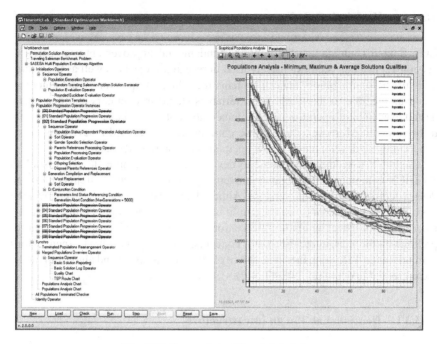

Fig. 10.3. Screenshot of HeuristicLab 2.x

algorithmic concepts, as an algorithm does not have to be implemented and compiled as a plugin.

In order to enable interrupting, saving and restoring of algorithms at any time during execution, a 2-phase commit strategy has been realized. If an algorithm is stopped during its execution, the actual program flow is located somewhere in the depths of the operator tree. As operators also may have local status variables, it has to be assured that the whole operator tree is left in a consistent state, so that execution can be continued, if the algorithm is restored or restarted again. The actual iteration that has not been finished yet has to be rolled back. Consequently, each operator has to keep a local copy of its status variables to be able to restore the last save state of the last completed (and committed) iteration. At the end of an iteration (i.e., a single execution of the top level algorithm of a workbench), a commit is propagated through the whole operator tree indicating that the actual state can be taken for granted.

All these concepts described above were implemented in the HeuristicLab 2.x prototype and were evaluated in several research projects showing the advantages of the new architecture in terms of code reuse, algorithm development time and flexibility. An overview of these applications can be found in Section 10.6 at the end of this chapter.

10.3.3 HeuristicLab 3.x

Even though HeuristicLab 2.x contains fundamental improvements compared to version 1.x and has been extensively used in research projects as well as in lectures on heuristic optimization, major problems regarding its operator model emerged: For example, due to the local status variables stored in operators and due to the nested execution of operators, the implementation of parallel algorithms turned out to be difficult. Moreover, the 2-phase commit strategy caused a severe overhead concerning the development of new operators and the required memory.

Therefore, it seemed reasonable to develop a new version called HeuristicLab 3.x (HL3) completely from scratch to overcome limitations due to architectural and design decisions of previous versions. Although this decision led to starting over the design and development process again, it offered the essential possibility to build a thoroughly consistent heuristic optimization software system by picking up ideas of preliminary projects, integrating novel concepts, and always keeping learned lessons in mind. In the following, the architecture of HL3 and its object and algorithm model are presented in detail (cf. [46, 47, 41]).

10.3.3.1 HeuristicLab Plugin Infrastructure

A plugin is a software module that adds new functionality to an existing application after the application has been compiled and deployed. In other words, plugins enable modularity not only at the level of source code but also at the level of object or byte code, as plugins can be developed, compiled, and deployed independently of the main application. Due to dynamic loading techniques offered in modern application frameworks such as Java^{TM} or Microsoft$^{®}$ .NET, a trend towards software systems can be observed in the last few years that use plugins as main architectural pattern. In these systems the main application is reduced to a basic plugin management infrastructure that provides plugin localization, on demand loading, and object instantiation. All other parts of an application are implemented as plugins. Communication and collaboration of plugins is based on extension points. An extension point can be used in a plugin to provide an interface for other plugins to integrate additional functionality. In other words, plugins can define extension points or provide extensions which leads to a hierarchical application structure. By this approach, very flexible applications can be built, as an application's functionality is determined just by its plugins. If a comprehensive set of plugins is available, a huge variety of applications can be developed easily by selecting, combining and deploying appropriate plugins.

In Section 10.2 genericity has been identified as an essential quality criterion of heuristic optimization software systems. In order to enable integration of custom optimization problems and algorithms, main parts of the system such as objective functions, operators, or solution encodings have to be exchangeable. A high degree of modularity is required and consequently a plugin-based architecture is also reasonable for a heuristic optimization software system [48].

Motivated by the benefits of plugin-based software systems especially in the case of heuristic optimization, plugins are used in HL3 as architectural paradigm as in the versions HeuristicLab 1.x and 2.x. In contrast to previous versions, a lightweight plugin concept is implemented in HL3 by keeping the coupling between plugins very simple: Collaboration between plugins is described by interfaces. The plugin management mechanism contains a discovery service that can be used to retrieve all types implementing an interface required by the developer. It takes care of locating all installed plugins, scanning for types, and instantiating objects. As a result, building extensible applications is just as easy as defining appropriate interfaces (contracts) and using the discovery service to retrieve all objects fulfilling a contract. Interfaces that can be used for plugin coupling do not have to be marked by any specific declaration.

Meta-information has to be provided by each plugin to supply the plugin management system with further details. For example, when installing, loading, updating, or removing a plugin, the plugin infrastructure has to know which files belong to the plugin and which dependencies of other plugins exist. With this information the plugin infrastructure can automatically install other required plugins, or disable and remove dependent plugins, if a base plugin is removed. Hence, it is guaranteed that the system is always in a consistent state. In the HL3 plugin infrastructure all plugin-related data is stored in the source files together with the plugin code. Plugin meta-information is expressed using code annotations instead of separate configuration files, keeping the configuration of the plugin system simple and clean (see Listing 10.1 for an example).

Additionally, a special kind of plugin is necessary: Some designated plugins, called application plugins, have to be able to take over the main application flow. Application plugins have to provide a main method and usually they also offer a GUI. Due to these application plugins the HeuristicLab plugin infrastructure leads to a flexible hierarchical system structure. It is possible to have several different front-ends (applications) within a single system.

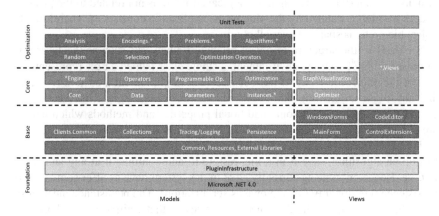

Fig. 10.4. Plugin structure of HL3

Figure 10.4 illustrates the plugin structure of HL3 and Listing 10.1 shows the meta-information of the HeuristicLab Optimizer application plugin as an example.

```
 1   using HeuristicLab.PluginInfrastructure;
 2
 3   namespace HeuristicLab.Optimizer {
 4     [Plugin("HeuristicLab.Optimizer", "3.3.6.7400")]
 5     [PluginFile("HeuristicLab.Optimizer-3.3.dll",
 6                     PluginFileType.Assembly)]
 7     [PluginDependency("HeuristicLab.Clients.Common", "3.3")]
 8     [PluginDependency("HeuristicLab.Collections", "3.3")]
 9     [PluginDependency("HeuristicLab.Common", "3.3")]
10     [PluginDependency("HeuristicLab.Common.Resources", "3.3")]
11     [PluginDependency("HeuristicLab.Core", "3.3")]
12     [PluginDependency("HeuristicLab.Core.Views", "3.3")]
13     [PluginDependency("HeuristicLab.MainForm", "3.3")]
14     [PluginDependency("HeuristicLab.MainForm.WindowsForms", "3.3")]
15     [PluginDependency("HeuristicLab.Optimization", "3.3")]
16     [PluginDependency("HeuristicLab.Persistence", "3.3")]
17     public class HeuristicLabOptimizerPlugin : PluginBase { }
18
19     [Application("Optimizer", "HeuristicLab Optimizer 3.3.6.7400")]
20     internal class HeuristicLabOptimizerApplication : ApplicationBase {
21       public override void Run() {
22         [...]
23       }
24     }
25   }
```

Listing 10.1. HeuristicLab.Optimizer application plugin

10.3.3.2 Object Model

Based on the plugin infrastructure, a generic object model is provided by HL3 which is described in detail in this section. Following the paradigm of object-oriented software development, the whole HL3 system is represented as a set of interacting objects. In a logical view these objects are structured in an object hierarchy using the principle of inheritance. Though, this logical structure is not related to the physical structure which is defined by the plugins the application consists of. Therefore, the object hierarchy is spanned over all plugins.

Similarly to the structure of object-oriented frameworks such as the Java^{TM} or Microsoft$^{®}$ .NET runtime environment, a single class named Item represents the root of the object hierarchy. It has been decided to implement a custom root class and not to use an existing one of the runtime environment, as it is necessary to extend the root class with some additional properties and methods which usually are not available in the root classes of common runtime environments.

Due to the requirement of replicability and persistence, HL3 offers functionality to stop an algorithm, save it in a file, restore it, and continue it at any later point in time. Hence, all objects currently alive in the system have to offer some kind of persistence functionality (also known as serialization). The persistence mechanism has to be able to handle circular object references, so that arbitrarily complex object graphs can be stored without any difficulties. To simplify communication with other systems and to enable integration into an existing IT environment, XML is used as a generic way of data representation for storing objects. Additionally, all objects should offer deep cloning functionality which is also determined by the replicability and persistence requirement. Seen from a technical point of view, cloning is very similar to persistence, as cloning an object graph is nothing else than persisting it in memory.

A graphical user interface is the second requirement which has to be considered in the design of the basic object model. The user should be able to view and access all objects currently available at any time, for example to inspect current quality values, to view some charts, or to take a look at a custom representation of the best solution found so far[2]. Nevertheless, the representation of data in a graphical user interface is a very time consuming task and therefore critical concerning performance. Consequently, visualizing all objects per default is not an option. The user has to be able to decide which objects should be presented in the GUI, and to open or close views of these objects as needed. Therefore, a strict separation and loose coupling between objects and views is required.

To realize this functionality, event-driven programming is a suitable approach. If the state of an objects changes and its representation has to be updated, events are used to notify all views. In object-oriented programming this design is known as the observer pattern [14] and is derived from model-view-controller. Model-view-controller (MVC) is an architectural pattern which was described by Trygve Reenskaug for the first time when working together with the Smalltalk group at Xerox PARC in 1979. It suggests the separation into a model (containing the data), views (representing the model) and a controller for handling user interactions (events). By this means, it is assured that the view is decoupled from the model and the control logic and that it can be exchanged easily. A comprehensive description of MVC is given by Krasner and Pope in [27].

In HL3 these concepts are used for object visualization. In addition, a strict decoupling of objects and views simplifies the development of complex views which represent compound objects. A view of a complex object containing several other objects can be composed by adding views of the contained objects. Figure 10.5 outlines this structure in a schematic way and Figure 10.6 shows a screenshot of the HL3 GUI.

As a summary, the basic properties of each HL3 object can be defined as follows:

- **Persistence**
  Objects can be persisted in order to save and restore themselves and all contained objects.
- **Deep Cloning**
  Objects provide a cloning mechanism to create deep copies of themselves.
- **Visualization**
  Views can be created for each object to represent it in a graphical user interface.

In Figure 10.7 the structure of HL3 objects defined by the basic object model is shown.

---

[2] Additionally, a graphical user interface is also very helpful for reducing the learning effort required when starting to work with the system.

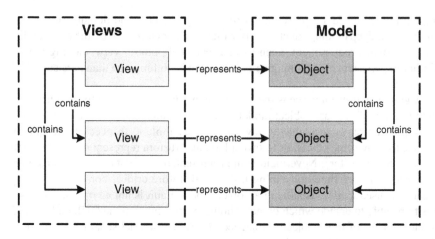

Fig. 10.5. Compound views representing complex objects

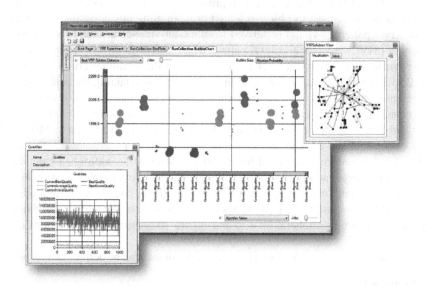

Fig. 10.6. Screenshot of HeuristicLab 3.x

10.3.3.3 Algorithm Model

In the previous section, the HL3 object model has been described. It is the basis for implementing arbitrary objects interacting in HL3. In this section the focus is now on using this object model to represent heuristic optimization algorithms and problems.

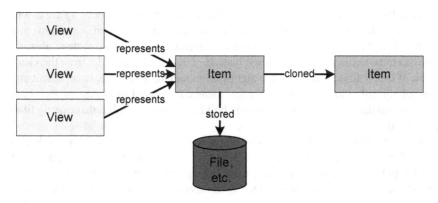

Fig. 10.7. HL3 object model

As stated in the requirements analysis in Section 10.2, the algorithm model of a heuristic optimization software system has to be very generic. Users have to be able to implement custom solution representations and objective functions and to realize individual optimization algorithms. However, the area of heuristic optimization algorithms and problems is very heterogeneous, as many different phenomena (e.g., evolution, hill climbing, foraging, or cooling of matter) were used as a source of inspiration leading to a broad spectrum of algorithms. These algorithms were applied in various problem domains including medicine, finance, engineering, economics, biology, chemistry, and many more.

Furthermore, due to the increase of computational power, many new problem domains are opened up for heuristic algorithms. Today, problems are solved for which using a (meta-)heuristic algorithm was unthinkable a few years ago because of unacceptable execution times. Every year new paradigms of heuristic optimization are introduced, new hybrid algorithms combining concepts of different optimization techniques are developed, and new optimization problems are solved. Therefore, diversity of algorithms and problems in the area of heuristic optimization is growing steadily and it can be expected that this trend will continue within the next years. Hence, developing an algorithm model capable of representing all these different cases is quite challenging.

For software engineers this reveals an interesting problem: On the one hand, a uniform and generic algorithm model is necessary to be able to represent all different optimization paradigms. Even more, the model has to be flexible enough to realize new algorithms that are not even known today and probably will be invented in the future. On the other hand, heterogeneity in the field of heuristic optimization makes it very difficult to develop such a generic model, as the different concepts and paradigms of heuristic optimization can hardly be unified. Although, some efforts were made in the scientific community to develop common models for subareas of heuristic optimization (e.g., for evolutionary algorithms as described in [9]), still there is no common theoretical model of heuristic optimization algorithms in general.

To solve this challenging task, radical abstraction and meta-modeling are essential success factors. As it is impossible to foresee which kinds of algorithms and problems will be implemented in a heuristic optimization software system, abstraction has to be shifted one level higher. Instead of developing a model from the viewpoint of heuristic optimization, the algorithm model has to be able to represent any kind of algorithm in any domain. By this means, the model turns into an algorithm meta-model that enables users to quickly build customized models that exactly fit to their needs.

In order to define such a generic and domain-independent algorithm model, an algorithm can be represented as an interaction of three parts: It is a sequence of *steps* (operations, instructions, statements) describing manipulation of *data* (input and output variables) that is executed by a *machine* (or human). Therefore, these three components (data, operators, and execution) are the core of the HL3 algorithm model and are described in detail in the following.

Data Model

Data values are represented as objects according to the HL3 object model. Therefore, each value can be persisted and viewed. Standard data types such as integers, doubles, strings, or arrays that do not offer these properties have to be wrapped in HL3 objects. In imperative programming, variables are used to represent data values that are manipulated in an algorithm. Variables link a data value with a (human readable) name and (optionally) a data type, so that they can be referenced in statements. Adapting this concept in the HL3 data model, a variable object stores a name, a description, and a value (an arbitrary HL3 object). The data type of a variable is not fixed explicitly but is given by the type of the contained value.

In a typical heuristic optimization algorithm a lot of different data values and therefore variables are used. Hence, in addition to data values and variables, another kind of objects called scopes is required to store an arbitrary number of variables. To access a variable in a scope, the variable name is used as an identifier. Thus, a variable name has to be unique in each scope the variable is contained.

Hierarchical structures are very common in heuristic optimization algorithms. For example, in an evolutionary algorithm an environment contains several populations, each population contains individuals (solutions) and these solutions may consist of different solution parts. Moreover, hierarchical structures are not only suitable for heuristic optimization. In many algorithms complex data structures (also called compound data types) are assembled from simple ones. As a result, it is reasonable to combine scopes in a hierarchical way to represent different layers of abstraction. Each scope may contain any number of sub-scopes which leads to an n-ary tree structure. For example, a scope representing a set of solutions (population) contains other scopes that represent a single solution each.

As operators are applied on scopes to access and manipulate data values or sub-scopes (as described in the next section), the hierarchical structure of scopes also has another benefit: The interface of operators does not have to be changed according to the layer an operator is applied on. For example, selection operators and manipulation operators are both applied on a single scope. However, in the case of selection this scope represents a set of solutions and in the case of manipulation it stands for a single solution. Therefore, users are able to create as many abstraction layers as required, but existing operators do not have to be modified. Especially in the case of parallel algorithms, this aspect is very helpful and will be discussed in detail later on.

The hierarchy of scopes is also taken into account when accessing variables. If a variable is not found in a scope, looking for the variable is dedicated to the parent scope of the current scope. The lookup is continued as long as the variable is not found and as long as there is another parent scope left (i.e., until the root scope is reached). Each variable is therefore "visible" in all sub-scopes of the scope that contains it. However, if another variable with the same name is added to one of the sub-scopes, the original variable is hidden due to the lookup procedure[3].

Operator Model

According to the definition of an algorithm, steps are the next part of algorithms that have to be considered. Each algorithm is a sequence of clearly defined, unambiguous and executable instructions. In the HL3 algorithm model these atomic building blocks of algorithms are called operators and are represented as objects. In analogy to imperative programming languages, operators can be seen as statements that represent instructions or procedure calls.

Operators fulfill two major tasks: On the one hand, they are applied on scopes to access and manipulate variables and sub-scopes. On the other hand, operators define which operators are executed next.

Regarding the manipulation of variables, the approach used in the HL3 operator model is similar to formal and actual parameters of procedures. Each operator contains parameters which contain a (formal) name, a description, and a data type and are used to access and manipulate values. In general, the HL3 operator model provides two major groups of parameters: value parameters and lookup parameters. In value parameters the parameter value is stored in the parameter object itself and can be manipulated directly by the user. They usually represent input parameters of an operator which have to be set by the user. Lookup parameters do not contain but refer to a value and are responsible for retrieving the current parameter value dynamically at runtime. For this purpose an actual name has to be defined in each lookup parameter which is used in an internal value lookup mechanism to fetch the current value from the variables stored in the scope tree.

By this means, operators are able to encapsulate functionality in an abstract way: For example, a simple increment operator contains a single parameter object indicating that the operator manipulates an integer variable. Inside the operator this

---

[3] This behavior is very similar to blocks in procedural programming languages. That is the reason why the name "scope" has been chosen.

parameter is used to implement the increment operation. After instantiating an increment operator and adding it to an algorithm at run time, the user has to define the concrete name of the value which should be incremented. This actual name is also set in the parameter. However, the original code of the operator (i.e., the increment operation) does not have to be modified, regardless which value is actually incremented. The implementation of the operator is decoupled from concrete variables. Therefore, the operator can be reused easily to increment any integer value.

Regarding their second major purpose, operators define the execution flow of an algorithm. Each operator may contain parameters which refer to other operators, which defines the static structure of an algorithm. When an operator is executed, it can decide which operators have to be executed next. Hence, complex algorithms are built by combining operators.

Control operators can be implemented that do not manipulate data, but dynamically define the execution flow. For example, branches can be realized by an operator that chooses a successor operator depending on the value of a variable in the scope the branch operator is applied on (cf. if- or switch-statements).

In contrast to scopes, operators are not combined hierarchically, but represent a graph. An operator used in an upper level of an algorithm can be added as a sub-operator in a lower level again. Thus, operator references may contain cycles. In combination with branches, these cycles can be used to build loops (see Section 10.4 for a detailed description of control operators).

As described above, classical concepts of programming such as sequences, branches, loops, or recursion can be represented in the operator model. Therefore, the HL3 algorithm model is capable of representing any algorithm that can be described in imperative programming languages.

Execution Model

The execution of algorithms is the last aspect which has to be defined in the HL3 algorithm model. Algorithms are represented as operator graphs and are executed step-by-step on virtual machines called engines. In each iteration an engine performs an operation, i.e., it applies an operator to a scope. Before executing an algorithm, each engine is initialized with a single operation containing the initial operator and an empty global scope.

As the execution flow of an algorithm is dynamically defined by its operators, each operator may return one or more operations that have to be executed next. Consequently, engines have to keep track of all operations that wait for execution. These pending operations are kept in a stack. In each iteration, an engine pops the next operation from the top of its stack, executes the operator on the scope, and pushes all returned successor operations back on the stack again in reversed order[4]. By this means, engines perform a depth-first expansion of operators. Listing 10.2 states the main loop of engines in pseudo-code.

---

[4] Reversing the order of operations is necessary to preserve the execution sequence, as a stack is a last-in-first-out queue.

```
1  clear global scope // remove all variables and sub-scopes
2  clear operations stack
3  push initial operation // initial operator and global scope
4
5  WHILE NOT operations stack is empty DO BEGIN
6      pop next operation
7      apply operator on scope
8      push successor operations // in reverse order
9  END WHILE
```

Listing 10.2. Main loop of HL3 engines

Finally, as a summary of the HL3 algorithm model and its three sub-models (data model, operator model, execution model), Figure 10.8 shows the structure of all components.

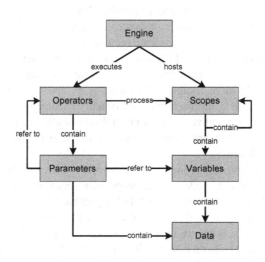

Fig. 10.8. HL3 algorithm model

10.3.3.4   Parallelism

As already pointed out in Section 10.2, short execution time is very important in many real-world applications of heuristic optimization. In order to increase performance, concepts of parallel and distributed computing are used frequently to utilize multiple cores or even computers and to distribute the work load. In the area of parallel heuristic optimization several models of parallelization have been developed which reflect different strategies. In general, there are two main approaches:

The first approach is to calculate the quality of solutions in parallel. In many optimization problems, evaluating a solution requires much more runtime than the execution of solution manipulation operators. As an example consider heuristic optimization for production planning or logistics: In that case, evaluating a solution is done by building a schedule of all jobs or vehicles (for example by using the Giffler-Thompson scheduling algorithm [15]), whereas a solution manipulation operator just needs to change permutations[5]. As another example, heuristic optimization of data representation or simulation models can be mentioned. In these applications, evaluating a solution means executing the whole model (i.e., performing the simulation or checking the quality of the model for all training data). In both examples (and there are many more), evaluating a single solution is independent of all other solutions. Therefore, quality calculation can be easily executed in parallel which is called *global parallelization* [2]. However, the heuristic algorithm performing the optimization is still executed sequentially.

The second approach is to parallelize heuristic optimization algorithms directly [2, 6]. By splitting solution candidates into distinct sets, an algorithm can work on these sets independently and in parallel. For example, parallel multi-start heuristics are simple representatives of that idea. In a parallel multi-start heuristic algorithm, multiple optimization runs are executed with different initial solutions in order to achieve larger coverage of the solution space. Nevertheless, no information is exchanged between these runs until all of them are finished and the best solution is determined. In more complex algorithms (e.g., coarse- or fine-grained parallel genetic algorithms), information is additionally exchanged from time to time to keep the search process alive and to support diversification of the search. Hence, population-based heuristic optimization algorithms are especially well suited for this kind of parallelization.

Consequently, a heuristic optimization software system should consider parallelization in its algorithm model. It has to provide sequential as well as parallel blocks, so that all different kinds of parallel algorithms can be represented. Furthermore, the definition of parallel parts in an algorithm has to be abstracted from parallel execution. By this means, users can focus on algorithm development and do not have to rack their brains on how parallelism is actually implemented.

In the HL3 algorithm model presented in the previous section, data, operators and algorithm execution have been separated strictly. Consequently, parallelism can be integrated by grouping operations into sets that might be executed in parallel. As an operator may return several operations that should be executed next, it can mark the successor operations as a parallel group. These operations are then considered to be independent of each other and the engine is able to decide which kind of parallel execution should be used. How this parallelization is actually done just depends on the engine and is not defined by the algorithm.

HL3 offers two engines which implement different parallelization concepts. The *ParallelEngine* is based on the Microsoft® .NET Task Parallel Library (TPL) and

---

[5] This depends on the solution encoding, but variations of permutation-based encoding are frequently used for combinatorial optimization problems and have been successfully applied in many applications.

enables parallel execution on multiple cores. The *HiveEngine* is based on HeuristicLab's parallel and distributed computing infrastructure called Hive and provides parallel execution of algorithms on multiple computers. By this means, the user can specify the parallelization concept used for executing parallel algorithms by choosing an appropriate engine. The definition of an algorithm is not influenced by that decision. Additionally, HL3 also contains a *SequentialEngine* and a *DebugEngine* which both do not consider parallelism at all and execute an algorithm sequentially in every case. These engines are especially helpful for testing algorithms before they are really executed in parallel.

Based on this parallelization concept, HL3 provides special control operators for parallel processing. Data partitioning is thereby enabled in an intuitive way due to the hierarchical structure of scopes. For example, the operator *SubScopesProcessor* can be used to apply different operators on the sub-scopes of the current scope in parallel. Therefore, parallelization can be applied on any level of scopes which enables the definition of global, fine- or coarse-grained parallel heuristic algorithms (for a detailed description of parallel control operators see Section 10.4).

10.3.3.5   Layers of User Interaction

Working with HL3 on the level of its generic algorithm model offers a high degree of flexibility. Therefore, this level of user interaction is very suitable for experts who focus on algorithm development. However, dealing with algorithms on such a low level of abstraction is not practical for practitioners or students. Practitioners require a comprehensive set of predefined algorithms which can be used as black box optimizers right away. Similarly, predefined algorithms and also predefined problems are equally important for students, so that they can easily start to experiment with different algorithms and problems and to learn about heuristic optimization. Consequently, HL3 offers several layers for user interaction that reflect different degrees of detail [3].

Predefined algorithms are provided that contain entire operator graphs representing algorithms such as evolutionary algorithms, simulated annealing, hill climbing, tabu search, or particle swarm optimization. Therefore, users can work with heuristic optimization algorithms right away and do not have to worry about how an algorithm is represented in detail. They only have to specify problem-specific parts (objective function, solution encoding, solution manipulation) or choose one of the predefined optimization problems. Additionally, custom views can be provided for these algorithms that reveal only the parameters and outputs the user is interested in. By this means, the complexity of applying heuristic optimization algorithms is significantly reduced.

Between the top layer of predefined solvers and the algorithm model, arbitrary other layers can be specified that offer algorithm building blocks in different degrees of detail. For example, generic models of specific heuristic optimization paradigms, as for example evolutionary algorithms or local search algorithms, are useful for experimenting with these algorithms. These models also hide the full complexity and flexibility of the algorithm model from users who therefore can solely concentrate on the optimization paradigm. Especially in that case the graphical user interface is very important to provide a suitable representation of algorithms.

In Figure 10.9 the layered structure of user interaction in HL3 is shown schematically. Users are free to decide how much flexibility they require and on which level of abstraction they want to work. However, as all layers are based on the algorithm model, users can decrease the level of abstraction step by step, if additional flexibility is necessary in order to modify an algorithm.

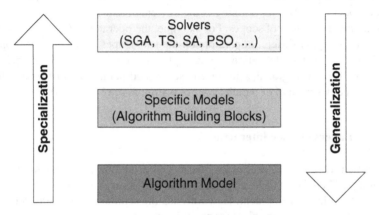

Fig. 10.9. Layers of user interaction in HL3

10.3.4 *Analysis and Comparison*

A comparison of the different versions of HeuristicLab concerning the requirements defined in Section 10.2 is shown in Table 10.1. A bullet (•) indicates that a requirement is fulfilled, a circle (∘) marks requirements which are somehow addressed but are not satisfactorily fulfilled, and a dot (·) shows requirements which are not considered.

HeuristicLab 1.x provides a paradigm-independent algorithm model which is very similar to the model of Templar [21, 22], although it lacks a clean separation of optimization problems and solution representations. Furthermore, algorithm runs cannot be interrupted, saved, and restarted. It provides a plugin mechanism that enables dynamic extension, web-based deployment, and clean integration into other applications. A broad spectrum of plugins for trajectory-based as well as population-based metaheuristics and many different optimization problems have been implemented. Furthermore, a bunch of plugins developed in the HeuristicLab Grid project enables parallel and distributed batch execution. Additionally, GUI components are integrated into the core framework and enable graphical and interactive configuration and analysis of algorithms.

Table 10.1. Evaluation and comparison of HeuristicLab

	HeuristicLab 1.x	HeuristicLab 2.x	HeuristicLab 3.x
Automation	●	○	●
Customizable Output	●	●	●
Generic Algorithm Model	●	●	●
Generic Operators	○	●	●
Generic Objective Functions	○	○	●
Generic Solution Representations	○	●	●
Graphical User Interface	●	●	●
Integration	●	●	●
Learning Effort	●	○	●
Parallelism	○	○	●
Parameter Management	○	○	●
Performance	○	○	○
Predefined Algorithms and Problems	●	○	●
Replicability and Persistence	·	●	●

HeuristicLab 2.x was designed to overcome some of the drawbacks of the previous version. The main purpose was to combine a generic and fine-grained algorithm model with a GUI to enable dynamic and interactive prototyping of algorithms. Algorithm developers can use the GUI to define, execute, and analyze arbitrary complex search strategies and do not have to be experienced programmers. Furthermore, HeuristicLab 2.x also offers support for replicability and persistence, as algorithm runs can be aborted, saved, loaded and continued. However, HeuristicLab 2.x is a research prototype and has never been officially released. Therefore, it still contains some drawbacks concerning performance, stability, usability, learning effort, and documentation. Additionally, the implementation of parallel metaheuristics is difficult due to the nested execution of operators and the local status variables which are used in many operators.

Finally, HL3 fulfills almost all requirements identified so far. Just performance remains as the last requirement that is still a somehow open issue. Even though HL3 enables parallel execution of algorithms and can therefore utilize multi-core CPUs or clusters, the runtime performance of sequential algorithms is worse compared to other frameworks. Reasons for this drawback are the very flexible design of HL3 and the dynamic representation of algorithms. As algorithms are defined as operator graphs, an engine has to traverse this graph and execute operators step by step. Of course, this requires more resources in terms of runtime and memory as is the case when algorithms are implemented as static blocks of code. However, if more effort has to be put on the evaluation of solutions, the overhead of the HeuristicLab

3.x algorithm model is less critical. This is in fact an important aspect, as most resources are consumed for the evaluation of the objective function in many real-world optimization applications. Therefore, the benefits of HeuristicLab 3.x in terms of flexibility, genericity and extensibility should outweigh the additional overhead. Furthermore, the efficiency of HeuristicLab 3.x can be easily improved by executing operators in parallel.

10.4   Algorithm Modeling

In HeuristicLab algorithms are modeled using the operator concept described in Section 10.3. HeuristicLab offers a wide range of operators that can be used to model any type of algorithm. Of course because HL is primarily used to implement metaheuristics, there are a lot of operators which are specially designed for this kind of algorithms. In the following section, operators are presented that are later used to build some typical examples of standard heuristic optimization algorithms. Furthermore, these operators serve as algorithm building blocks for successively defining more complex parallel and hybrid metaheuristics, showing the flexibility and genericity of the framework.

10.4.1   Operators

First of all, simple operators are discussed that perform typical tasks required in every kind of algorithm.

EmptyOperator

The *EmptyOperator* is the most simple form of an operator. It represents an operator that does nothing and can be compared to an empty statement in classical programming languages.

To get an idea of how an operator implementation looks like in HL, Listing 10.3 shows its implementation. Each operator is inherited from the abstract base class *SingleSuccessorOperator*. This base class takes care of aspects that are identical for all operators (e.g., storing of sub-operators and variable information, persistence and cloning, and events to propagate changes). Note that the most important method of each operator is *Apply* which is called by an engine to execute the operator. Apply may return an object implementing *IOperation* which represents the successor operations. If the operator has no successor operations, null is returned.

```
1   public sealed class EmptyOperator : SingleSuccessorOperator {
2     public EmptyOperator() : base() { }
3
4     public override IOperation Apply() {
5       return base.Apply();
6     }
7   }
```

Listing 10.3. Source code of EmptyOperator

Random

As most heuristic optimization algorithms are stochastic processes, uniformly distributed high quality pseudo-random numbers are required. For creating random numbers a single pseudo-random number generator (PRNG) should be used to enable replicability of runs. By setting the PRNG's random seed the produced random number sequence is always identical for each run; it is therefore useful to create a single PRNG into the global scope. Although PRNGs are also represented as data objects and consequently can be stored in variables, it has been decided to implement a custom operator *RandomCreator* for this task. Reasons are that it can be specified for a RandomCreator whether to initialize the PRNG with a fixed random seed to replay a run or to use an arbitrary seed to get varying runs.

Counter

IntCounter is a basic operator for incrementing integer variables. It increases the value of a variable in a scope by a specified value. As an example, this operator can be used for counting the number of evaluated solutions or to increment the actual iteration number.

As the Counter operator is the first operator that changes a variable, Listing 10.4 shows its implementation to give an impression how the manipulation of variables in a scope is done. In the constructor two parameters are defined for looking up the values needed for executing the increment operation. First of all a parameter for looking up the *Value* to increment is added as well as a parameter that contains the value with which the number will be incremented. If there is no *Increment* variable defined in the scope a default value of 1 is used. In the *Apply* method the actual value of *Value* and *Increment* are used. *ActualValue* automatically translates the formal name into the actual name of the parameter (which e.g. has been specified by the user in the GUI) and does the parameter lookup on the scope. The *ValueLookup-Parameter* uses the locally defined value if it can't find the name on the scope. If the *ValueParameter* can't find a value for the given name, a new value is created in the *Apply* method. Finally, the value is incremented and the successor operation is executed.

```
1   public class Counter : SingleSuccessorOperator {
2     public ILookupParameter<IntValue> ValueParameter {
3       get {
4         return (ILookupParameter<IntValue>)Parameters["Value"];
5       }
6     }
7     public IValueLookupParameter<IntValue> IncrementParameter {
8       get {
9         return (IValueLookupParameter<IntValue>)Parameters["Increment"];
10      }
11    }
12
13    public Counter() {
14      Parameters.Add(new LookupParameter<IntValue>("Value",
15        "The value which should be incremented."));
16      Parameters.Add(new ValueLookupParameter<IntValue>(
17        "Increment",
18        "The increment which is added to the value.",
19        new IntValue(1)));
20    }
21
22    public override IOperation Apply() {
23      if (ValueParameter.ActualValue == null)
24        ValueParameter.ActualValue = new IntValue();
25      ValueParameter.ActualValue.Value +=
26        IncrementParameter.ActualValue.Value;
27      return base.Apply();
28    }
29  }
```

Listing 10.4. Source code of the Counter

Comparator

The *Comparator* operator is responsible for comparing the values of two variables. It expects two input variables which should be compared and a comparison operation specifying which type of comparison should be applied (e.g., less, equal, greater or equal). After retrieving both variable values and comparing them, Comparator creates a new Boolean variable containing the result of the comparison and writes it back into the scope.

ConditionalBranch

The operator *ConditionalBranch* can be used to model simple binary branches. It retrieves a Boolean input variable from the scope tree. Depending on the value of this variable, an operation collection containing the successor operation and additionally either the first (true branch) or the second sub-operator (false branch) is returned. Note that do-while or repeat-until loops can be constructed without any other specific loop operator by combining the operators ConditionalBranch and any sequence of operations to execute in the body of the loop. Figure 10.10 shows an operator graph for these two loop structures.

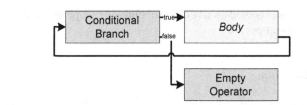

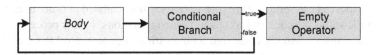

Fig. 10.10. Operator graphs representing a while and a do-while loop

StochasticBranch

In heuristic optimization algorithms it is a common pattern to execute operations with a certain probability (for example mutation of individuals in evolutionary algorithms or post-optimization heuristics in hybrid algorithms). Of course, this could be realized by using an operator creating a random number into a scope in combination with the Comparer and ConditionalBranch operators. However, for convenience reasons the *StochasticBranch* operator performs this task in one step. It expects a double variable as an input specifying the probability and a PRNG. When applying the operator a new random number between 0 and 1 is generated and compared with the probability value. If the random number is smaller, the true branch (first sub-operator) or otherwise the false branch (second sub-operator) is chosen.

UniformSubScopesProcessor

HeuristicLab offers operators to navigate through the hierarchy levels of the scope tree (i.e., to apply an operator on each sub-scope of the current scope). The *Uniform-SubScopesProcessor* fulfils this task by returning an operation for each sub-scope and its first sub-operator. Additionally this operator has a parameter for enabling parallel processing of all sub-scopes if the engine supports it. This leads to a single program multiple data style of parallel processing.

SubScopesProcessor

As a generalization of the UniformSubScopesProcessor is the *SubScopesProcessor* that returns an operation not only for the first sub-operator and every sub-scope, but pair sub-operators and sub-scopes together. For each sub-scope there has to be a sub-operator which is executed on its corresponding sub-scope to enable individual processing of all sub-scopes.

Selection and Reduction

Seen from an abstract point of view, a large group of heuristic optimization algorithms called improvement heuristics follows a common strategy: In an initialization step one or more solutions are generated either randomly or using construction heuristics. These solutions are then iteratively manipulated in order to navigate through the solution space and to reach promising regions. In this process manipulated solutions are usually compared with existing ones to control the movement in the solution space depending on solution qualities. Selection splits solutions into different groups either by copying or moving them from one group to another; replacement merges solutions into a single group again and overwrites the ones that should not be considered anymore.

In the HL algorithm model each solution is represented as a scope and scopes are organized in a hierarchical structure. Therefore, these two operations, selection and replacement, can be realized in a straight forward way: Selection operators split sub-scopes of a scope into two groups by introducing a new hierarchical layer of two sub-scopes in between, one representing the group of remaining solutions and one holding the selected ones as shown in Figure 10.11a. Thereby solutions are either copied or moved depending on the type of the selection operator. Reduction operators represent the reverse operation. A reduction operator removes the two sub-scopes again and reunites the contained sub-scopes as shown in Figure 10.11b. Depending on the type of the reduction operator this reunification step may also include elimination of some sub-scopes that are no longer required.

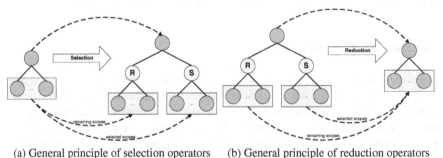

(a) General principle of selection operators   (b) General principle of reduction operators

Fig. 10.11. Selection and reduction operators

Following this simple principle of selection and reduction of solutions, HL provides a set of predefined selection and reduction operators that can be used as a basis for realizing complex selection and replacement schemes.

Selection Operators

The most trivial form of selection operators are the two operators *LeftSelector* and *RightSelector* which select sub-scopes either starting from the leftmost or the rightmost sub-scope. If the sub-scopes are ordered for example with respect to solution quality, these operators can be used to select the best or the worst solutions of a group. If random selection of sub-scopes is required, *RandomSelector* can be used which additionally expects a PRNG as an input variable.

In order to realize more sophisticated ways of selection, *ConditionalSelector* can be used which selects sub-scopes depending on the value of a Boolean variable contained in each sub-scope. This operator can be combined with a selection preprocessing step to create this Boolean variable into each scope depending on some other conditions.

Furthermore, HL also offers a set of classical quality-based selection schemes well-known from the area of evolutionary algorithms, as for example fitness proportional selection optionally supporting windowing (*ProportionalSelector*), linear rank selection (*LinearRankSelector*), or tournament selection with variable tournament group sizes (*TournamentSelector*). Additionally, other individual selection schemes can be integrated easily by implementing custom selection operators.

Reduction Operators

Corresponding reverse operations to LeftSelector and RightSelector are provided by the two reduction operators *LeftReducer* and *RightReducer*. Both operators do not reunite sub-scopes but discard either the scope group containing the selected or the group containing the remaining scopes. LeftReducer performs a reduction to the left and picks the scopes contained in the left sub-scope (remaining scopes) and RightReducer does the same with the right sub-scopes (selected scopes). Additionally, another reduction operator called *MergingReducer* is implemented that reunites both scope groups by merging all sub-scopes.

The following sections show how the described operators are used for designing a genetic algorithm and simulated annealing.

10.4.2   *Modeling Genetic Algorithms*

Genetic algorithms [19] are population-based metaheuristics which apply processes from natural evolution to a population of solution candidates. This includes natural selection, crossover, mutation and a replacement scheme for updating the old population with the newly generated one. Figure 10.12 gives and overview of a genetic algorithm.

The genetic algorithm starts with generating a population of solution candidates. The selection operation then selects individuals based on a certain scheme (e.g., randomly or according to their qualities). In the next two steps the selected individuals

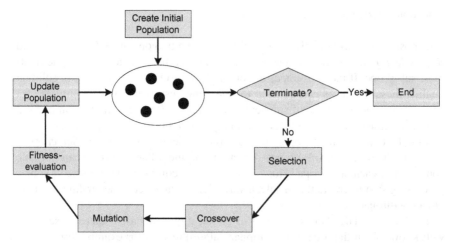

Fig. 10.12. Overview of a genetic algorithm

are crossed and mutated leading to a new population of solution candidates that then get evaluated. In the last step the old population gets replaced with the newly generated one. This procedure is repeated until a termination criterion (e.g., maximum number of generations) is reached.

In the following it will be outlined how a genetic algorithm can be implemented using HeuristicLab's operator graph. Based on the above description, the GA needs at least the following operators:

- Operator for creating the initial population
- Crossover operator
- Mutation operator
- Fitness evaluation operator
- Selection operator
- Population update method

The first four operators are problem- or encoding-specific operators. As the genetic algorithm should work with all types of encodings and problems which HeuristicLab offers, place-holders are used for these operators so that the user can configure which operator should be used. The same concept holds for the selection operator, though this operator is not specific to a problem or encoding. HL offers several common selection strategies, crossover and mutation operators which can be configured by the user and be executed by the placeholders. Before discussing the population update method, the representation of individuals and populations is covered in more detail.

Individuals in HL are represented as scopes. Each individual is a subscope containing at least it's genotype and optionally additional values describing the solution candidate (e.g. the quality). The selection operators create two subscopes from the population. The first subscope contains the old population while the second subscope contains the selected individuals. Crossover operators in HeuristicLab assume that they are applied to a subscope containing the two indivduals to be crossed. Therefore a *ChildrenCreator* is used to divide the selected subscope into groups of two individuals. To these subscopes the crossover and mutation operators can then be easily applied. Additionally the two parents have to be removed after a new individual was generated as they are not needed any more. The population update method has to delete the subscope of the old population and extract the newly generated individuals from the second (selected) subscope. Figure 10.13 outlines how the subscopes change over the course of the selection, crossover, mutation and population update methods.

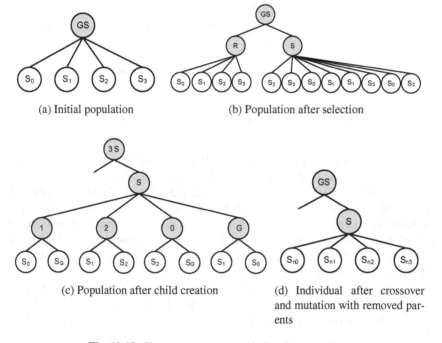

Fig. 10.13. Changes on scope tree during GA execution

Figure 10.13a shows the initial population which is created beneath the root of the scope tree (Global Scope - GS). Figure 10.13b depicts the scope tree after applying a selection operator. It shows that the population is now divided into the remaining (R) and the selected (S) population. The *ChildrenCreator* operator then introduces subscopes in the selected subscope that contain two individuals (Figure 10.13c). After crossover and mutation, the parents are removed with the *SubScopesRemover* and the remaining scopes contain the new solutions (Figure 10.13d). After having generated a new population the old population is removed and the new population moved back into the global scope to produce the original structure shown in Figure 10.13a. This task is accomplished with the previously described *RightReducer*.

Figure 10.14 shows the operator graph for generating the offspring in the genetic algorithm.

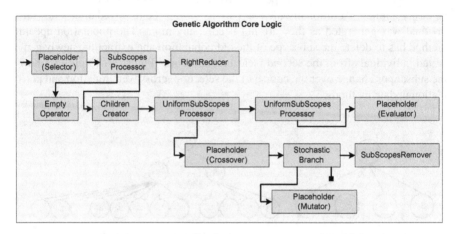

Fig. 10.14. Offspring generation in the genetic algorithm

After the selection operation a *SubScopesProcessor* is used to apply the *Children-Creator* to the selected individuals. An *UniformSubScopesProcessor* is used to apply the crossover to each group of parents. After the crossover the *StochasticBranch* chooses based on the mutation rate, if the mutation operator should be applied to the newly generated offspring. After the mutation the parents are removed with the help of the *SubScopesRemover* operator. Next a quality is assigned to the offspring using the *Evaluator*. The last step of the core logic is to remove the old population and move the offspring to the global scope.

The algorithm so far represents the steps that one iteration of the GA is made up of. Figure 10.15 depicts the loop that applies the offspring generation method until the maximum number of generations is reached as well as the initialization of the algorithm.

After initializing the random number generator, a variable *Generations* is created and added to the results collection. It is incremented by the *IntCounter* every generation and should be displayed on the results page of HL to show the progress of the algorithm. After incrementing *Generations*, a *Comparator* is used to check if *Generations* has reached it's allowed maximum. If *MaximumGenerations* is reached, the variable *Abort* is set to true and the *ConditionalBranch* executes the true branch, which is empty and therefore the algorithm stops.

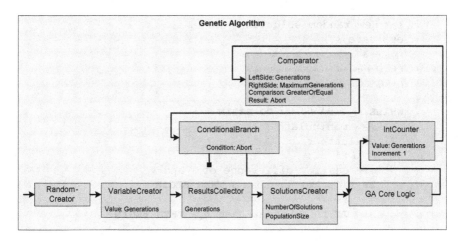

Fig. 10.15. Initialization and main loop of the genetic algorithm

10.4.3   Modeling Simulated Annealing

As stated in [4], simulated annealing (SA) [24] is commonly said to be the oldest among the metaheuristics and one of the first algorithms that contained an explicit strategy to escape from local optima. As greedy local search algorithms head for the optimum located in the attraction basin of the initial solution, they severely suffer from the problem of getting stuck in a local optimum. To overcome this problem many heuristic optimization algorithms use additional strategies to support diversification of the search. One of these algorithms is simulated annealing which additionally introduces an acceptance probability. If a worse solution is selected in the solution space, it is accepted with some probability depending on the quality difference of the actual (better) and the new (worse) solution and on a parameter called temperature; to be more precise, the higher the temperature and the smaller the quality difference, the more likely it is that a worse solution is accepted. In each iteration the temperature is continuously decreased leading to a lower and lower acceptance probability so that the algorithm converges to an optimum in the end. A detailed description of SA is given in Listing 10.5.

SA starts with an initial solution s which can be created randomly or using some heuristic construction rule. A solution s' is randomly selected from the neighborhood of the current solution in each iteration. If this solution is better, it is accepted and replaces the current solution. However, if s' is worse, it is not discarded immediately but is also accepted with a probability depending on the actual temperature parameter t and the quality difference. The way how the temperature is decreased over time is defined by the cooling scheme. Due to this stochastic acceptance criterion the temperature parameter t can be used to balance diversification and intensification of the search.

```
1    s ← new random solution // starting point
2    evaluate s
3    s_best ← s  // best solution found so far
4    i ← 1  // number of evaluated solutions
5    t ← t_i  // initial temperature
6
7    WHILE i ≤ maxSolutions DO BEGIN
8         s' ← manipulate s  // get solution from neighborhood
9         evaluate s'
10        i ← i + 1
11        q ← quality difference of s and s'
12        IF s' is better than s THEN BEGIN
13             s ← s'
14             IF s is better than s_best THEN BEGIN
15                  s_best ← s
16             END IF
17        END
18        ELSE IF Random(0, 1) < e^{-\frac{|q|}{t}} BEGIN
19             s ← s'
20        END IF
21        t ← t_i  // calculate next temperature
22   END WHILE
23
24   RETURN s_best
```

Listing 10.5. Simulated annealing

Figure 10.16 shows the operator graph for generating solutions in simulated annealing.

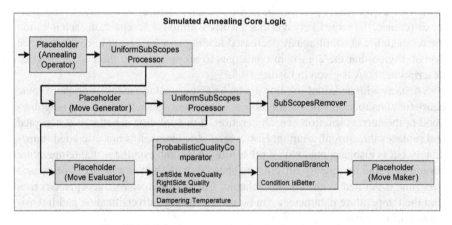

Fig. 10.16. Solution generation in simulated annealing

The annealing operator defines the cooling scheme and is modeled with a place-holder as HL offers different cooling schemes. Concrete operators for placeholders can be provided by plugins and selected by the user through the GUI. In contrast to the GA the global scope of the algorithm contains not a population of solution candidates but only one solution. A *UniformSubScopesProcessor* is used to apply a *Move Generator* to this solution. Move generators are operators that generate a con-figurable amount of solution candidates for a given solution. A placeholder is used for the *Move Generator* again, so that the user can decide how to generate moves. Each new solution candidate is then evaluated with the *Move Evaluator*. The *Proba-bilisticQualityComparator* calculates whether the current quality has gained a better quality than the so far found best solution. The operator additionally considers the current temperature as well as a random number (depicted in Listing 10.5). The following *ConditionalBranch* decides based on the result of the *ProbabilisticQuali-tyComparator* if the global solution has to be updated. If that is the case the update is performed by the *Move Maker*. If all generated solution candidates have been evaluated, a *SubScopesRemover* is used to dispose them.

As with genetic algorithms, simulated annealing's solution generation method is applied repeatedly and the number of iterations can be defined by the user. Figure 10.17 shows the main loop of the SA algorithm and the initialization operators.

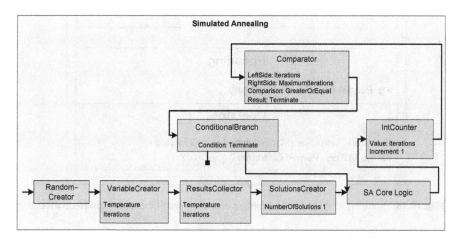

Fig. 10.17. Initialization and main loop of simulated annealing

After initializing the random number generator, two variables, *Temperature* and *Iterations*, are created and added to the results collection. *Temperature* is used in the core algorithm to save the result of the cooling scheme operator as well as for deciding if the quality of solution candidates is better than the global best solution. *Iterations* is incremented after each solution generation phase and is compared to the maximum number of allowed iterations. If *MaximumIterations* is reached, the algorithm is stopped. *SolutionsCreator* creates only one solution which is used for generating moves and is continuously improved over the course of the algorithm execution.

10.5 Problem Modeling

After having defined a set of generic operators in Section 10.4 which build the basis for every heuristic optimization algorithm, in the next sections the focus is on problem-specific aspects such as solution encoding, quality evaluation and manipulation operators. Figure 10.18 gives an overview of how problems and encodings are organized in HL.

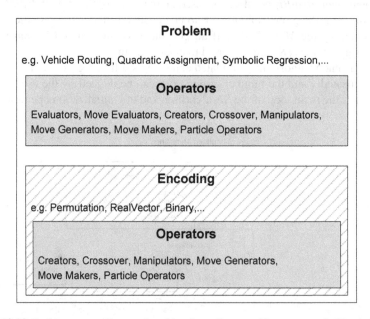

Fig. 10.18. Problems, encodings and problem/encoding-specific operators in HeuristicLab

HeuristicLab offers a range of already implemented problems. These problems use encodings to represent solution candidates. Problems and encodings offer various operators that a user can choose from. As seen in Section 10.4 algorithms usually are modeled to allow configuration of certain operators. These placeholders can be filled with encoding- or problem-specific operators. In addition to the representation of a solution, encodings usually offer operators for creating solution candidates, crossover and mutation (*Manipulator*) operators as well as operators for trajectory-based metaheuristics (*Move Generators* and *Move Makers*) or particle swarm optimization. Problems hold the problem information (e.g. the distance matrix of a TSP) and offer problem-specific operators. Primarily these operators are the *Evaluators* and *Move Evaluators* for assigning a fitness value to a solution candidate. Of course problems can also offer the same operators as the encodings if there is a need for incorporating problem-specific information.

In the following, three concrete problem implementations in HeuristicLab are presented in more detail, showing how HL can be used to represent the quadratic assignment problem, to optimize simulations, or to do genetic programming.

10.5.1 Quadratic Assignment Problem

The Quadratic Assignment Problem (QAP) was introduced in [26] and is a well-known problem in the field of operations research. It is the topic of many studies, treating the improvement of optimization methods as well as reporting successful application to practical problems in keyboard design, facility layout planning and re-planning as well as in circuit design [18, 7]. The problem is NP hard in general and, thus, the best solution cannot easily be computed in polynomial time. Many different optimization methods have been tried, among them popular metaheuristics such as tabu search [36] and genetic algorithms [10].

The problem can be described as finding the best assignment for a set of facilities to a set of locations so that each facility is assigned to exactly one location which in turn houses only this facility. The problem can also model situations in which there are a greater number of locations available. By introducing and assigning facilities with no flows a solution indicates which locations are to be selected. Likewise the problem can also model situations that contain a larger set of facilities by introducing locations with very high distances. Generally, an assignment is considered better than another when the flows between the assigned facilities run along smaller distances.

More formally the problem can be described by an $N \times N$ matrix W with elements w_{ik} denoting the weights between facilities i and k and an $N \times N$ matrix D with elements d_{xy} denoting the distances between locations x and y. The goal is to find a permutation π with $\pi(i)$ denoting the location that facility i is assigned to so that the following objective is achieved:

$$\min \sum_{i=1}^{N} \sum_{k=1}^{N} w_{ik} \cdot d_{\pi(i)\pi(k)} \tag{10.1}$$

A permutation is restricted to contain every number just once, hence, it satisfies the constraint of a one-to-one assignment between facilities and locations:

$$\forall_{i,k} i \neq k \Leftrightarrow \pi(i) \neq \pi(k) \tag{10.2}$$

The complexity of evaluating the quality of an assignment according to Eq. (10.1) is $O(N^2)$, however several optimization algorithms move from one solution to another through small changes, such as by swapping two elements in the permutation. These moves allow to reduce the evaluation complexity to $O(N)$ and even $O(1)$ if the previous qualities are memorized [36]. Despite changing the solution in small steps iteratively, these algorithms can, nevertheless, explore the solution space and interesting parts thereof quickly. The complete enumeration of such a "swap" neighborhood contains $N * (N - 1)/2$ moves and, therefore, grows quickly with the problem size. This poses a challenge for solving larger instances of the QAP.

HeuristicLab offers an implementation of the QAP in the form of the *QuadraticAssignmentProblem* plugin that is based on the *PermutationEncoding* plugin. The *Permutation* is used as a representation to the QAP. Besides the solution vector a QAP solution additionally holds the quality assigned by the *QAPEvaluator* operator.

HeuristicLab also provides access to all instances of the Quadratic Assignment Problem Library (QAPLIB) [5] which is a collection of benchmark instances from different contributors. According to the QAPLIB website[6], it originated at the Graz University of Technology and is now maintained by the University of Pennsylvania, School of Engineering and Applied Science. It includes the instance descriptions in a common format, as well as optimal and best-known solutions or lower bounds and consists of a total of 137 instances from 15 contributing sources which cover real-world as well as random instances. The sizes range from 10 to 256 although smaller instances are more frequent. Despite their small size these instances are often hard to solve so a number of different algorithms have emerged [36, 35, 10].

The *QuadraticAssignmentProblem* class in HeuristicLab extends from the *SingleObjectiveHeuristicOptimizationProblem* class provided by the HeuristicLab 3.3 framework. Besides support for single and multi objective problems HeuristicLab also distinguishes between problems and heuristic optimization problems. *IProblem* specifies that each problem has to hold a list of operators that can be queried. Problems are therefore responsible for discovering encoding and problem specific operators that algorithms can retrieve and use. *IHeuristicOptimizationProblem* defines that a heuristic optimization problem has to contain at least parameters for an evaluator and a solution creator that an algorithm can use to create and evaluate solution candidates. *ISingleObjectiveHeuristicOptimizationProblem* further specifies that the evaluator has to be an *ISingleObjectiveEvaluator* in contrast to an *IMultiObjectiveHeuristicOptimizationProblem* that requires an *IMultiObjectiveEvaluator* which generates multiple fitness values.

The QAP itself holds the matrices for the distances between the locations and the weights between the facilities. These are parameters that are used by the QAPEvaluator to compute the fitness value of a solution candidate. The plugin also provides move evaluators such as the *QAPSwap2MoveEvaluator* that evaluates the quality of a swap move. This move evaluator has a special method that allows to compute the move quality in $O(1)$ as described in [36]. It requires that the complete swap neighborhood is evaluated each iteration and that the move qualities from the previous iteration, as well as the previously selected move are remembered. The *QuadraticAssigmentProblem* class also implements interfaces which allow to parameterize it with benchmark instances from various libraries. As described above the QAPLIB is

---

already integrated in HeuristicLab, but as the QAP is a generalization of the Traveling Salesman Problem (TSP) it can be parameterized with problems from the according benchmark instance library TSPLIB[33]. For this purpose the problem class acts as an *IProblemInstanceConsumer* for *QAPData* and *TSPData*. Figure 10.19 shows a screenshot of the QAP implementation in HL with the chr12a problem instance loaded from the QAPLIB.

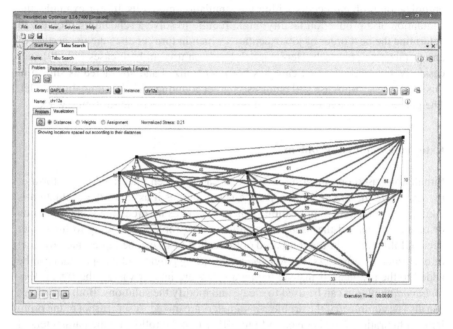

Fig. 10.19. QAP in HeuristicLab showing a QAPLIB instance. The layout of the locations is obtained by performing multi-dimensional scaling on the distances matrix.

10.5.2   Simulation-Based Optimization

The optimization of simulation parameters has been discussed before [12, 11], however there is still a lack of widely available ready-to-use optimization systems that can be employed to perform the optimization task. Some simulation frameworks come with optimizers such as OptQuest [16] included, but even in this case the included optimizer might not be well suited to solve the given problem. The cause for this can be found in the *no free lunch theorem* [50] which states that no algorithm can outperform any other algorithm on all problems. Due to the high heterogeneity of the actual simulation models and the meaning of their parameters, the problems can be of many different natures. In one case there might be only a single optimal solution that can be obtained through a basic local search algorithm, but in other cases the model response might be more complex and different strategies are required with different levels of diversification and intensification. In any case, the more solutions that need to be evaluated, the longer the optimization process will last. The selection of a suited optimization method and suited algorithm parameters is therefore crucial to find good parameters for the given simulation model.

Generally, two different cases of simulation-based optimization can be identified. In the first case the simulation model acts as a fitness function, it will take a number of parameters and calculate the resulting fitness value. In the second case, the optimization problem occurs within the simulation model itself. For example, the simulation of a production facility might require to solve a scheduling problem to determine the best order of job executions which in turn requires the integration of an optimization approach. HeuristicLab has been used in both cases [32, 39] successfully, but while the first case has been generalized and abstracted as shown in this work, the second case still requires a tighter coupling with the simulation model. The generalization and abstraction of the second case is a topic for future work.

External Evaluation Problem

Simulation-based optimization in HeuristicLab has been integrated in the form of the *ExternalEvaluationProblem*. As the name implies it assumes the evaluation is taking place in another application. This problem has no predefined representation or operators, instead the user can customize the problem according to his or her needs. If the actual simulation-based optimization tasks can be represented by a set of real-valued parameters, the *RealVectorEncoding* plugin and its operators can be added to the problem. If instead the parameters are integer values, the *IntegerVectorEncoding* plugin can be used to create and modify the solutions. Both encodings can also be combined if the problem parameters are mixed. A screenshot of the problem configuration view is given in Figure 10.20. In the following the parameters of this problem are explained.

BestKnownQuality and *BestKnownSolution:* These parameters are used and updated by certain analyzers and remember the best quality that has been found so far as well as the corresponding best solution.
Cache: Together with the appropriate evaluation operator this parameter can be used to add an evaluation cache. The cache stores already seen configurations and their corresponding quality so that these need not be simulated again.
Clients: Contains a list of clients in the form of communication channels. At least one must be specified, but the list can contain multiple channels if the simulation model is run on multiple machines.
Evaluator: This operator is used to collect the required variables, packs them into a message, and transmits them to one of the clients. If the cached evaluator is used the cache will be filled and the quality of previously seen configurations will be taken directly from the cache.
Maximization: This parameter determines if the received quality values should be maximized or minimized.

Operators: This list holds all operators that can modify and process solutions such as for example crossover and mutation operators. Any operator added to the list can be used in an algorithm and certain algorithms might require certain operators to work.

SolutionCreator: This operator is used to create the initial solution, typically it randomly initializes a vector of a certain length and within certain bounds.

Interoperability

The representation of solutions as scope objects which contain an arbitrary number of variables and the organization of scopes in a tree provides an opportunity for integrating a generic data exchange mechanism. Typically, an evaluator is applied on the solution scope and calculates the quality based on some variables that it would expect therein. The evaluator in the ExternalEvaluationProblem will however collect a user specified set of variables in the solution scope, and adds them to the *Solution-Message*. This message is then transmitted to the external application for evaluation. The evaluation operator then waits for a reply in form of the *QualityMessage* which contains the quality value and which can be inserted into the solution scope again. This allows to use any algorithm that can optimize single-objective problems in general to optimize the ExternalEvaluationProblem. The messages in this case are protocol buffers[7] which are defined in a .proto file. The structure of these messages is shown in Listing 10.6.

The protocol buffer specification in form of the message definitions is used by a specific implementation to generate a tailored serializer and deserializer class for each message. The format is designed for very compact serialized files that do not impose a large communication overhead and the serialization process is quick due to the efficiency of the specific serialization classes. Implementations of protocol buffers are provided by Google for Java[TM], C++, and Python, but many developers have provided open source ports for other languages such as C#, Clojure, Objective C, R, and many others[8]. The solution message buffer is a so called "union type", that means it provides fields for many different data types, but not all of them need to be used. In particular there are fields for storing Boolean, integers, doubles, and strings, as well as arrays of these types, and there is also a field for storing bytes. Which data type is stored in which field is again customizable. HeuristicLab uses a *SolutionMessageBuilder* class to convert the variables in the scope to variables in the solution message. This message builder is flexible and can be extended to use custom converters, so if the user adds a special representation to HeuristicLab a converter can be provided to store that representation in a message. By default, if the designer of an ExternalEvaluationProblem would use an integer vector, it would be stored in an integer array variable in the solution message. The simulation model can then extract the variable and use it to set its parameters. The protocol buffer

---

[7] http://code.google.com/p/protobuf

[8] http://code.google.com/p/protobuf/wiki/ThirdPartyAddOns

is also extensible in that new optional fields may be added at a later date. Finally, transmission to the client is also abstracted in the form of *channels*. The default channel is based on the transmission control protocol (TCP) which will start a connection to a network socket that is opened by the simulation software. The messages are then exchanged over this channel.

Parallelization and Caching

If the required time to execute a simulation model becomes very long, users might want to parallelize the simulation by running the model on multiple computers. In HeuristicLab this is easily possible through the use of the *parallel engine*. The parallel engine allows multiple evaluation operators to be executed concurrently which in turn can make use of multiple channels defined in the *Clients* parameter. It makes use of the *TaskParallelLibrary* in Microsoft® .NET which manages the available threads for efficient operations. To further speed up the optimization the user can add aforementioned *EvaluationCache* and the respective evaluator. The cache can be persisted to a file or exported as a comma-separated-values (CSV) file for later analysis [32].

10.5.3 Genetic Programming

HeuristicLab includes an implementation of tree-based genetic programming and and extensive support for symbolic regression and classification. This implementation will be described in the following sections.

10.5.3.1 Symbolic Expression Tree Encoding

The central part of our implementation of tree-based GP is the symbolic expression tree encoding. It defines the structure of individuals and provides problem-independent classes that can be reused for GP problems. All standard methods for tree creation, manipulation, crossover and compilation are located in the encoding and therefore a newly implemented problem just has to provide concrete symbols and methods to evaluate solution candidates. For example, in the case of a symbolic regression problem, the evaluator calculates the error of the model predictions and uses an interpreter to calculate the output of the formula for each row of the dataset.

Any algorithm that uses recombination and mutation operators to generate new solution candidates, for instance a genetic algorithm (GA), can be used to solve any problem using the symbolic expression tree encoding for instance a symbolic regression problem. A specialized algorithm for genetic programming with reproduction and crossover probability is not yet provided but is planned to be added soon.

```
1    message SolutionMessage {
2      message IntegerVariable {
3        required string name = 1;
4        optional int32 data = 2;
5      }
6      message IntegerArrayVariable {
7        required string name = 1;
8        repeated int32 data = 2;
9        optional int32 length = 3;
10     }
11     //... similar sub-messages omitted for brevity ...
12     message RawVariable {
13       required string name = 1;
14       optional bytes data = 2;
15     }
16     required int32 solutionId = 1;
17     repeated IntegerVariable integerVars = 2;
18     repeated IntegerArrayVariable integerArrayVars = 3;
19     repeated DoubleVariable doubleVars = 4;
20     repeated DoubleArrayVariable doubleArrayVars = 5;
21     repeated BoolVariable boolVars = 6;
22     repeated BoolArrayVariable boolArrayVars = 7;
23     repeated StringVariable stringVars = 8;
24     repeated StringArrayVariable stringArrayVars = 9;
25     repeated RawVariable rawVars = 10;
26   }
27
28   message QualityMessage {
29     required int32 solutionId = 1;
30     required double quality = 2;
31   }
```

Listing 10.6. Definition of the generic interface messages

10.5.3.2   Symbolic Expression Trees

The most important interfaces of the symbolic expression tree encoding are: *ISymbolicExpressionTree*, *ISymbolicExpressionTreeNode*, and *ISymbol*.

The structure of a tree is defined by linked nodes, and the semantic is defined by symbols attached to these nodes. The *SymbolicExpressionTree* represents trees and provides properties for accessing the root node, getting its length and depth, and iterating all tree nodes. Every node of a tree can be reached beginning with the root node as the *ISymbolicExpressionTreeNode* provides properties and methods to manipulate its parent, its subtrees and its symbol. A GP solution candidate is

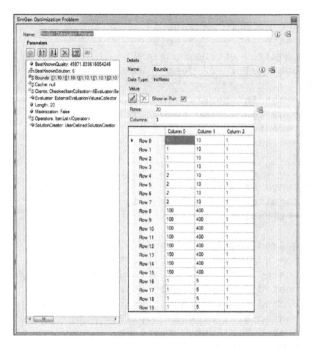

Fig. 10.20. Example of a simulation-based optimization problem configuration in Heuristic-Lab 3.3 with multiple bounds shown for each dimension of an integer parameter vector. The first column denotes the minimum value, the second column the maximum value, and the step size can be given in the third column.

therefore simply created by building a tree by linking tree nodes beginning with the root node.

The root node of a tree always contains the *ProgramRootSymbol* and must have a child node with the *StartSymbol*. This convention is necessary to support ADFs which are defined as additional sub-trees of the root node. The root node of ADF definitions contains a *DefunSymbol* (cf. Section 10.5.3.4).

10.5.3.3 Symbols and Grammars

In addition to the structure of a tree, symbols and grammars are necessary for individual creation. A symbol defines the semantic of a tree node (how it is interpreted) and specifies a minimum and maximum arity; terminal symbols have a minimum and maximum arity of zero. The set of available symbols must be defined with a grammar from which the set of valid and well-formed trees can be derived. We have chosen to implement this by defining which *Symbols* are allowed as child symbol of other symbols, and at which position they are allowed. For example, the first child of a conditional symbol must either be a comparison symbol or a Boolean function symbol.

Problem-specific grammars should derive from the base class for grammars which is provided by the encoding and define the rules for allowed tree structures. The initialization of an arithmetic expression grammar is shown in Listing 10.8. The statements in lines 1–10 create the symbols and add them to lists for easier handling. Afterwards the created symbols are added to the grammar (lines 12–13) and the number of allowed subtrees is set to two for all function symbols (lines 14–15). Terminal symbols do not have to be configured, because the number of allowed subtrees is automatically set to zero. The last lines define which symbols are allowed at which position in the tree. Below the *StartSymbol* all symbols are allowed (lines 16 and 17) and in addition, every symbol is allowed under a function symbol (Lines 18–21).

```
1  <expr> := <expr> <op> <expr> | <terminal>
2  <op>   := + | - | / | *
3  <terminal>  := variable | constant
```

Listing 10.7. Backus-Naur Form of an arithmetic grammar defining symbolic expression trees to solve a regression problem.

Default grammars are implemented and pre-configured for every problem which can be solved by GP. These grammars can be modified within the GUI to change the arity of symbols or to enable and disable specific symbols.

A typical operator for tree creation first adds the necessary nodes with the *Root-Symbol* and the *StartSymbol* and afterwards uses one of the allowed symbols returned by the *Grammar* as the starting point for the result producing branch. This procedure is recursively applied to extend the tree until the desired size is reached. In addition, the crossover and mutation operators also adhere to the rules defined by the grammar so during the whole algorithm run only valid and well-formed trees are produced.

10.5.3.4 Automatically Defined Functions

The GP implementation of HeuristicLab also supports automatically defined functions (ADFs). ADFs are program subroutines that provide code encapsulation and reuse. They are not shared between individuals but have to be evolved separately in individuals, either by crossover or mutation events and are numbered according to their position below the tree root. The *Defun* tree node and symbol define a new

```
1   var add = new Addition();
2   var sub = new Subtraction();
3   var mul = new Multiplication();
4   var div = new Division();
5   var constant = new Constant();
6   var variableSymbol = new Variable();
7   var allSymbols = new List<Symbol>()
8     {add,sub, mul,div,constant,variableSymbol};
9   var funSymbols = new List<Symbol>()
10    {add,sub,mul,div};
11
12  foreach (var symb in allSymbols)
13     AddSymbol(symb);
14  foreach (var funSymb in funSymbols)
15     SetSubtreeCount(funSymb, 2, 2);
16  foreach (var symb in allSymbols)
17     AddAllowedChildSymbol(StartSymbol, symb);
18  foreach (var parent in funSymbols) {
19     foreach (var child in allSymbols)
20        AddAllowedChildSymbol(parent, child);
21  }
```

Listing 10.8. Source code for the configuration of the *ArithmeticGrammar* formally defined
in Listing 10.7.

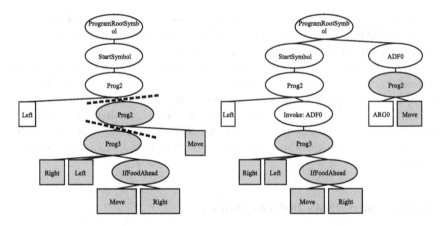

Fig. 10.21. Creation of an ADF in an artificial ant program. The dashed lines indicate the cut
points in the tree from which the ADF is created.

subroutine and are used next to the *StartSymbol* directly below the *ProgramRoot-
Symbol*. ADFs can be called through the *InvokeFunction* symbol from point in the
symbolic expression tree, except from ADFs with a lower index to prevent infinite
recursions and non-stopping programs.

ADFs are created during algorithm execution either by the subroutine creator or by the subroutine duplicator. The subroutine creator moves a subtree of an individual into a subroutine and inserts an *InvokeFunctionTreeNode* instead of the original subtree. Furthermore, ADFs can have an arbitrary number of arguments that are used to parameterize the subroutines. An example for creating an ADF with one argument is shown in Figure 10.21. On the left hand side the original tree describing an artificial ant program is displayed. Additionally, two cut points for the ADF extraction are indicated by dashed lines. The subtree between the cut points is added beneath a *DefunTreeNode* displayed as the newly defined ADF *ADF0* and as a replacement an *InvokeFunctionTreeNode* is inserted. The subtree below the second cut point is left unmodified and during interpretation its result is passed to *ADF0* as the value of *ARG0*.

Architecture altering operators for subroutine and argument creation, duplication, and deletion are provided by the framework. All of these work by moving parts of the tree to another location, either in the standard program execution part (below the *StartTreeNode*) or into an ADF (below the *DefunTreeNode*). For example, the subroutine deletion operator replaces all tree nodes invoking the affected subroutine by the body of the subroutine itself and afterwards deletes the subroutine from the tree by removing the *DefunTreeNode*. All architecture altering operators can be called in place of mutation operators as described by Koza, however in contrast to mutation operators, architecture altering operators preserve the semantics of the altered solution.

The combination of architecture altering operators and tree structure restrictions with grammars is non-trivial as grammars must be dynamically adapted over time. Newly defined ADFs must be added to the grammar; however, the grammar of each single tree must be updated independently because ADFs are specific to trees. This has led to a design where tree-specific grammars contain dynamically extended rules and extend the initially defined static grammar. The combination of the tree-specific grammar and the static grammar defines the valid tree structures for each solution and also for its child solutions because grammars must be inherited by child solutions. If ADFs are not allowed the tree-specific grammar is always empty because no symbols are dynamically added during the run.

Architecture manipulating operators automatically update the tree-specific grammar correctly by altering the allowed symbols and their restrictions. This mechanism allows to implement crossover and mutation operators without special cases for ADFs.

10.5.3.5   Symbolic Regression

In this section the implementation for evaluating symbolic regression models represented as symbolic expression trees is described. Symbolic regression is frequently used as a GP benchmark task for testing new algorithmic concepts and ideas. If symbolic regression is applied to large real-world datasets with several thousand data rows, performance as well as memory efficiency becomes an important issue.

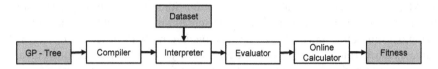

Fig. 10.22. Workflow for calculating the fitness of symbolic regression models

The main concepts for the symbolic regression evaluation in HeuristicLab are streaming and lazy evaluation provided by constructs of the Microsoft® .NET framework. Figure 10.22 depicts how the symbolic expression trees are passed through different operators for fitness value calculation, which is explained in the following sections.

Interpretation of Trees

Evaluators for symbolic regression calculate the error of the predicted values of the model and the actual target values. To prevent the allocation of large double arrays we implemented an interpreter for symbolic regression models that yields a lazy sequence of predicted values for a given model, a dataset and a lazy sequence of row indexes. As a preparatory step the interpreter first compiles the model represented as a symbolic expression tree down to an array of instructions. This preparation can be done in a single pass over all nodes of the tree so the costs are rather small and the linear instruction sequence can be evaluated much faster. First of all the nodes of the original tree are scattered on the heap, while the instruction array is stored in a continuous block of memory. Additionally, the instructions have a small memory footprint as they consist of a single byte for the operation code (opcode), a byte for the number of arguments of the function and an object reference which can hold additional data for the instruction. As a result the instruction array is much more cache friendly and the number of cache misses of tree interpretation can be reduced. Another benefit of the compilation step is that simple static optimizations for instance constant folding can be applied.

The interpretation of the instruction array is implemented with a simple recursive *Evaluate* method containing a large switch statement with handlers for each opcode. Listing 10.9 shows an excerpt of the evaluation method with the handlers for the opcodes for addition, division and variable symbols. The recursive evaluation method is rather monolithic and contains all necessary code for symbol evaluation. This goes against fundamental OO design principles, however, the implementation as a single monolithic switch loop with recursive calls is very efficient as no virtual calls are necessary, the switch statement can be compiled down to a relative jump instruction, and the arguments are passed on the runtime stack which again reduces cache misses.

```
1   double Evaluate(Dataset ds, State state) {
2     var curInstr = state.NextInstruction();
3     switch (curInstr.opCode) {
4       case OpCodes.Add: {
5         double s = Evaluate(dataset, state);
6         for (int i = 1; i < curInstr.nArgs; i++) {
7           s += Evaluate(dataset, state);
8         }
9         return s;
10      }
11      // [...]
12      case OpCodes.Div: {
13        double p = Evaluate(dataset, state);
14        for (int i = 1; i < curInstr.nArgs; i++) {
15          p /= Evaluate(dataset, state);
16        }
17        if (curInstr.nArgs == 1) p = 1.0 / p;
18        return p;
19      }
20      // [...]
21      case OpCodes.Variable: {
22        if (state.row < 0 || state.row >= dataset.Rows)
23          return double.NaN;
24        var varNode = (VariableTreeNode)curInstr.dynamicNode;
25        var values = ((IList<double>)curInstr.iArg0)
26        return values[state.row];
27      }
28    }
29  }
```

Listing 10.9. Excerpt of the evaluation method for symbolic regression models showing handlers for the addition, division and variable opcodes.

An alternative design would be to implement a specific evaluation method in each symbol class. This would be the preferable way regarding readability and maintainability. However, with this alternative design a costly indirect virtual call would be necessary for each node of the tree and for each evaluated row of the dataset.

In addition to the recursive interpreter, HeuristicLab also provides an interpreter implementation that compiles symbolic expression trees to linear code in intermediate language (IL) using the *Reflection.Emit* framework which can be executed directly by the Microsoft® .NET CLR. This interpreter is useful for large datasets with more than 10,000 rows as the generated IL code is further optimized and subsequently compiled to native code by the framework JIT-compiler. The drawback is that the JIT-compiler is invoked for each evaluated tree and these costs can be amortized only when the dataset has a large number of rows.

Interpretation of ADFs

The interpreter must be able to evaluate trees with ADFs with a variable number of arguments. The instruction for calling ADF uses the *Call* opcode and contains the index of the called ADF and the number of arguments of the ADF. The code fragment for the interpretation of ADFs and function arguments is shown in Listing 10.10. First the interpreter evaluates the subtrees of the *Call* opcode and stores the results. Next the interpreter creates a stackframe which holds the current program counter and the argument values. Stackframes are necessary to support ADFs that subsequently call other ADFs and recusive ADFs. After the stackframe has been created the interpreter jumps to the first instruction of the ADF. When an argument symbol is encountered while interpreting the ADF instructions the interpreter accesses the previously calculated argument values which are stored in the top-most stackframe and returns the appropriate value. This approach to ADF interpretation using precalculated argument values is only possible because symbolic regression expressions do not have side effects. Otherwise, the interpreter would have to jump back to the subtrees of the call symbol for each encountered *ARG* opcode. At the end of the ADF definition the interpreter deletes the top-most stackframe with *RemoveStackFrame* and continues interpretation at the point after the subtrees of the just evaluated *Call* opcode.

Online Evaluation of Programs

The first step for the evaluation of programs is to obtain the dataset on which the trees have to be evaluated on and to calculate the rows that should be used for fitness evaluation. If all samples are to be used, the rows are streamed as an *Enumerable* beginning with the start of the training partition until its end. Otherwise, the row indices to evaluate the tree on, are calculated and yielded by the selection sampling technique [25].

The row indices, together with the dataset and the individual are passed to the interpreter that in fact returns a sequence of numbers. Until now no memory is allocated (except the space required for the iterators) due to the streaming capabilities of the interpreter and the way of calculating row indices. But the whole streaming approach would by pointless if the estimated values of the interpreter were stored in a data structure for fitness calculation. Therefore, all fitness values must be calculated on the fly which is done by *OnlineCalculators*. Such calculators are provided for the mean and the variance of a sequence of numbers and for calculation metrics between two sequences such as the covariance and the Pearson's R^2 coefficient. Further error measures are the mean absolute and squared error, as well as scaled ones, the mean absolute relative error and the normalized mean squared error. *OnlineCalculators* can be nested; for example the *MeanSquaredErrorOnlineCalculator* just calculates the squared error between the original and estimated values and then passes

```
1   // [...]
2   case OpCodes.Call: {
3       // evaluate subtrees
4       var argValues = new double[curInstr.nArgs];
5       for (int i = 0; i < curInstr.nArgs; i++) {
6         argValues[i] = Evaluate(dataset, state);
7       }
8       // push on argument values on stack
9       state.CreateStackFrame(argValues);
10
11      // save the pc
12      int savedPc = state.ProgramCounter;
13      // set pc to start of function
14      state.PC = (ushort)curInstr.iArg0;
15      // evaluate the function
16      double v = Evaluate(dataset, state);
17
18      // delete the stack frame
19      state.RemoveStackFrame();
20
21      // restore the pc => evaluation will
22      // continue at point after my subtrees
23      state.PC = savedPc;
24      return v;
25  }
26  case OpCodes.Arg: {
27      return state.GetStackFrameValue(curInstr.iArg0);
28  }
29  // [...]
```

Listing 10.10. Code fragment for the interpretation of ADFs and function arguments

the result to the *MeanAndVarianceOnlineCalculator*. The code of the *MeanAnd-VarianceOnlineCalculator* is presented in Listing 10.11 and in the *Add* method it can be seen how the mean and variance are updated, when new values are added.

The source for calculating the mean squared error of an individual is shown in Listing 10.12, where all the parts described are combined. First the row indices for fitness calculation are generated and the estimated and original values obtained (lines 1-3). Afterwards these values are enumerated and passed to the *OnlineMeanSquaredErrorEvaluator* that in turn calculates the actual fitness.

```
1  public class OnlineMeanAndVarianceCalculator {
2    private double oldM, newM, oldS, newS;
3    private int n;
4
5    public int Count { get { return n; } }
6    public double Mean {
7      get { return (n > 0) ? newM : 0.0; }
8    }
9    public double Variance {
10     get { return (n > 1) ? newS / (n-1) : 0.0; }
11   }
12
13   public void Reset() { n = 0; }
14   public void Add(double x) {
15     n++;
16     if(n == 1) {
17       oldM = newM = x;
18       oldS = newS = 0.0;
19     } else {
20       newM = oldM + (x - oldM) / n;
21       newS = oldS + (x - oldM) * (x - newM);
22
23       oldM = newM;
24       oldS = newS;
25     }
26   }
27 }
```

Listing 10.11. Source code of the *MeanAndVarianceOnlineCalculator.*

```
1  var rows = Enumerable.Range(0,trainingEnd);
2  var estimated = interpreter.GetExpressionValues(tree,
3      dataset, rows).GetEnumerator();
4  var original = dataset.GetDoubleValues(targetVariable,
5      rows).GetEnumerator();
6  var calculator = new OnlineMSECalculator();
7
8  while(original.MoveNext() & estimated.MoveNext()) {
9    calculator.Add(original.Current, estimated.Current);
10 }
11 double meanSquaredError = calculator.MeanSquaredError;
```

Listing 10.12. Source code for calculating the mean squared error between the original values and the estimated values of an individual.

10.6   Conclusion

The main goal of this chapter was to describe the architecture and design of the HeuristicLab optimization environment which aims at fulfilling the requirements of three heterogeneous user groups, namely practitioners, heuristic optimization experts, and students. Three versions of HeuristicLab, referred to as HeuristicLab 1.x, HeuristicLab 2.x, and HeuristicLab 3.x, have been implemented by the authors since 2002 which were discussed in this chapter. By incorporating beneficial features of existing frameworks as well as several novel concepts, especially the most recent version, HeuristicLab 3.x, represents a powerful and mature framework which can be used for the development, analysis, comparison, and productive application of heuristic optimization algorithms. The key innovations of HeuristicLab can be summarized as follows:

- **Plugin-Based Architecture**
  The concept of plugins is used as the main architectural pattern in HeuristicLab. In contrast to other monolithic frameworks, the HeuristicLab main application just provides a lightweight plugin infrastructure. All other parts are implemented as plugins and are loaded dynamically at runtime. This architecture offers a high degree of flexibility. Users can easily integrate custom extensions such as new optimization algorithms or problems by developing new plugins. They do not need to have access to all the source code of HeuristicLab or to recompile the whole application. Furthermore, the modular nature of the plugin-based architecture simplifies the integration into existing software environments, as only the plugins required in a specific optimization scenario have to be deployed.
- **Generic Algorithm Model**
  As there is no unified model for all different heuristic optimization techniques in general, a generic algorithm model is implemented in HeuristicLab that is not restricted to a specific heuristic optimization paradigm. Any kind of algorithm can be represented. To achieve this level of flexibility, algorithms are not implemented as static blocks of code but are defined as operator graphs which are assembled dynamically at runtime. Users can define custom algorithms by combining basic operators for different solution encodings or optimization problems provided by several plugins, or they can add custom operators to integrate specific functionality. Consequently, not only standard trajectory-based or population-based heuristic optimization algorithms but also generic or problem-specific extensions as well as hybrid algorithms can be easily realized.
- **Graphical User Interface**
  As practitioners, students, and also in some cases heuristic optimization experts might not have comprehensive programming skills, a suitable user interface is required to define, execute, and analyze algorithms. Consequently, a graphical user interface (GUI) is integrated in HeuristicLab. According to the model-view-controller pattern, each HeuristicLab object (e.g., operators, variables, or data values) can provide a view to present itself to the user. However, as graphical visualization of objects usually is a performance critical task, these views are

shown and updated on demand. Furthermore, the GUI reduces the required learning effort significantly. Similarly to standard software products, it enables users to apply heuristic optimization algorithms immediately.

- **Parallelism**
  Last but not least parallel execution of algorithms is also respected in HeuristicLab. Dedicated control operators can be used to define parts of an algorithm that should be executed in parallel. These operators can be used anywhere in an algorithm which enables the definition of parallel heuristic optimization methods, as for example global, coarse-grained, or fine-grained parallel GAs. However, how parallelization is actually done does not depend on the operators but is defined when executing an algorithm by choosing an appropriate execution engine. Several engines are provided in HeuristicLab to execute parallel algorithms for example using multiple threads on a multi-core CPU or multiple computers connected in a network.

Since 2002 all versions of HeuristicLab have been extensively used in the research group "Heuristic and Evolutionary Algorithms Laboratory (HEAL)" of Michael Affenzeller for the development of enhanced evolutionary algorithms as well as in several research projects and lectures. The broad spectrum of these applications is documented in numerous publications and highlights the flexibility and suitability of HeuristicLab for the analysis, development, test, and productive use of metaheuristics. A comprehensive description of the research activities of the group can also be found in the book "Genetic Algorithms and Genetic Programming - Modern Concepts and Practical Applications" [1].

However, the development process of HeuristicLab has not come to an end so far. As it is very easy to apply and compare different algorithms with HeuristicLab, it can be quickly identified which heuristic algorithms and which corresponding parameter settings are effective for a certain optimization problem. In order to systematically analyze this information, the authors plan within the scope of the research laboratory "Josef Ressel-Centre for Heuristic Optimization (Heureka!)"[9] to store the results of all algorithm runs executed in HeuristicLab in a large database. The ultimate goal of this optimization knowledge base is to identify correlations between heuristic optimization algorithms and solution space characteristics of optimization problems. This information will provide essential clues for the selection of appropriate algorithms and will also encourage the development of new enhanced and hybrid heuristic optimization algorithms in order to solve problems for which no suitable algorithms are known yet.

Acknowledgements. The work described in this chapter was done within the Josef Ressel-Centre HEUREKA! for Heuristic Optimization sponsored by the Austrian Research Promotion Agency (FFG). HeuristicLab is developed by the Heuristic and Evolutionary Algorithm Laboratory (HEAL)[10] of the University of Applied Sciences Upper Austria. It can be

---

[9] http://heureka.heuristiclab.com

[10] http://heal.heuristiclab.com/

downloaded from the HeuristicLab homepage[11] and is licensed under the GNU General Public License.

References

1. Affenzeller, M., Winkler, S., Wagner, S., Beham, A.: Genetic Algorithms and Genetic Programming - Modern Concepts and Practical Applications. In: Numerical Insights. CRC Press (2009)
2. Alba, E. (ed.): Parallel Metaheuristics: A New Class of Algorithms. Wiley Series on Parallel and Distributed Computing. Wiley (2005)
3. Arenas, M.G., Collet, P., Eiben, A.E., Jelasity, M., Merelo, J.J., Paechter, B., Preuß, M., Schoenauer, M.: A framework for distributed evolutionary algorithms. In: Guervós, J.J.M., Adamidis, P.A., Beyer, H.-G., Fernández-Villacañas, J.-L., Schwefel, H.-P. (eds.) PPSN 2002. LNCS, vol. 2439, pp. 665–675. Springer, Heidelberg (2002)
4. Blum, C., Roli, A., Alba, E.: An introduction to metaheuristic techniques. In: Alba, E. (ed.) Parallel Metaheuristics: A New Class of Algorithms, Wiley Series on Parallel and Distributed Computing, ch. 1, pp. 3–42. Wiley (2005)
5. Burkard, R.E., Karisch, S.E., Rendl, F.: QAPLIB – A quadratic assignment problem library. Journal of Global Optimization 10(4), 391–403 (1997), http://www.opt.math.tu-graz.ac.at/qaplib/
6. Cantu-Paz, E.: Efficient and Accurate Parallel Genetic Algorithms. Kluwer (2001)
7. de Carvalho Jr., S.A., Rahmann, S.: Microarray layout as quadratic assignment problem. In: Proceedings of the German Conference on Bioinformatics (GCB). Lecture Notes in Informatics, vol. P-83 (2006)
8. Cox, B.J.: Planning the software industrial revolution. IEEE Software 7(6), 25–33 (1990), http://www.virtualschool.edu/cox/pub/PSIR/
9. DeJong, K.A.: Evolutionary Computation: A Unified Approach. In: Bradford Books. MIT Press (2006)
10. Drezner, Z.: Extensive experiments with hybrid genetic algorithms for the solution of the quadratic assignment problem. Computers & Operations Research 35(3), 717–736 (2008), Part Special Issue: New Trends in Locational Analysis, http://www.sciencedirect.com/science/article/pii/S0305054806001341, doi:10.1016/j.cor.2006.05.004
11. Fu, M., Glover, F., April, J.: Simulation optimization: A review, new developments, and applications. In: Proceedings of the 2005 Winter Simulation Conference, pp. 83–95 (2005)
12. Fu, M.C.: Optimization for simulation: Theory vs. practice. Informs J. on Computing 14(3), 192–215 (2002), http://www.rhsmith.umd.edu/faculty/mfu/fu_files/fu02.pdf
13. Gagné, C., Parizeau, M.: Genericity in evolutionary computation software tools: Principles and case-study. International Journal on Artificial Intelligence Tools 15(2), 173–194 (2006)
14. Gamma, E., Helm, R., Johnson, R., Vlissides, J.: Design Patterns: Elements of Reusable Object-Oriented Software. Addison-Wesley (1995)
15. Giffler, B., Thompson, G.L.: Algorithms for solving production-scheduling problems. Operations Research 8(4), 487–503 (1960)

---

[11] http://dev.heuristiclab.com/

16. Glover, F., Kelly, J.P., Laguna, M.: New advances for wedding optimization and simulation. In: Farrington, P.A., Nembhard, H.B., Sturrock, D.T., Evans, G.W. (eds.) Proceedings of the 1999 Winter Simulation Conference, pp. 255–260 (1999), http://citeseer.ist.psu.edu/glover99new.html
17. Greenfield, J., Short, K.: Software Factories: Assembling Applications with Patterns, Models, Frameworks, and Tools. Wiley (2004)
18. Hahn, P.M., Krarup, J.: A hospital facility layout problem finally solved. Journal of Intelligent Manufacturing 12, 487–496 (2001)
19. Holland, J.H.: Adaption in Natural and Artificial Systems. University of Michigan Press (1975)
20. Johnson, R., Foote, B.: Designing reusable classes. Journal of Object-Oriented Programming 1(2), 22–35 (1988)
21. Jones, M.S.: An object-oriented framework for the implementation of search techniques. Ph.D. thesis, University of East Anglia (2000)
22. Jones, M.S., McKeown, G.P., Rayward-Smith, V.J.: Distribution, cooperation, and hybridization for combinatorial optimization. In: Voß, S., Woodruff, D.L. (eds.) Optimization Software Class Libraries. Operations Research/Computer Science Interfaces Series, vol. 18, ch. 2, pp. 25–58. Kluwer (2002)
23. Keijzer, M., Merelo, J.J., Romero, G., Schoenauer, M.: Evolving Objects: A general purpose evolutionary computation library. In: EA 2001, Evolution Artificielle, 5th International Concerence in Evolutionary Algorithms, pp. 231–242 (2001)
24. Kirkpatrick, S., Gelatt, C.D., Vecchi, M.P.: Optimization by simulated annealing. Science 220, 671–680 (1983)
25. Knuth, D.E.: The Art of Computer Programming, 3rd edn. Seminumerical Algorithms, vol. 2. Addison-Wesley (1997)
26. Koopmans, T.C., Beckmann, M.: Assignment problems and the location of economic activities. Econometrica, Journal of the Econometric Society 25(1), 53–76 (1957), http://cowles.econ.yale.edu/P/cp/p01a/p0108.pdf
27. Krasner, G.E., Pope, S.T.: A cookbook for using the model-view-controller user interface paradigm in Smalltalk-80. Journal of Object-Oriented Programming 1(3), 26–49 (1988)
28. Lenaerts, T., Manderick, B.: Building a genetic programming framework: The added-value of design patterns. In: Banzhaf, W., Poli, R., Schoenauer, M., Fogarty, T.C. (eds.) EuroGP 1998. LNCS, vol. 1391, pp. 196–208. Springer, Heidelberg (1998)
29. McIlroy, M.D.: Mass produced software components. In: Naur, P., Randell, B. (eds.) Software Engineering: Report of a conference sponsored by the NATO Science Committee, pp. 138–155 (1969)
30. Nievergelt, J.: Complexity, algorithms, programs, systems: The shifting focus. Journal of Symbolic Computation 17(4), 297–310 (1994)
31. Parejo, J.A., Ruiz-Cortes, A., Lozano, S., Fernandez, P.: Metaheuristic optimization frameworks: A survey and benchmarking. Soft Computing 16(3), 527–561 (2012)
32. Pitzer, E., Beham, A., Affenzeller, M., Heiss, H., Vorderwinkler, M.: Production fine planning using a solution archive of priority rules. In: Proceedings of the IEEE 3rd International Symposium on Logistics and Industrial Informatics (Lindi 2011), pp. 111–116 (2011)
33. Reinelt, G.: TSPLIB - A traveling salesman problem library. ORSA Journal on Computing 3, 376–384 (1991)
34. Ribeiro Filho, J.L., Treleaven, P.C., Alippi, C.: Genetic-algorithm programming environments. IEEE Computer 27(6), 28–43 (1994)
35. Stützle, T.: Iterated local search for the quadratic assignment problem. European Journal of Operational Research 174, 1519–1539 (2006)

36. Taillard, E.D.: Robust taboo search for the quadratic assignment problem. Parallel Computing 17, 443–455 (1991)
37. Voß, S., Woodruff, D.L.: Optimization software class libraries. In: Voß, S., Woodruff, D.L. (eds.) Optimization Software Class Libraries. Operations Research/Computer Science Interfaces Series, vol. 18, ch. 1, pp. 1–24. Kluwer (2002)
38. Voß, S., Woodruff, D.L. (eds.): Optimization Software Class Libraries. Operations Research/Computer Science Interfaces Series, vol. 18. Kluwer (2002)
39. Vonolfen, S., Affenzeller, M., Beham, A., Wagner, S., Lengauer, E.: Simulation-based evolution of municipal glass-waste collection strategies utilizing electric trucks. In: Proceedings of the IEEE 3rd International Symposium on Logistics and Industrial Informatics (Lindi 2011), pp. 177–182 (2011)
40. Wagner, S.: Looking Inside Genetic Algorithms. Schriften der Johannes Kepler Universität Linz, Reihe C: Technik und Naturwissenschaften. Universitätsverlag Rudolf Trauner (2004)
41. Wagner, S.: Heuristic optimization software systems - Modeling of heuristic optimization algorithms in the HeuristicLab software environment. Ph.D. thesis, Johannes Kepler University, Linz, Austria (2009)
42. Wagner, S., Affenzeller, M.: HeuristicLab Grid - A flexible and extensible environment for parallel heuristic optimization. In: Bubnicki, Z., Grzech, A. (eds.) Proceedings of the 15th International Conference on Systems Science, vol. 1, pp. 289–296. Oficyna Wydawnicza Politechniki Wroclawskiej (2004)
43. Wagner, S., Affenzeller, M.: HeuristicLab Grid. - A flexible and extensible environment for parallel heuristic optimization 30(4), 103–110 (2004)
44. Wagner, S., Affenzeller, M.: HeuristicLab: A generic and extensible optimization environment. In: Ribeiro, B., Albrecht, R.F., Dobnikar, A., Pearson, D.W., Steele, N.C. (eds.) Adaptive and Natural Computing Algorithms, pp. 538–541. Springer, Heidelberg (2005)
45. Wagner, S., Affenzeller, M.: SexualGA: Gender-specific selection for genetic algorithms. In: Callaos, N., Lesso, W., Hansen, E. (eds.) Proceedings of the 9th World Multi-Conference on Systemics, Cybernetics and Informatics (WMSCI 2005), vol. 4, pp. 76–81. International Institute of Informatics and Systemics (2005)
46. Wagner, S., Kronberger, G., Beham, A., Winkler, S., Affenzeller, M.: Modeling of heuristic optimization algorithms. In: Bruzzone, A., Longo, F., Piera, M.A., Aguilar, R.M., Frydman, C. (eds.) Proceedings of the 20th European Modeling and Simulation Symposium, pp. 106–111. DIPTEM University of Genova (2008)
47. Wagner, S., Kronberger, G., Beham, A., Winkler, S., Affenzeller, M.: Model driven rapid prototyping of heuristic optimization algorithms. In: Quesada-Arencibia, A., Rodrígue, J.C., Moreno-Diaz Jr., R., Moreno-Diaz, R. (eds.) 12th International Conference on Computer Aided Systems Theory EUROCAST 2009, vol. 2009, pp. 250–251. IUCTC Universidad de Las Palmas de Gran Canaria (2009)
48. Wagner, S., Winkler, S., Pitzer, E., Kronberger, G., Beham, A., Braune, R., Affenzeller, M.: Benefits of plugin-based heuristic optimization software systems. In: Moreno Díaz, R., Pichler, F., Quesada Arencibia, A. (eds.) EUROCAST 2007. LNCS, vol. 4739, pp. 747–754. Springer, Heidelberg (2007)
49. Wilson, G.C., McIntyre, A., Heywood, M.I.: Resource review: Three open source systems for evolving programs - Lilgp, ECJ and Grammatical Evolution. Genetic Programming and Evolvable Machines 5(1), 103–105 (2004)
50. Wolpert, D.H., Macready, W.G.: No free lunch theorems for optimization. IEEE Transactions on Evolutionary Computation 1(1), 67–82 (1997)

Part II
Network Management Essential Problems

Chapter 11
A Biomimetic SANET Middleware Infrastructure for Guiding and Maneuvering Autonomous Land-Yacht Vessels

Christopher Chiu and Zenon Chaczko

Abstract. This chapter elaborates an approach of guiding land-yachts according to a predefined maneuvering strategy. The simulation of the path as well as the controller scheme is simulated for a sailing strategy (tacking), where the scheme of the resultant path and the sailing mechanism is driven using Sensor-Actor Network (SANET) middleware infrastructure. The addition of obstacle avoidance and detection heuristics aids in the guiding process. By incorporating SANETs in the sailing craft, a range of sensory mechanisms can be employed to monitor and handle local obstacles in an effective manner, while data collated from the sensory environment can be transmitted to a base station node for monitoring conditions from a holistic dimension.

Keywords: Biomimetic Methodologies, Spring Tensor Analysis, Software Middleware, Autonomous Sailing Systems, Wireless Sensor Networks (WSN), Sensor Actor Networks (SANET).

11.1 Introduction

Autonomous vehicles using flow-based propulsion have been examined by Jouffroy [8] as a means to provide efficient navigation and control in variable conditions. A particular concern is that the control strategies necessary to assess stability and performance [8] cannot be solved using traditional Artificial-Intelligence techniques. Given the progression of technology advances in Sensor Actor Network (SANET) technologies [12], a distributed control approach using SANETS combined with sailing kinematics for path coordination and neural networks for obstacle avoidance can be embedded within the embedded system for monitoring and control [2].

In this chapter, a multi-paradigm approach is employed to monitor the environmental surrounds of the vessel, while control aspects assist in sailing maneuvers and

Christopher Chiu · Zenon Chaczko
Faculty of Engineering & IT, University of Technology, Sydney, Australia
e-mail: {Christopher.Chiu,Zenon.Chaczko}@uts.edu.au

R. Klempous et al. (eds.), *Advanced Methods and Applications in Computational Intelligence*,   265
Topics in Intelligent Engineering and Informatics 6,
DOI: 10.1007/978-3-319-01436-4_11, © Springer International Publishing Switzerland 2014

trajectory planning. The use of a unified middleware framework to implement the controller functionality enables a uniform methodology to coordinate multiple sailing vessels; as well as provide localized awareness of vessels in the vicinity of a localized regional zone [2]. As the vessel navigates to its desired destination, real-time feedback of obstacles allows the SANET infrastructure to map the global terrain for the benefit of all vessels active in the environment [5]. This distributed approach to obtain a complete environmental map of the vessel's surroundings, provides redundancy in the situation where Global Positioning System (GPS) navigation signals are unavailable, or in the instance of outdated or unobtainable terrain maps of the region [2].

11.2 A Sailing Vessel as a Actor-Based Process

The actor-based paradigm, as elaborated by Rao and Georgeff [13], has been employed to incorporate the rule-sets for the sailing path maneuvers, and the tensor-analysis heuristics to incorporate obstacle avoidance [11]. In this system design, each software actor is associated with the sensors incorporated into the sailing craft (including the anemometer, GPS navigation sensor and ultrasonic detection sensors), and the motor controller mechanisms (to regulate steering and manipulate sail orientation). A core actor will represent the craft itself, such that a hierarchical structure of actor responsibilities exists within the sailing vessel model. The actor-based characteristics defined in the model are elaborated as follows [2]:

- **Centralized gateway:** To aggregate statistical data from all sailing crafts in a predefined region, and coordinate scheduled activities by the administrator or coordinator user:

  - **Weather forecasting:** Weather monitoring data collected from all sailing craft actors, along with meteorological data imported from external data sources; assist in forecasting potential low-pressure formations that would affect trajectory path planning.
  - **Terrain mapping:** Obstacle data accumulated from the crafts can be geo-tagged in a global map, thus allowing for the new mapping of undefined terrain or refreshing existing terrain maps for future expeditions.

- **Sailing vessel:** To incorporate various sensor data via a multi-modal approach [8, 9], communicate with neighboring vessels in a peer-to-peer fashion, and centralized gateway exchanges for collaborative data exchange [2]:

  - **Anemometer and barometer:** Wind direction and Speed distribution monitoring, along with atmospheric pressure measurement will monitor localized weather patterns and predict short-term weather forecasts.
  - **Global positioning system:** Satellite Navigation system to provide accurate positioning of vessels relative on Earth's coordinates; with sailing vessels incorporating GPS sensors are designated as anchor node vessels.

- **Ultrasonic and infra-red sensors:** For close-proximity obstacle detection so preemptive path navigation can be established; additionally vessels can incorporate digital video capture with embedded machine vision algorithms for enhanced obstacle recognition.

Harnessing the land yacht control dynamics as elaborated by Jouffroy [8], the purpose of the chapter is to examine the coordination of multiple craft irrespective of the environmental terrain. The coordination and control of autonomous systems have its origins in autonomous robotic control [12, 14], with the customization in place to account for land yacht kinematic parameters. The main study of investigation is to determine what optimum method can be used to steer a fleet of craft in uncertain conditions.

11.3   Heuristic Analysis for Autonomous Sailing Craft

Land-yachts by their nature need to perform complex maneuvers due in part to their main method of propulsion being a sail [6, 15]. This is of particular consequence especially when the craft needs to travel in the direction of the wind, because the vehicle quickly loses propulsion [9]. This no-sailing angular zone is resolved by maneuvering in a zig-zagged direction of the wind, where each angular turn of direction in the zig-zag formation is known as a tack [7]. The motion planning required for a tacking maneuver is important, because of the potential for competing interests between sailing in a safe manner to maintain propulsion, and trying to maintain a minimum trajectory [2].

The reason is because to ensure that a direct sailing path is made to the destination when traveling in the direction of the wind, the tack's angle should be made at the angle closest to the no-sailing zone [10, 16]. In other words, the zig-zag sailing formation should be as narrow as possible, but can never achieve an absolute linear path. However, tacking closer to the no-sailing zone angle also leads to a higher likelihood of propulsion loss, and thus stalling the craft. Therefore, an optimum balance must be made between achieving an optimum sailing trajectory, while simplifying the number of tacking maneuvers [8].

SANET middleware-based heuristics can be applied when a fleet of autonomously controlled craft need to be coordinated effectively, with each craft aware of its surroundings and neighboring vessels utilizing wireless radio-frequency communication [5, 10]. To develop an effective mechanism for trajectory mapping while simplifying tacking maneuvers, a trajectory mapping for sailing has been developed to linearize the trajectory itself. As shown in Figure 18.1 (*Left*), traditional trajectory mapping methods rely on local perception to maneuver around its neighbours, such as neural networks. Such a trajectory path places emphasis on intelligence at the source, without realizing the bigger picture: that the path can be potentially complex and meandering. By taking a SANET-based approach to the solution, we treat the environment as a global perception in Figure 18.1 (*Right*). This means that the

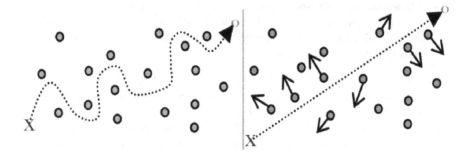

Fig. 11.1. Trajectory Mapping of the Destination Point using: (*Left*) Localized Awareness; (*Right*) Global Perception

neighboring vessels coordinate together to ensure a linearized path can be established by the sailing vessel, and final trajectory is executed as a result.

11.3.1 Application of Tensor Analysis for Trajectory Mapping

The method to determine a globalized trajectory mapping method has been inspired by protein fluctuation dynamics. As discussed in depth by Lin and Song [11], the premise of the spring tensor model is to determine conformational changes in proteins using second-order partial derivatives (Hessian matrices). Conformational change is the transition of macro-molecular structures in proteins as a result in a change of acidity, temperature, voltages and so forth. The spring tensor model is an enhancement of anisotropic modeling and Gaussian modeling methods, as while the former determines fluctuations of an atom's direction, the latter is better at determining the prediction of magnitudes of direction [11]. Thus by combining the two methodologies, the spring tensor model can be applied to a coordinated land-yacht system as follows:

- **Anisotropic modeling:**
  The determination of conformational variation or fluctuation in *direction* between elements.

  - *Adaptation:* This is suitable for determining how the interactions between neighboring land-yachts will result in the degree of directional fluctuation. The variation of potential direction will indicate what possible directions a vessel can travel if it is in proximity with a neighboring vessel.
  - *Meaning:* Smaller anisotropic values indicate a smaller potential to alter the direction, while larger values indicate a larger potential to alter the direction.

- **Gaussian modeling:**
  The determination of conformational variation or fluctuation in *magnitude* between elements.

  – *Adaptation:* This is appropriate to ascertain how interactions between land-yachts will result in the magnitude or total range of the fluctuation. The variation of potential magnitude indicates the possible maximum range the vessel can travel towards if it is in proximity with a neighboring vessel.
  – *Meaning:* Smaller magnitudes values indicate a smaller potential to alter the distance, while larger magnitudes indicate a larger potential to alter the distance of the vessel.

In the application of the Spring Tensor Model (STEM) by Lin and Song, the Go-like potential [2] is considered to take the non-native and native conformations as the input data [11]; for this instance these values are the difference in the land-yacht's Cartesian coordinates between time n and time $n+1$. These terms are divided into four terms as shown in the expression in Figure 18.2 (*Left*). The sum of the first term, V_1, determines the radius of connectivity; the sum of the second term, V_2, determines the bond angle; the sum of the third term, V_3, determines the torsional interaction. What is of interest is examining the final term, V_4, that determines non-local or global interactions.

$Y(X,Y_0) = \sum_{V1\,Bonds} + \sum_{V2\,Angles}$ $+ \sum_{V3\,Dihedral} + \sum_{V4\,NonLocal}$	(1.) The $V(X,Y\_0)$ values are sum of radial lengths, bonding angles and dihedral angles of consecutive objects for i and j. The non-local values are used.
$V_4 = \varepsilon \left[5\left(\frac{r_{0,ij}}{r_{ij}}\right)^{12} - 6\left(\frac{r_{0,ij}}{r_{ij}}\right)^{10} \right]$	(2.) The final non-local contact term is derived from the Go-like potential as discussed in Lin and Song's theoretical work.
$V_4 = -\varepsilon + \frac{120\varepsilon}{r_{o,ij}^2}\left(r_{ij} - r_{0,ij}\right)^2$	(3.) The Taylor expansion of the initial non-local contact term yields the following equation, where r_ij and r_0,ij are the consecutive long-term values for objects i and j.
$\frac{\delta^2 V_4}{\delta X_i \delta Y_j} = -\frac{240\varepsilon}{r_{o,ij}^2}\left(X_j - X_i\right)\left(Y_j - Y_i\right)/r_{ij}^2$	(4.) As the focus is on the equilibrium fluctuations, r_ij is equal or approximately equal to r_0,ij at equilibrium; thus the derivatives of V_4 can be further simplified as shown.

Fig. 11.2. The Go-like Potential Expression and its Taylor Expansion of the Non-Local Contact Term [11]

The STEM model's fourth term is of interest as it examines the global interactions of the elements. The final term, examined in Figure 18.2 (*Right*), is shown with its Taylor expansion form. The final non-local derivation is adopted from Lin and Song's calculations, which is used as a point of reference in this research project [11]. Using the parameters stated by Clementi [2], the value of epsilon (ε) adopted is

0.36 as per conformation observations of macro-molecular protein structures using X-ray crystallography.

Although beyond the scope of this current research, the future task is to obtain a unique value of epsilon and Taylor expansion parameters suitable for a wireless SANET environment in land-yacht sailing contexts. These values are obtained through experimental observation of the sailing vessels as they interact in the physical environment, and determining the maximum thresholds of the land-yacht's direction and magnitude to make a tacking maneuver in different environmental conditions (i.e. finding the difference between calm versus stormy weather forecasts). It is noted that there is no fixed parameter values that can be used for all land-yachts, although a close approximation can be made for classes or category types of land-yachts that is sufficient for the majority of results. To give an example of this property, Class 2 land-yachts will have a different tacking angle to Class 5 land-yachts due to their differentiation in size, weight and the total surface area of the sail).

11.3.2   Developmental Approach and Methodology

For the design prototype, the STEM Hessian is calculated using the experimental methodology depicted in Figure 18.3; the excerpt source code is provided in Figure 18.4. The purpose of the experimental approach is to examine how the Tensor Analysis heuristic can be used in conjunction with swarming algorithms to optimize global trajectory mapping for the sailing craft.

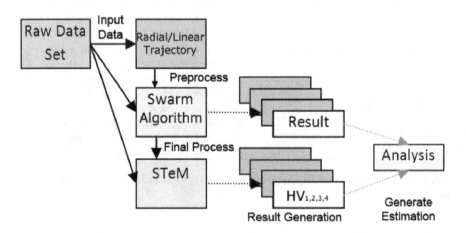

Fig. 11.3. Methodology of Analyzing Data Set using Tensor Analysis under Radial/Linear Trajectories and Particle Swarm Optimization

```
function preRunStemFunction()
global gHandles
clearOutput();
preProcessParamInput();
handles         = gHandles.handles;
dataMapMode     = get(handles.popup_DataMap,'Value');
maxIteration    = str2double(get(handles.input_Iterations,'String'));
maxIteration    = setMaxIteration(dataMapMode, maxIteration);
waitTime        = str2double(get(handles.input_WaitTime,'String'));
singMode        = 1;
diffMode        = 2;
inputData       = zeros(1,1);
for i=1:maxIteration
    switch dataMapMode
        case singMode
            [inputData, errInput] = getInputData(inputData, i);
            errTraj = getTrajData(i);
            if isDataInvalid(errInput, errTraj); return; end
            inputData = manipulateData(inputData, i, maxIteration);
            inputData = trajectoryEvent(inputData, i);
            runStemFunction(inputData, i, maxIteration);
        case diffMode
            diff = cell(1,2); k = 1;
            for j=i:i+1
                [inputData, errInput] = getInputData(inputData, j);
                errTraj = getTrajData(j);
                if isDataInvalid(errInput, errTraj); return; end
                inputData = manipulateData(inputData, j, maxIteration+1);
                inputData = trajectoryEvent(inputData, j);
                diff{k}   = inputData; k = k + 1;
            end
            runStemFunction(diff{2}-diff{1}, i, maxIteration, diff{2});
    end
end
```

Run Single Data Mode:
1. Manipulate Data
2. Add Trajectory Event
Run Heuristics

Run Dataset Difference:
1. Manipulate Data
2. Add Trajectory Event

Run Tensor Analysis

```
function hessian=fourthTerm(hessian,caArray,distance,numOfResidues,Epsilon,L_term)
% derive the hessian of the first term (off diagonal)
for i=1:numOfResidues;
    for j=1:numOfResidues;
        if abs(i-j)>3
            bx=caArray(i,1)  - caArray(j,1);
            by=caArray(i,2)  - caArray(j,2);
            bz=caArray(i,3)  - caArray(j,3);
            distijsqr=distance(i,j)^4;
            % diagonals of off-diagonal super elements (1st term)
            hessian(3*i-2,3*j-2)    = hessian(3*i-2,3*j-2)-L_term*Epsilon*bx*bx/distijsqr;
            hessian(3*i-1,3*j-1)    = hessian(3*i-1,3*j-1)-L_term*Epsilon*by*by/distijsqr;
            hessian(3*i,3*j)        = hessian(3*i,3*j)-L_term*Epsilon*bz*bz/distijsqr;
            % off-diagonals of off-diagonal super elements (1st term)
            hessian(3*i-2,3*j-1)    = hessian(3*i-2,3*j-1)-L_term*Epsilon*bx*by/distijsqr;
            hessian(3*i-2,3*j)      = hessian(3*i-2,3*j)-L_term*Epsilon*bx*bz/distijsqr;
            hessian(3*i-1,3*j-2)    = hessian(3*i-1,3*j-2)-L_term*Epsilon*by*bx/distijsqr;
            hessian(3*i-1,3*j)      = hessian(3*i-1,3*j)-L_term*Epsilon*by*bz/distijsqr;
            hessian(3*i,3*j-2)      = hessian(3*i,3*j-2)-L_term*Epsilon*bz*bx/distijsqr;
            hessian(3*i,3*j-1)      = hessian(3*i,3*j-1)-L_term*Epsilon*bz*by/distijsqr;
            % Hii: update the diagonals of diagonal super elements
            hessian(3*i-2,3*i-2)    = hessian(3*i-2,3*i-2)+L_term*Epsilon*bx*bx/distijsqr;
            hessian(3*i-1,3*i-1)    = hessian(3*i-1,3*i-1)+L_term*Epsilon*by*by/distijsqr;
            hessian(3*i,3*i)        = hessian(3*i,3*i)+L_term*Epsilon*bz*bz/distijsqr;
            % update the off-diagonals of diagonal super elements
            hessian(3*i-2,3*i-1)    = hessian(3*i-2,3*i-1)+L_term*Epsilon*bx*by/distijsqr;
            hessian(3*i-2,3*i)      = hessian(3*i-2,3*i)+L_term*Epsilon*bx*bz/distijsqr;
            hessian(3*i-1,3*i-2)    = hessian(3*i-1,3*i-2)+L_term*Epsilon*by*bx/distijsqr;
            hessian(3*i-1,3*i)      = hessian(3*i-1,3*i)+L_term*Epsilon*by*bz/distijsqr;
            hessian(3*i,3*i-2)      = hessian(3*i,3*i-2)+L_term*Epsilon*bz*bx/distijsqr;
            hessian(3*i,3*i-1)      = hessian(3*i,3*i-1)+L_term*Epsilon*bz*by/distijsqr;
        end
    end
end
end
```

Calculation of Forth
Term of Tensor Analysis
Code (From Lin & Song)

Fig. 11.4. Source Code for Data Preprocessing and Sample of STEM Hessian Code

The experimental approach conducted is categorized according into the dataset processing method used. This includes pseudo-randomized data, grid-lattice structured data and a Particle Swarm Optimization approach:

- **Random data generation**
  The data is randomly generated using a Pseudo-Random algorithm (Mersenne-twister method) as a control baseline to evaluate the other dataset layouts. The random data is subject to the limits of variation among the total population size using a pre-determined threshold value. Furthermore, dataset randomization limits ensure the environment is chaotic within reasonable bounds to control the experimental structure.

- **Grid-lattice layout**
  The data is generated according to a grid lattice plane, with each element equidistant from each other. This data structure is used to evaluate how radial and linear trajectories can impact on geometrically layered structures, and its impact on network structures that are evenly distributed in the environment. A layered lattice structure emulates the crystal structure found in carbon allotropes, such as graphite mineral.

- **Particle swarm optimization**
  The data is subject to Particle Swarm Optimization (PSO) as described by Clerc [3, 4], with random, grid-lattice datasets and future data sources being processed by PSO algorithms. The optimization of the data to obtain a global minimum or maximum value is limited by the total number of iterations that can be executed, the objection function used and the thresholds values for inertia and correction factors.

Implementation of the trajectory analytical system was developed by building upon the original STEM source code provided by Lin and Song [11]. The user frontend environment is primarily utilitarian: with the main panel in Figure 18.5 (*Left*) used to configure the experimental scenarios, and the visual display panel in Figure 18.5 (*Right*) showing the direction and magnitude of each element's long-term Hessian value. Apart from the provisioning of data manipulation and simulation of path trajectory functions, the analytical system allows the user to customize the Tensor Analysis parameters to evaluate a range of variations in epsilon and its eventual impact on the Hessian long-term values.

The prototype is developed in a step-wise fashion, where the execution of the simulation environment is based on the parameters defined in these steps. The environment is configured as follows:

- *Environment configuration:* Importation of the sail craft positions and their relative coordinates with one another;
- *External influences:* Introduction of environmental interference to the network structure, such as the initial detection of obstacles or foreign craft.
- *Internal influences:* Setup of the path trajectory to determine a craft's behavior as it encounters an obstacle. Execute results once final step is configured.

Additional data representation is made using surface map plots of the Hessian matrices as captured in Figure 18.6 (*Above*) for 20 elements with 3 coordinates

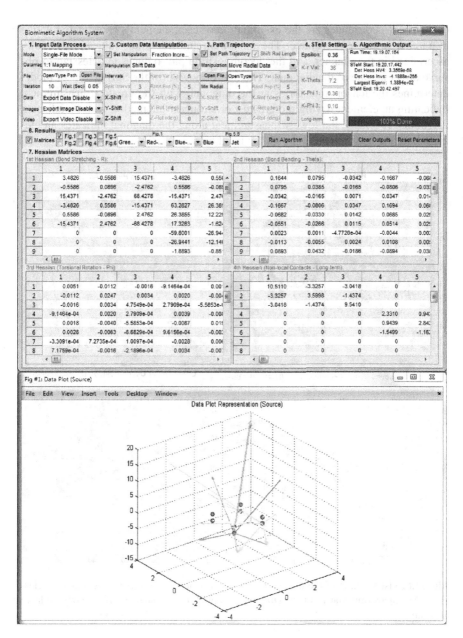

Fig. 11.5. (*Above*) Tensor Analysis Application Frontend; (*Below*) An example of Trajectory Analysis Visualization

each. The x-coordinates are the examination of each element's coordinate relative to another in the y-coordinate. The maximum and minimum values are visually ascertained in the manner, and the decomposition plot reduces the Hessian to observe only one side of the diagonal. The figure shows how the Spring Tensor model determines the tensor trajectories at the diagonal, with the greatest measurement indicating an element having the greatest or strongest interaction with its adjacent neighbors. Inversely, the lowest measurement along the diagonal indicates an element have the least interaction with the neighbors.

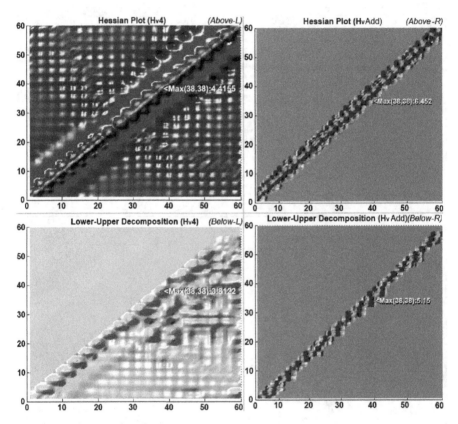

Fig. 11.6. Visualization of the (*Above-L*) Long-Term Hessian and (*Above-R*) Additive Hessian; Decomposition Matrix of the (*Below-L*) Long-Term Hessian and (*Below-R*) Additive Hessian.

Therefore, a strong interaction by one element can emphasize its significance relative to other elements; thus making it easier to identify hotspots such as when a land-yacht is in distress. Resource calculations are multiplied with the number of elements being processed; so using the example of 20 elements consisting of 3 coordinates each, a 60x60 long-term Hessian matrix is determined. With larger element values, the information space can magnify significantly resulting in resource constraints. As shown in Figure 18.6 (*Below*), lower-upper decomposition is used to find the maxima which results in a minor loss of information fidelity, but preserve memory resources by up to 50%.

11.4 Evaluation of the Tensor Analysis Framework

The experiment of the prototype framework examines the usability of STEM to determine projections with a range of different data sets. It is noted that the accuracy of the projections in land-yacht scenarios is not examined in this chapter, as this will be a future research task. The main objective of the prototype is to determine suitability, with reliability and efficiency being considered at a later stage of development.

11.4.1 Experiment of Heuristics

The following experimental parameters are established as follows:

1. The software platform consists of a Windows 7 Enterprise operating system running MathWorks MATLAB Release 2012a (Version 7.14). The MATLAB process runs on a C-based processing engine, with a Java UI frontend.
2. The hardware specification is a desktop computer with an Intel Pentium Dual-core Processor (2.10Ghz) and with 4GB of memory.

The experimental procedure executing the Tensor Analysis platform is elaborated below:

1. The data structures examined for the experimental case study:

   a. The first is using randomized data structures generated pseudo-randomly using the Mersenne-twister method. 50% of the data structure is subject to randomization at each iteration;
   b. The second is using a crystal lattice grid structure generated with each note equidistant amongst each neighbor. The lattice is shifted in a positive linear direction at each iteration;
   c. The third is a random data structure being processed with a Particle Swarm Optimization algorithm according to the spherical objective function at its maxima: $x^2 + y^2 + z^2 = r$

2. The experiment is run for 50 iterations for each agent population, with the following environmental bounds:

   a. The virtual land area is scaled down to 1m x 1m (Approximately scaled to 1:1000)
   b. Each land-yacht occupies an area of 10cm x 5cm (Craft occupies 0.5% of the total land mass)

  c. A trajectory covers the diagonal path from the bottom left to the top right
     corner in iterative steps.

3. The population consist of the total number of land-yachts in the environment:
   10, 20, 30, 40, 50, 60, 80 and 100 agents are examined in total.

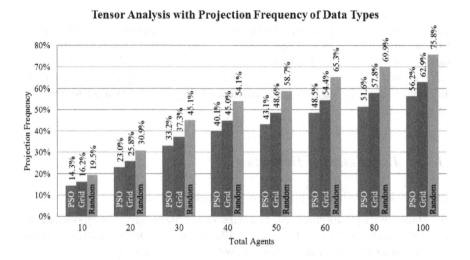

Fig. 11.7. Tensor Analysis with Projection Frequency of Various Data Processing Types

As observed in Figure 18.7, the purpose of the experiment is to measure the
total number of projections made using the Tensor Analysis method. The results
show that as the number of agents exists in the environment, the number of tensor
projections increase. That is expected because the increase in population density
will result in a larger number of interactions between land-yachts. However, it is
important to take note of the projection frequencies between random data sets: grid
lattice structures having an average 15% lower difference of projections compared
to random data, and PSO with an average 18% difference of projections compared
to random data. As the purpose of utilizing the STEM heuristic is to minimize the
number of interactions of land-yachts as they cross through a SANET-connected
field, the results demonstrate that a PSO method is suitable to extend the heuristic
process further for environmental global awareness.

11.4.2   Analysis and Further Work

The difference in the observations about the data processing methods used in the
experiment along with the final results, thus leads to the analysis of how structure
impacts on the global perspective of a SANET-connected network:

- **Pseudo-random data**

  - Random coordinate structures are the least optimum method to distribute a network by its chaotic nature. Although randomized data will result in an evenly structured distribution of nodes in the network, making local observations on the data structure shows some elements cluster more tightly than others - though the global structure remains even.
  - Thus, the purpose of using a randomized data set is to provide experimental control. This is because it is a known data structure that is the least efficient in ensuring optimum resource distribution, considering multi-dimensional factors beyond physical geometry.

- **Grid-lattice structure**

  - From the physical observation, the grid lattice's equidistant distribution between nodes ensures points are evenly organized in a structured manner. However, equidistant lattices can only be organized evenly to its closest neighbors, and the distribution is linearly structured. As a result, the lattice structure loses its distributive quality when looking from a global perspective, because the observation of equidistance occurs only at the absolute latitude and longitude planes.
  - As an example, if a land-yacht traveling in the horizontal longitude plane needs to make a tacking maneuver by $25°$ clockwise, the lattice arrangement of neighboring land-yachts craft is lost with respect to the traveling yacht's perspective. In addition, equidistance is only a measure in the physical plane and does not translate to the non-physical plane - an example is the equitable distribution of resources and capability amongst all land-yachts.

- **Particle swarm optimized data**

  - PSO of the input data is a heuristic method to determine global minima or maxima values by adopting the biological collective behavior of swarming. The mathematical model devised to model this collective behavior is based on following the three main rules [3, 4]: to move in the direction as your neighbors, remain close as possible, but also avoid any collisions.
  - Particle swarming in robotic contexts is based on the principle that the autonomous actuators all move in a particular direction, but for the context of land-yacht trajectory control this is not the desired behavior. The current experimental model demonstrates how particle swarming can assist in global perception of land-yachts.
  - However, further process needs to be developed so that the swarming is required; only when a land-yacht needs to execute a path that is obscured by neighboring land-yachts. Further work needs to be developed to build a model that can balance the needs of the individual land-yacht agent, while harnessing the cooperative capability of SANET infrastructure space.

From the experimental observations, the results demonstrate the combination of Particle Swarm Optimization with Tensor Analysis can provide a two-fold effect on the global perception of the SANET field:

- **Tensor analysis**
  General summary: To determine competitive elements; obtaining the magnitude and direction of change.

  - Using STEM is an effective method to analyze how the effect of nearby land-yachts can interfere with the general trajectory of another land-yacht. In particular with further research, it will consider the effect of prioritizing a land-yacht's trajectory over the rest of the field. The dynamic between competitive and cooperative behavior is important especially for SANET-based infrastructures, as a quantitative measure can be established as to what optimum movement can be made by one land-yacht without affecting the whole field of yachts.
  - The increase in the projection frequency using a grid-lattice structure for the experiment is a good example of how qualitative measures of optimization (i.e. such as rational distribution of elements on a physical geographic basis) may not be compatible with global interests (i.e. maintaining stability of a global optima value or an objective function). In effect, tensor analysis provides a useful metric to determine how competitive behavior to establish a land-yacht trajectory. Thus, it eventually impacts on the globally-connected network of SANET-connected yachts in a positive or negative manner.

- **Particle swarm optimization**
  General summary: To establish cooperative behavior; finding how elements can coordinate activity.

  - Using PSO with STEM compliments the balance between competition and cooperation, as PSO seeks to coordinate a swarm towards a common goal or objective point. If one is to consider the point of achieving a global trajectory by a land-yacht, it is important to consider how the system would react. This is especially if all yachts moved synchronously as a result of one yacht's need to reach a direction. Furthermore, it needs to be considered how some yachts will need to perform complex tacking maneuvers, if they have no choice but to sail in the wind.
  - For this problem, future work will be the development solution to close the feedback loop between competition and cooperation. An example to apply this feedback loop is elaborated: when a land-yacht will move in one direction, it runs a PSO routine to make neighboring nodes guide a path away from the land-yacht intended trajectory path. Using STEM analysis, if an analysis of the potential change of magnitude and direction indicates a yacht may need to make a tight tack to avoid a nearby vessel. This will trigger a halting condition of the PSO routine until the changes of potential values reach satisfactory thresholds.

Table 11.1. Summary of Local Awareness and Global Perspective of a SANET-connected environment

	Local Awareness of Environment	**Global Perception of Environment**
Competitive Behavior	• *Main Reasoning:* Local-competitive awareness is the essence of competitive behavior, as it requires the land-yacht to be awareness of its surrounds at all times while it traverses through the SANET field. Such a behavior is considered to be preservation of self, because the existence of the agent requires it to be aware for it to look after itself. • *Example of Occurrence:* When a land-yacht needs to perform a tacking maneuver to avoid an on-coming collision by a neighboring land-yacht or an obstacle that is suddenly detected without warning.	• *Main Reasoning:* Global-competitive perspective is required when a group of land-yachts need to perform a singular task. It is the unintended consequence when the behavior of one group competes with the actions of another group executing a predefined action. This behavior is considered to be the preservation of the team. • *Example of Occurrence:* When one team of land-yachts need to approach a goal point while another team of land-yachts is traveling on a separate or alternate trajectory that is opposing the first team.
Cooperative Behavior	• *Main Reasoning:* Local-cooperative awareness from a cooperative perspective typically occurs when neighboring land-yachts cooperate to perform a common goal. This is considered to be local cooperation because each land-yacht assists its nearest neighbor to execute a common task on behalf of the primary land-yacht. • *Example of Occurrence:* When one land-yacht requires the assistance of a neighboring land-yacht to provide safe passage in a terrain field that is uncharted, undefined or populated with obstacles in the environment.	• *Main Reasoning:* Global-cooperative perspective is achieving awareness on a distributed scale. It requires a balance between competing local interests and global objectives. Such a behavior is considered global cooperation because each land-yacht's existence is dependent upon ensuring all yachts cooperate for stability. • *Example of Occurrence:* When a land-yacht seeks to travel in a trajectory, the neighboring land-yachts cooperate to move away from the trajectory path; thus resulting in the safe passage of the particular land-yacht.

An elaboration of the global and local concerns exhibited by a SANET land-yacht environment is described in Table 18.1. The goal to design a globally-aware system that performs cooperative behavior is the main objective in designing a land-yacht cooperative heuristic agent. Combining Spring Tensor Analysis with Particle Swarm Optimization endeavors to balance the need for cooperative behavior to reach an objective goal, with the metrics necessary to ensure global perception of the environment [11, 12]. This is because behavior that is excessively competitive results in individual agents leveraging over the needs of the group, while excessively cooperative behavior results in group needs overriding the actions of the individual.

The following recommendations are suggested for further examination to investigate the dynamics between competitive and cooperative behavior:

- **Alternate data reduction methods**

  - The use of the Spring Tensor Model is a computationally dependent activity, relying on memory to store the multi-dimensional matrices. Simplifying the matrices with other linear or non-linear approaches can assist in the simplification of the output Hessian matrices.
  - Using either non-linear or linear reduction approaches will result in a loss of data precision, so it is important to obtain a reduction method that achieves preserves data precision as practically as possible.

- **Integrate prototype framework in physical environment**

  - Once a suitable global perspective metric and cooperative behavior heuristic is determined for a land-yacht environment, the deployment of the prototype to real-life scenarios should be considered. The integration will need to consider the system architecture, the middleware infrastructure and the connectivity framework to interface the heuristic service with the physical world.
  - An organized integration strategy is important because the parameters can be easily modified to match theoretical calculations with observed behavior (such as obtaining the epsilon value for STEM or the objective function for PSO).
  - Furthermore, optimization heuristics can be evaluated to determine how uncontrolled variables affect the system environment. Hence, this resolves the practical implications of using a heuristic and the effect it makes on the physical integrity of the land-yacht.

- **Alternate processing methods for the spring tensor model**

  - The use of Hessians to determine the STEM model is an effective method by Lin and Song to determine anisotropic and Gaussian factors for protein models [11]. For a SANET land-yacht environment, a future recommendation is to investigate a method that enhances the projection process.
  - The use of other techniques such as Lagrangians or Single Value Decomposition should be considered to obtain metrics for global perspectives; with comparisons made between the information accuracy of using alternate methods.

- This accuracy will matter depending on the level of precision is required in a land-yacht guidance system. Finally in certain circumstances, one method may be advantageous over another given different environmental conditions placed on the system.
- Examples of such a condition would be the operation of a land-yacht in a high-altitude or a high wind area and the effect on the projection process.

The future work recommended for the STEM heuristic analysis can be applied to any context where a SANET infrastructure space is suitable. This is because the physical environment should be integrated using uniform standards, in terms of protocols and data structures used. The benefit of such an approach will mean the land-yacht controller can be harmonized with external systems that are value-adding for the system. An example would be the integration of a base emergency warning system, such that the detection of threats like a forest fire will dispatch a fleet of land-yachts to the heat source. Then, the STEM heuristic will be executed to determine the yacht maneuvers establishing possible magnitudes and directions to generate potential routes to the destination.

In summary, the results demonstrate that the data processing method used to determine a trajectory path makes an impact on the global environment. As a consequence, this impact will determine the potential for future interactions with neighboring vessels in the physical space. Heuristic approaches that model methods in biological organization are ideal for land-yacht trajectory mapping, because they provide a template for initiating the local awareness for the neighboring SANET-enabled land-yachts. Coupling with STEM allows a global analysis of the impacting effects that an organizational method has on the entire network.

11.4.3   Providing Representation and Context to Land-Yacht Systems

The STEM Code has been used in a land-yacht context to demonstrate how tensor analysis provides a global representation of the SANET network space. This is especially for the case in the magnitude and direction of elements, as a result of change in the physical environment. The visualization of the Long-Term Hessian shown as a 3D surface map in Figure 18.6 represents the Hessian values generated from the STEM Code; being interpreted by referencing the physical coordinate values and the Hessian projection values as a visualized directional plot in Figure 18.5. It is this visualization of the projection values that adds value to land-yacht control systems by assisting the coordinator in terms of coordinating and facilitating the actions of land-yacht agents. This assistance, either human controlled or by heuristic-algorithm, will respond to local impacts in the SANET infrastructure - such as the priority trajectory manipulation of a land-yacht due to the occurrence of a critical event.

Inspired by the conceptual behavior of social systems, the goal is to design a fleet of land-yachts that cooperatively organize when a prioritized traffic path needs to be established. Path prioritization is organized on a centralized level is the direct equivalent of having global perception with competitive behavior. An example would be one land-yacht's path trajectory leading to the halting of all other land-yacht's processes in its path. Thus to improve the way in which agent-based systems cooperate, Tensor Analysis provides an efficient and uniform approach to sharing the information space from a global perspective. Therefore only selective land-yachts are controlled if they are in the direct path of a land-yacht's trajectory.

11.5 Conclusion

In this chapter, the use of the Spring Tensor Model with Particle Swarm Optimization seeks to achieve the balance between global optimality and local environmental awareness in the development of a SANET land-yacht coordination system. The use of Spring Tensors captures the metric values necessary for projecting the magnitude and direction of change in the network, while Particle Swarm Optimization provides the navigational context to guide a land-yacht in the environment. This achieves the desire of achieving cooperative behavior from a global context, as complete visibility of the network structure is maintained throughout the optimization process to generate a trajectory in the SANET field. Although further research work is required to obtain parameter values that are suitable for analyzing cooperative land-yachts from a physical perspective, the current design prototype demonstrates the potential of utilizing Tensor Analysis for systems-engineering domains. These domains include systems requiring cooperative behavior in distributed, networked environments, such as cloud computing and traffic management systems.

References

1. Chiu, C., Chaczko, Z.: Steering a Swarm of Autonomous Sailing Vehicles implementing Biomimetic SANET Middleware. In: Proceedings from the Australian Conference on the Applications of Systems Engineering, pp. 85–86 (2012)
2. Clementi, C., Nymeyer, H., Onuchic, J.N.: Topological and Energetic Factors: What Determines the Structural Details of the Transition State Ensemble and En-route Intermediates for Protein Folding? An Investigation for Small Globular Proteins. Journal of Molecular Biology 298, 937–953 (2000)
3. Clerc, M.: Discrete Particle Swarm Optimization, illustrated by the Traveling Salesman Problem, New Optimization Techniques in Engineering, pp. 219–239. Springer (2004)
4. Clerc, M.: Particle Swarm Optimization, International Scientific and Technical Encyclopedia. Wiley ISTE (2006)
5. Clerc, M., Kennedy, J.: The Particle Swarm - Explosion, Stability, and Convergence in a Multi-dimensional Complex Space. IEEE Transactions on Evolutionary Computation 6(1), 58–73 (2002)
6. Fossen, T.I.: Marine Control Systems: Guidance, Navigation and Control of Ships, Rigs and Underwater Vehicles, Marine Cybernetics (2002)
7. Harland, J.: Seamanship in the Age of Sail. Naval Institute Press (1984)

8. Jouffroy, J.: A Control Strategy for Steering an Autonomous Surface Vehicle in a Tacking Maneuver System, Systems, Man and Cybernetics - SMC Conference (2009)

9. Kanayama, Y., Hartman, B.I.: Smooth Local Path Planning for Autonomous Vehicles. In: Proceedings of the IEEE Conference on Robotics and Automation, Scottsdale, Arizona, USA (1989)

10. Kiencke, U., Nielsen, L.: Automotive Control Systems. Springer Publishing, Germany (2000)

11. Lin, T.L., Song, G.: Generalized Spring Tensor Models for Protein Fluctuation Dynamics and Conformation Changes. In: Computational Structural Bioinformatics Workshop, Washington D.C., USA (2009)

12. Low, K.H., et al.: Task Allocation via Self-organizing Swarm Coalitions in Distributed Sensor Networks. In: 19th Artificial Intelligence Conference, pp. 28–33 (2004)

13. Rao, M., Georgeff, P.: BDI-agents: From Theory to Practice. In: Proceedings of the First International Conference on Multi-agent Systems - ICMAS (1995)

14. Ritter, H., Schulten, K., Denker, J.S.: Topology Conserving Mappings for Learning Motor Tasks: Neural Networks for Computing, American Institute of Physics. In: 151st Conference Publication Proceedings, Snowbird, Utah, pp. 376–380 (1986)

15. Xiao, C.M., Austin, P.C.: Yacht Modeling and Adaptive Control. In: Proceedings of the IFAC Conference on Manoeuvering and Control of Marine Craft, Aalborg, Denmark (2000)

16. Yeh, E.C., Bin, J.C.: Fuzzy Control for Self-steering of a Sailboat. In: Proceedings of the Singapore International Conference on Intelligent Control and Instrumentation, Singapore (1992)

Chapter 12
Improvement of Spatial Routing in WSN Based on LQI or RSSI Indicator

Jan Nikodem, Ryszard Klempous, Maciej Nikodem, and Zenon Chaczko

Abstract. The chapter presents theoretical work and simulation results of spatial routing in wireless sensor network (WSN). We propose a new framework of the routing in WSN that is based on the set theory and on three elementary relations: subordination (π), tolerance (ϑ) and collision ($\varkappa$) necessary to describe behaviour of nodes in distributed system. The essence of using relations is decentralization of decision making process. Relations introduce the ability to delegate decission making process to nodes, which base their actions on situation in the vicinity, information that is avaliabe to them and capabilities they have. Since relations, support local elements to make decisions, they are a better instrument compared to functions which does not facilitate such choices. Spatial routing algorithm, during network organization, builds a framework - space for globally oriented communication towards base station. Within this framework, during the network lifetime, each node, decides if the routing path should be changed in case of local interferences, neighbor nodes failure or changes in environmental conditions. Therefore, the proposed approach combines existing features of the spatial routing and LQI or RSSI indicators to aide in route selection within a neighborhood.

12.1 Introduction

The wireless sensor network (WSN) is a relay system with a large number of nodes communicating among each other using wireless channels. On the one hand, each

Jan Nikodem · Ryszard Klempous · Maciej Nikodem
Wrocław University of Technology, Institute of Computer Engineering,
Control and Robotics, Wybrzeże Wyspiańskiego 27, 50-370 Wrocław, Poland
e-mail: {jan.nikodem,ryszard.klempous,maciej.nikodem}@pwr.wroc.pl

Zenon Chaczko
Faculty of Engineering and IT, University of Technology Sydney, Blvd.1, Broadway,
Ultimo, 2007 NSW, Australia
e-mail: zenon.chaczko@uts.edu.au

R. Klempous et al. (eds.), *Advanced Methods and Applications in Computational Intelligence*,   285
Topics in Intelligent Engineering and Informatics 6,
DOI: 10.1007/978-3-319-01436-4_12, © Springer International Publishing Switzerland 2014

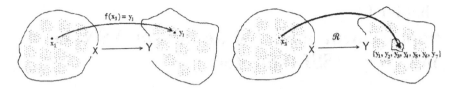

Fig. 12.1. Functional vs. relational mapping

sensor works in its vicinity autonomously, interacting with environmental stimuli. On the other hand, sensors must communicate with each other, therefore communication channels are crucial elements of the WSN architecture.

There is a considerable body of research related to routing algorithms for WSNs that allow sending information between nodes and the base staion (BS) [2, 3]. The aim of so called flat algorithms, e.g. flooding, is to ensure that every piece of information finally reaches the BS, as long as there is a communication path between every node and the BS. Such communication protocols use broadcast communication and basically flood the whole WSN with data, causing collisions in the communication channels and packet retransmissions. The flooding approach, besides being a real waste of resources, also lowers the bandwidth, limits communication speed and causes a degradation of quality of communication by simply cutting off some parts of the network. These disadvantages are minimized in gossip type protocols [4] where the broadcast communication concerns only neighboring nodes.

The routing protocols such as Data Centric Routing, Sensor Protocol for Information via Negotiation (SPIN) aim to orient communication towards time, query, event or information requirements. The Directed Diffusion protocol enables further minimisation of energy consumption at the execution of the routing further, but at the cost of lower efficiency due to the large amounts of queries. Rumor routing [2] presents another attempt that draws upon a software agents and a mechanism of directed communication. Positive results are obtained for a rather limited number of scenarios when costs of additional agents are negligible compared to the overall costs.

Hierarchical protocols, such as LEACH, TEEN and PEGASIS [8, 12] address directed communication that is achieved through creating hierarchal topologies using clustering or divisions into zones. The work presented in this chapter proposes to use a routing protocol based on spatial or chain structures. All aforementioned protocols construct a single multi-hop communication path between every node and the BS. An established communication path is utilized as long as it is energetically justified [5, 7].

12.2 Relation Based Spatial Routing

The relational attempt is based on the set theory and considers three relations: subordination (π), tolerance (ϑ) and collision $(\varkappa)$ that are necessary to describe behaviour of various objects in distributed systems. Based on these relations it is possible to initially organize WSN and manage message routing during its operation. What is so special about these three relations and why were these three relations chosen over others? In distributed systems, decision making is often undertaken locally by a single node. Globally defined principles allow for making the decision locally, while preserving the centrally established operational aims. In literature [1, 2, 5, 8], these principles are defined using specific functions. Why then, when describing behaviour of various components in distributed systems the use of mathematical functions is insufficient and applying relations seems a better choice? The essence of decentralized decision making process introduces the ability to delegate decission making process to nodes, which base their action on situation in their vicinity, informathion that is avaliabe to them and capabilities they have. All these aspects need to ensure an effective selection process regarding each component of the system. Application of $f(x)$ function specifies explicitly the action $f(x_1) = y_1$ and leaves the node x_1 with no decision nor choice. Relational approach overcome this difficulty by defining relations $\mathscr{R}(x)$ that specify sets of possible actions each node can take indepedently. Globally, delegating decission making process to node x_1, we determine a set of allowed actions $\mathscr{R}(x_1) = \{y_1, y_2, \ldots, y_k\}$. Node x_1 uses relation $\mathscr{R}(x_1) = \{y_1, y_2, \ldots, y_k\}$ to make an independent decision which action to take, based on local situation. For example on Fig. 12.1 node x_1 can choose any suitable, non-empty subset from the set $\{y_1, y_2, \ldots, y_k\}$ of the actions having $2^k - 1$ different options ($k = 7$).

Since relations, support local elements to make independent, locally determinded, decisions, they are a better instrument than functions which did not facilitate such choices. So, why were these specific relations chosen? Let us remind [9, 10] that:

- subordination (π) is transitive (enables construction of chains) and non-symetric (prevents creation of loops in the chains),
- tolerance (ϑ) is symmetric (opposite to subordination) and reflexive,
- collision ($\varkappa$) is symmetric (opposite to subordination) and irreflexive (opposite to tolerance).

The above set of relations is exhaustive and non-redundant and well suited to define various phenomena that take place in WSNs. These relations can help in modelling the communication activity area within WSNs. The communication activity can be best modelled using the subordination (π) relation. Various limitations and issues of communication among nodes in distributed systems are modelled using the relation of collision ($\varkappa$). Remaining phenomena, that do not disturb, but frequently even facilitate the communication, are modelled using the relation of tolerance (ϑ).

Using a traditional (functional) approach, at the initialisation of WSN, a routing tree is constructed. Figure 12.2 shows initial routing tree as thin, dashed lines coupling regular nodes (dots) and cluster heads (stars). During network lifetime, for each selected point K, only one path can be formed in the routing tree (dashed line on Fig. 12.2). When a given routing path cannot be used any longer, the network must be reorganized again, and a new path must be selected.

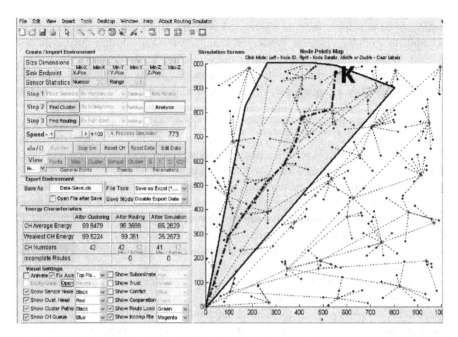

Fig. 12.2. Spatial (grey) vs. hierarchical (dashed line), based on tree structure, routing

Using relations we can also perform initial organization, but there is no necessity to repeat the reorganization process during network lifetime. It is so, since each node selects an element of the routing path, based on a local balance between (π), (ϑ) and $(\varkappa)$, so the final result is not specified by a concrete function but is governed by relational principles and situation in node's vicinity.

Making a local criteria ranking, nodes use deterministic and random elements. Deterministic, since intensity quotients for each relation π, ϑ, $\varkappa$ are determined globally during the initial organization and remain unchanged throughout the cycle of network operation. Random, because the ordering of the set of neighbour nodes to which packet can be transmitted is done locally and ad-hoc using current propagation conditions in the environment. The disturbances in the environment may occur at random thus the propagation conditions can be very dynamic.

Relational approach is a method to create an individual decision space for each node of the network. Relations allow nodes to take different actions depending on the situation in node's vicinity. This approach can be used to select a node to which packet will be routed. Since routing is decided based on random and deterministic elements the resulting path cannot be predicted globally, but still selected routing path is the best one. It is so, since nodes took into consideration actual local conditions while deciding to which node retransmit the packet.

During the initialisation of WSN, using the relational approach, applying subordination (π), tolerance (ϑ) and collision $(\varkappa)$ relations we can shape the area of the network in which routing is intended to take place (dotted area on Fig. 12.2). Every routing path element is selected independently, each time packet is retransmitted. Spatial routing belongs to a category of reactive protocols but when a route is needed, it does not attempt to flood the WSN with route request packets. Hence, no additional overheads occur in WSN by clogging up communication channels with route request information.

It is typical for WSN reactive protocols that a record of all routes available in a WSN is not kept. In realtion based spatial routing there is neither a central nor local storage of route tables; hence there is no need to calculate best path scenarios. When a node K requires to send a packet to its neighbors it fills the address field with many destination addresses rather than a single address. The list of addresses included in the packet is defined based on the subordination (π), tolerance (ϑ) and collision ($\varkappa$) relations (global decision). Order of addresses however, is determined by the actual situation in the neighborhood (local decision). The first receiver in the address list is the best and preferred destination of retransmission. If the preferred node is not able to receive and forward the packet, the next from the address list takes its turn. If a node that can retransmit the packet is reached, it sends back a message that confirms its availability. The first reply received determines the route element, within the neighborhood, to be used.

Application of relations allows to organise the network very early, i.e. during its initialisation. At that time, it is possible to define the area in which packet routing can be realised (from a nominated point of the network). That is why routing protocol that uses relations is called the spatial routing protocol [9, 10]. This research work aims to address the issues related to so called *spatial communication* that outdo path based communication and aims to construct multiple possible communication paths towards the base station (BS), through which information can be sent reliably. Construction of these routing paths is based on local information collected by a node that received information or an event. Based on this local knowledge the node decides where to send the information. The approach allows to share communication costs among multiple routing paths and nodes. During network initialisation the algorithm constructs a framework - space for globally oriented communication towards BS. Within this framework each node decides if the routing path should be changed in case of local interferences, neighbor nodes failure or changes in environmental conditions. Since, the structure of the network also plays an important role in its behaviour, the collection of above described three relations π, ϑ, $\varkappa$ is aided by the concept of neighborhood $\mathcal{N}$, which describes the structure of the network from the local point of view.

12.3   Constructing Neighborhoods in WSN

No node is an lonely island within WSN. Nodes form an interconnected network in which the neighborhood is an essential concept, especially when we concider a *multi-hop* WSN. Most sensor nodes do not communicate directly with the BS

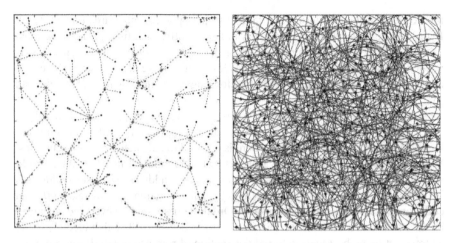

Fig. 12.3. Clusters and routing trees (left) vs. neighborhoods (right)

because it is out of a node's radio communication range. The WSN is a distributed system where nodes provide necessary computations, communicate with each other within some neighborhood; and maintain the state of the unified (whole) network infrastructure. There are quite a few WSN algorithms which take advantage of some concept of a neighborhood, e.g. multi-hop, reliable, bi-directional or geographic. To define neighbourhood we need auxiliary definitions. Let $Map(X;Y)$ be a set of mapping functions from X onto Y (surjection), where $Sub(X)$ is defined as a *family* of all X subsets.

The neighborhood $\mathcal{N}$ can be mathematically expressed as follows:

$$\mathcal{N} \in Map(Nodes, Sub(Nodes)). \tag{12.1}$$

Hence, $\mathcal{N}(k)$ is the neighborhood of node k, and $\mathcal{N}(C)$ is the neighborhood of C (set of nodes) defined as:

$$\mathcal{N}(k)_{|k \in Nodes} := \{ y \in Nodes \mid y \; \mathcal{R}_{\mathcal{N}} \, k \}, \tag{12.2}$$

$$\mathcal{N}(C)_{|C \subset Nodes} := \{ y \in Nodes \mid (\exists x \in C)(y \; \mathcal{R}_{\mathcal{N}} \, x) \}. \tag{12.3}$$

where $\mathcal{R}_{\mathcal{N}}$ is *'to be a neighbor'* relation.

There are various different definitions of locality that can be found in literature [1]. It is the native locality that has been pinpointed as the one to work with, when trying to establish the neighborhood abstraction. Native locality is determined by obvious constrains (mostly technical) such as radio link range or anisotropic propagation. Daunted by the neighborhood complexity (see Fig. 12.3), many authors [1, 8, 12] have choosen to study cluster techniques. Clustering is a kind of simplification, that facilitates computation while at the same time limiting the set of possible solutions. As an end result, the native neighborhood is seen as the most suitable from the local point of view. Native neighborhoods within sensor network are an indexed family of sets $\aleph = \{\mathcal{N}_i \mid i \in I\}$, which is characterised by the following properties:

$$(\forall i \in I)(\mathcal{N}_i \neq \emptyset) \wedge \bigcup \mathcal{N}_i = Nodes \tag{12.4}$$

$$(\exists^{max} i, j \in I \mid i \neq j)(\mathcal{N}_i \bigcap \mathcal{N}_j \neq \emptyset), \tag{12.5}$$

which relate to local (at every node) condition:

$$(\forall y \in Nodes)(\exists^{max} i \in I \mid y \in \bigcap \mathcal{N}_i \neq \emptyset), \tag{12.6}$$

where $\exists^{max}$ can be translated as *'there are as many instances as the structure of the network allows'*.

The formulas (12.5), (12.6) denote that the native neighborhoods do not partition the set of the WSN nodes into mutually exclusive subsets.

Neighborhoods are much more densely packed than the clusters themselves, as we see on Fig. 12.3. These neighborhoods are interwoven like strands, allowing one node to cooperate with many others. For clarity, Fig. 12.3 depicts neighborhoods of every third node in the network, only. Plotting all neighborhoods will result in an unreadable diagram that would be entirely black due to a huge density of all boundary lines. Studying such complex system may seem almost futile. Even as depicted at Fig. 12.3, on the right, the neighborhoods diagram looks rather confusing as it is rather hard to see anything interesting in the jungle of circles that delimit boundaries of the neighborhoods. This aspect may not necessarily interfere in our studies. Considering WSN as a distributed system, decision making can be relocated to the neighborhood. Solving the tasks at the local level node is not aware of multiple of neighbourhood as it operates in local vicinity. Its perspective is limited to its neighborhood with superimposed relations set that need to be realised.

For example, node's *"whole world"*, neighborhood $\mathcal{N}(k)$ shown on Fig. 12.4 on the top, is composed of 10 nodes. This is simple example, but in real networks there might be 20-30 nodes in the neighborhood. Nevertheless the number of neighbors is a small fraction of the total number of elements in the WSN.

Figure 12.4 represents WSN model seen from the perspective of the whole network (in the middle) as well as the neighborhood $\mathcal{N}(k)$ of node k (at the bottom). In order to correctly define the structure of the whole network, 4 matrices are needed. We can use the 2-dimensional matrix **N** to represent the neighborhoods in WSN. Each *"1"* in cell n of row k in the binary matrix **N** represents a membership n in $\mathcal{N}(k)$. In order to model $[\pi, \vartheta, \varkappa]$ space we exploit three additional, 2-dimensional $[n \times n]$ matrices presented on Fig. 12.4, in the middle. The real number in any cell of these matrices expresses an intensity quotient of relation. Elements $r_{k,k}$ represent the required intensity quotient within $\mathcal{N}(k)$, while $r_{i,j}$ for $i \neq j$ denotes interaction intensity between nodes. The size of these tables is indeed large as they dimensions are equal to the number of nodes in the WSN network; however these are sparse matrices with only minimal non-zero elements.

To model the neighborhood locally, from the node perspective, Fig. 12.4 (at the bottom), only 2 vectors (for each node) are needed, each composed of $[1, \ldots, e_{max}]$

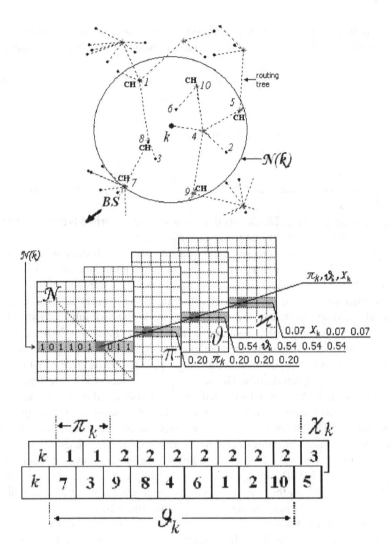

Fig. 12.4. Neighborhood $\mathcal{N}(k)$ (on the top) and its global (in the middle) and local (at the bottom) representations

elements, where e_{max} represents a maximum number of neighbors for a given node in WSN network, as follows:

$$e_{max} = Max\{Card\,(\mathcal{N}(y)) \mid y \in Node\}. \tag{12.7}$$

In our tests, for densely compacted network, this number rarely exceeded 25, and more likely it was between 8 to 16. Vectors defined for each node, contain indices

of their neighbors (at the front) as well as indices of relationship (at the back) of a
each neighbor with one of the 3 relations (1 indicates π, 2-ϑ and 3-$\varkappa$).

There are two advantages of using this type of modeling:

- it requires less memory, since vectors need $n \cdot 2\,e_{max}$ (where $e_{max} \ll n$) packed
  very densely locations, whereas four, 2-dimensional, $[n \times n]$ matrices need $4 \cdot n^2$
  locations in memory and they are remarkedly sparse as shown in Fig. 12.4,
- it allows for quick and simple calculations at nodes as calculations are performed
  using two vectors of small size only.

12.4 Ordering the Neighborhood Using LQI or RSSI Indicator

The relations $\pi, \vartheta, \varkappa$ allow us to make a projection of communication requirements
from global level, in reference to the desired direction of packet flow to the base
station, onto the area of local activity in the neighborhoods. This guarantees that
local operations, governed at the local level, will be compatible with global goal(s)
of the network as a whole. Mapping relations to each neighborhood results in a
collection of possible actions to take.

While investigating the packet routing paths in the WSN network, we obtain
related data collections that contain identifiers of neighboring nodes which then can
be used as relay nodes. In extreme cases, this set might be empty, which would
indicate that communication in such a neighborhood cannot be realized (e.g. due to
interferences or no neighboring nodes in communication range). The set can be a
singleton, consist of a single node only, in which case there is only one relay node
available. However, most often, the set of possible relay nodes consists of multiple
elements. In such situations, node needs to choose which node to select as the next
retransmitter.

In order to choose retransmitter from among many neighbors, node needs to rank
potential relays based on certain predefined criteria. Using the theory of sets language, this operation is called *set ordering* or *sequencing*. The collection of neighbors can make up; a preordered set, which only requires reflexivity and transitivity;
a partially ordered set, which additionally requires antisymmetry or a totally or-
der set, which for any pairs x, y requires $x\mathcal{R}y$ or $y\mathcal{R}x$. On a local level node can
use Link Quality Indicator (LQI) or Received Signal Strength Indicator (RSSI) that
are parameters often used in wireless communication. These indicators characterize
efficiency of radio transmissions and allow to evaluate the quality of radio signals
in a given area. This will help to identify the best retransmitter in the neighborhood.
The RSSI is used to show the loss of strength in a received radio signal, whereas the
LQI estimates the level of connection faults or the signal-to-noise ratio (SNR). Both
of these indicators are often used for various testing and are frequently discussed in
related literature [6, 11] as being the most valuable in defining the Packet Reception
Rate (PRR) coefficients.

The aim of this chapter is to introduce a new way of thinking about the relationship between communication nodes within a neighborhood. First, considering

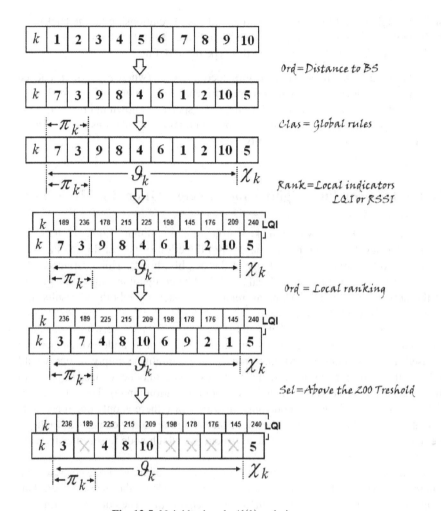

Fig. 12.5. Neighborhood $\mathcal{N}(k)$ ordering sequence

the distance from the base station (BS) to a node, all neighbors within $\mathcal{N}(k)$ are globally ordered ($Ord = Distance\, to\, Base\, Station$), Fig. 12.5 (on the top).

Second, the performance of routing algorithm is impacted by the selection of nodes which eventually transmit the data. For this purpose, we apply three relations π, ϑ and $\varkappa$ are applied in order to map the routing space in the WSN. At the local level it reflects a partition of the neighborhood into classes, according to a globally defined goal of routing ($Clas = Global\, rules$). Following relational approach, we are able to obtain four subsets of $\mathcal{N}(k)$ set. Three of them π_k, ϑ_k and $\varkappa_k$ are depicted in the central part of the Fig. 12.5. The fourth one (not shown in the figure) consists of all of the neighbors that are not in any of the 3 relations. The created subsets are not disjoint (as π_k and ϑ_k) but they cover the whole neighborhood and therefore the partition into subsets is complete.

Third, after completing the classification that meets the global requirements, the next stage is to involve which involves actions that consider the local conditions. The RSSI or LQI indicators form the foundation of this part of our algorithm. After one of the indicators (LQI or RSSI) is selected, we allocate a related *Rank = Local indicator* value for each of the neighboring nodes. With rankings of all neighbors at our disposal, within any of subsets π_k, ϑ_k and $\varkappa_k$ we reorder elements according to current *LQI* or *RSSI* value (*Ord = Locall ranking*, Fig. 12.5).

Finally, a selection *Sel = Above the xx Threshold* step is performed, where only those neighbors are selected for which ranking shows a higher value than the minimal threshold, i.e. on Fig. 12.5 it is the 200 value, that guarantees the success of the packet retransmission.

The first two, ordering *Ord = Distance to Base Station* and *Clas = Global rules* (Fig. 12.5) operations are completed at the initialization stage of the WSN infrastructure and there is no need to repeat these operations again during the lifetime of WSN as long as there is no change to its structure or the goal of its operation.

Sensor nodes monitor the transmissions in the communication channel all the time. The registration of transmission parameters and measurement of the current values of RSSI or LQI occurs at the node, even if packets are not addressed to them specifically. If, there is a packet in node k which should be retransmitted to the base station, it is only necessary to reorder elements of each of the subsets π_k, ϑ_k and $\varkappa_k$ and choose those for which the indicators exceed a given value of the threshold (*Sel = Above the 200 Threshold*, Fig. 12.5).

12.5   Conclusion

We have been investigating methods to enhance the route selection within neighborhoods in WSN. Specifically, an adaptive, *multi-hop* algorithm based on a *spatial routing* concept was proposed. However, the performance of a spatial routing algorithm will be impacted by the selection of many routes over which data is eventually transmitted. It has been shown that spatial routing algorithm, which distributes within neighborhood the energy consumption, not lead to adaptive network performance in wireless environments. Therefore, the proposed approach combines existing features of the spatial routing and LQI or RSSI indicators to aide in route selection within a neighborhood. We use the *IEEE 802.15.4* standard which enables the measurement of link quality between neighboring nodes in the network. This measurement is in the form of a LQI value that is reported with each received packet. Additionally, we perform experiments to utilise the RSSI parameters for ordering nodes in the WSN neighborhood. By ordering the neighbors according to the value of indicators (LQI or RSSI), adjustments of quality connection to the noise level can be successfully achieved. This approach allows circumventing areas of higher interference and executing a successful transmission even in adverse conditions.

The aim of applying the LQI and RSSI indicators was to increase adaptive characteristics of investigated algorithms. However, the adaptivity of WSN largely depends on the frequency of updates of these indicators. Low intensity of packet transmissions results in unacceptably low update rates of the LQI and RSSI indicators, thus also adversely affecting the values of PRR (packet reception rate) indicators. The LQI indicator requires a time window for its adjustment, which additionally could slow down the dynamics of node reactions on various disturbances. Application of the proposed technique in our tests, shows satisfying results for most slow-changing disturbances when the LQI or RSSI indicators can be accurately determined for all neighbors. This is valid for cases when the indicators reflect the actual propagation conditions in the network. Recently, our objective has been to develop further enhancements that increase the overall adaptability skills of the network by investigating methods of real-time cooperation among nodes within a neighborhood and to upturn the selection of relay nodes.

Acknowledgements. This chapter has been partially supported by the project entitled: *Detectors and sensors for measuring factors hazardous to environment – modeling and monitoring of threats*. The project financed by the European Union via the European Regional Development Fund and the Polish state budget, within the framework of the Operational Programme Innovative Economy 2007–2013. The contract for refinancing No. POIG.01.03.01-02-002/08-00.

References

1. Boukerche, A. (ed.): Algorithms and Protocols for Wireless Sensor Networks. Wiley & Sons, Inc., New Jersey (2009)
2. Braginsky, D., Estrin, D.: RumorRouting Algorithm for Sensor Networks. In: Proc. of the 1st Workshop on Sensor Networks and Applications, Atlanta, GA (October 2002)
3. Briesemeister, L., Hommel, G.: Localized Group Membership Services forAd Hoc Networks. In: Int. Workshop on Ad Hoc Networking (IWAHN), pp. 94–100 (August 2002)
4. Datta, A., Quarteroni, S., Aberer, K.: Autonomous Gossiping: A Self-Organizing Epidemic Algorithm for Selective Information Dissemination in Wireless Mobile Ad-Hoc Networks. In: Bouzeghoub, M., Goble, C.A., Kashyap, V., Spaccapietra, S. (eds.) ICSNW 2004. LNCS, vol. 3226, pp. 126–143. Springer, Heidelberg (2004)
5. Fang, Q., Gao, J., Guibas, L.J., de Silva, V., Zhang, L.: GLIDER: gradient landmark-based distributed routing for sensor networks. In: INFOCOM 2005, 24th Annual Joint Conference of the IEEE Computer and Communications Societies, vol. 1, pp. 339–350 (2005)
6. Holland, M., Aures, R., Heinzelman, W.: Experimental investigation of radio performance in wireless sensor networks. In: 2nd IEEE Workshop on Wireless Mesh Networks, WiMesh 2006, pp. 140–150 (2006)
7. Lim, A.: Distributed Services for Information Dissemination in Self-Organizing Sensor Networks, Distributed Sensor Networks for Real-Time Systems with Adaptive Reconfiguration. Journal of Franklin Institute 338, 707–727 (2001)

8. Manjeshwar, A., Agrawal, D.P.: TEEN: A Routing Protocol for Enhanced Efficiency in Wireless Sensor Networks. In: Parallel and Distributed Processing Symposium (IPDPS 2001) Workshops, vol. 3 (2001)

9. Nikodem, J., Klempous, R., Nikodem, M., Woda, M., Chaczko, Z.: Multihop Communication in Wireless Sensors Network Based on Directed Cooperation. In: 4th International Conference on Broadband Communication, Information Technology & Biomedical Applications, BroadBandCom 2009, Wrocław, Poland (2009)

10. Nikodem, J., Klempous, R., Nikodem, M., Chaczko, Z.: Enhanced Performance of Spatial Routing in WSN Based on LQI Metrics. In: Proceedings of 1st Australian Conference on the Applications of Systems Engineering, ACASE 2012, Sydney, Australia, pp. 108–114 (2012)

11. Srinivasan, K., Levis, P.: RSSI is Under Appreciated. In: Proceedings of the Third Workshop on Embedded Networked Sensors, EmNets (2006)

12. Sung-Min, J., Young-Ju, H., Tai-Myoung, C.: The Concentric Clustering Scheme for Efficient Energy Consumption in the PEGASIS. In: The 9th International Conference on Advanced Communication Technology, vol. 1, pp. 260–265 (2007)

Chapter 13
Centralized and Distributed CRRM in Heterogeneous Wireless Networks

Abdallah AL Sabbagh, Robin Braun, and Mehran Abolhasan

Abstract. The evolution of wireless networks has led to the deployment of different Radio Access Technologies (RATs) such as GSM/EDGE Radio Access Network (GERAN), UMTS Terrestrial Radio Access Network (UTRAN) and Long Term Evolution (LTE). Next Generation Wireless Networks (NGWNs) are predicted to interconnect various Third Generation Partnership Project (3GPP) Access Networks with Wireless Local Area Network (WLAN) and Mobile Worldwide Interoperability for Microwave Access (WiMAX). A major challenge is how to allocate users to the most suitable RAT for them. An intelligent solution will lead to efficient radio resource utilization, maximization of network operator's revenue and increasing in the users' satisfactions. Common Radio Resource Management (CRRM) was proposed to manage radio resource utilization in heterogeneous wireless networks. This paper discusses the need of CRRM for NGWN. Then, the paper presents a comparison between implementing or not the CRRM in heterogeneous wireless networks. After that, the interaction between RRM and CRRM entities is discussed. Then, different approaches for the distribution of RRM and CRRM entities among the Core Network (CN), RATs and User Terminals (UTs) will be presented. Finally, a comparison between distributed and centralized algorithms is presented.

13.1 Introduction

The integration of different Radio Access Technologies (RATs) led the Next Generation Wireless Network (NGWN) to be a heterogeneous wireless network and to be an Internet Protocol (IP) based networks. This will allow the interconnection between the Third Generation Partnership Project (3GPP) Access Networks such

Abdallah AL Sabbagh · Robin Braun · Mehran Abolhasan
Centre for Real-Time Information Networks (CRIN),
University of Technology Sydney (UTS), 15 Broadway, Ultimo, NSW 2007 Australia
e-mail: {abdallah.alsabbagh,robin.braun,
        mehran.abolhasan}@uts.edu.au

R. Klempous et al. (eds.), *Advanced Methods and Applications in Computational Intelligence,*   299
Topics in Intelligent Engineering and Informatics 6,
DOI: 10.1007/978-3-319-01436-4_13, © Springer International Publishing Switzerland 2014

as GSM/EDGE Radio Access Network (GERAN), UMTS Terrestrial Radio Access Network (UTRAN) and Long Term Evolution (LTE), IP networks (Internet) and the non 3GPP wireless access networks such as Wireless Local Area Network (WLAN) and Mobile Worldwide Interoperability for Microwave Access (WiMAX) [11, 17, 22].

The motivation in NGWN comes out from the fact that no single RAT could support widespread coverage and provide continuous high QoS levels over multiple hotspot areas, e.g. office, cafe, public smart areas, etc. [15]. In this case, multiple access networks that come from different technologies are spread in the same geographical space. The Third Generation Partnership Project (3GPP) has proposed different interconnected heterogeneous wireless network architectures such as Beyond 3G (B3G), Long Term Evolution (LTE) and LTE Advanced. B3G interconnects GERAN, UTRAN and WLAN through a common platform. LTE interconnects with all 3GPP wireless access networks such as GERAN and UTRAN and all non 3GPP wireless access networks such as WLAN and Mobile Worldwide Interoperability for Microwave Access (WiMAX). One of the key challenges need to be addressed in heterogeneous wireless network is how to allocate users to the most suitable RAT for them. An optimized solution will lead to efficient radio resource utilization Radio resource utilization and at the same time, it will increase the users' satisfactions. Common Radio Resource Management (CRRM) was proposed to manage radio resource utilization in heterogeneous wireless networks and to guarantee required Quality of Service (QoS) for users [4, 18].

A number of CRRM algorithms have been proposed in the literature for heterogeneous wireless networks. References [7, 5] present these algorithms. These algorithms can be categorized into centralized such as load balancing algorithm and policy based algorithm or distributed algorithms such as service based algorithm. Each one has its benefits and limitations. A comparison between some of these algorithms is presented in [8].

Centralized RAT selection algorithms have the benefit of considering more criteria during the making decision process. However, centralized algorithms do not guarantee required QoS for all admitted calls. In addition, they reduce network capacity as a result of the introduced signaling overheads or delay resulted by the communication between the network entities.

On the other hand, distributed algorithms have the benefit of considering users' preferences. A number of different distributed algorithms are proposed in [3, 9, 16]. These algorithms allow User Terminal (UT) to select the most efficient RAT that maximizes its satisfaction which is based on its preference such as best QoS or cheapest cost. However, distributed algorithms do not take into account the network benefits and policies. This may lead to inefficient radio resource utilization and it may create network bottlenecks. Radio resource utilization

The remainder of the paper is organized as follows. Sect. 13.2 discusses the need of CRRM in heterogeneous wireless networks. In Sect. 13.3, a comparison between implementing or not the CRRM in heterogeneous wireless networks is studied. The interaction between RRM and CRRM entities is presented in Sect. 13.4. In Sect. 13.5, different approaches for the distribution of RRM and CRRM entities among the Core Network (CN), RATs and UTs are studied. Distributed and centralized algorithms are compared in Sect. 13.6. Finally, this paper is concluded in Sect. 13.7.

13.2 Need for CRRM

In a homogeneous wireless network, each RAT has a local RRM entity that is responsible for admission control, congestion control, power control, packet scheduling, initial RAT selection algorithm, horizontal handover (HO) and vertical HO (VHO). It is implemented efficiently for the RAT that it is developed for. However, it is not suitable for a heterogeneous wireless network. Implementing RRM entity separately for each RAT will not have the ability to achieve efficient radio resource utilization and provide QoS for each call in a heterogeneousRadio resource utilization wireless network. Therefore, there is a need for CRRM to manage radio resource utilization and to guarantee the required QoS for users [12].

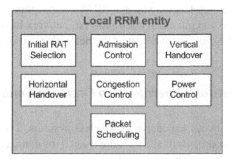

Fig. 13.1. Local RRM entity

CRRM is essential to guarantee the required QoS across different RATs, improve network reliability and stability, consider users' preferences and increase their satisfactions, allow network operators to gain maximum revenue, reduce blocking and dropping probability, and efficient utilization of radio resources. These factors are described in the following subsections.

13.2.1 Efficient Utilization of Radio Resources

There is an increasing demand for efficient utilization of radio resources in heterogeneous wireless networks. CRRM will distribute users load among the different available RATs. This will increase the efficiency of radio resources utilization.

Making the best use of radio resources is one of the most important objectives of CRRM which will enable the provision of QoS for users in heterogeneous wireless networks.

13.2.2   Reduce Blocking and Dropping Probability

Implementing CRRM will improve the utilization of radio resources in heterogeneous wireless networks and will provide requested new or VHO calls an opportunity to be allocated to another RAT if the selected RAT is unable to serve these calls. This will reduce both blocking and dropping probability.

13.2.3   Improve Network Reliability and Stability

An efficient utilization of radio resources will maximize the load capacity of each RAT in heterogeneous wireless networks. CRRM was proposed to manage radio resources and to support an equivalent distribution of network load between available RATs. This will lead to an improvement of the network reliability and stability.

13.2.4   Allow Network Operators' to Gain Maximum Revenue

Radio resources are always expensive in wireless systems. Therefore, an efficient utilization of radio resources will allows more users to be connected simultaneously. In addition, CRRM will allow users to have the ability to use multiple types of services such as voice call, web browsing, and video call, etc. As a result, CRRM has the potential to increase the network operator's revenue. Reference [24] proposed a CRRM algorithm which maximizes the network operator's revenue and compares it against non CRRM algorithm. Results show that the proposed CRRM algorithm has increased the network operator's revenue.

13.2.5   Guarantee Required QoS across Different RATs

Guaranteeing the required QoS for all users and their different types of request is one of the challenging tasks in a heterogeneous wireless network. When there is limitation in the availability of radio resources, congestion happens. CRRM will allocate user calls to the suitable RATs in heterogeneous wireless networks to achieve the required QoS. This will lead to a reduction in blocking and dropping probability, and improvement in the level of QoS.

13.2.6   Consider Users' Preferences and Increase Their Satisfactions

Different types of users have different preferences. Some may require best Quality of Service (QoS) and others may require cheapest cost of service or a RAT which offers a longer battery life time for their devices which will allow users to be connected for longer time when they are running out of battery or may require a RAT that has higher coverage area (because they are in high mobility) which will reduce the unnecessary VHO. CRRM can be used to achieve the users objectives based on their preferences, and therefore will lead to increasing the users' satisfactions.

13.3   Heterogeneous Wireless Networks with and without CRRM Algorithm

CRRM was proposed to manage radio resource utilizationRadio resource utilization in heterogeneous wireless networks and to improve RATs performances. RAT selection algorithms are part of the CRRM algorithms. Simply, their role is to verify if an incoming call will be suitable to fit into a heterogeneous wireless network, and to decide which of the available RATs is most suitable to fit the need of the incoming call and admit it. The flowchart of RAT selection process in a heterogeneous wireless network is presented in Fig. 13.2. It is composed into two main directions: with implementing CRRM algorithm and without implementing CRRM algorithm. In the first direction, when a new or VHO call arrives, firstly RATs will be sorted according to the RAT selection algorithm. Then, the RAT with highest satisfaction will be selected to serve the call. Therefore, radio resources will be allocated for the requested call. If the selected RAT is unable to serve the call, the RAT with second highest satisfaction will be selected. If none of the available RAT is able to serve the requested call, the call will be rejected. In the second direction, when a new or VHO call arrives, the RAT will be selected according to the selection algorithm results. If the selected RAT is unable to serve the requested call, the call will be rejected.

As shown in Fig. 13.2, it is clear that CRRM algorithms minimize the blocking and dropping probability. References [23, 1, 21, 13] present the CRRM environment. They compare the performance of the heterogeneous wireless networks against non CRRM algorithm. In [23, 1], results show that the use of CRRM algorithm outperforms the non CRRM algorithm in term of blocking probability. Fig. 13.3 shows the result of reference [23]. In [21], results show that admission control algorithms for CRRM schemes can reach 60-80% traffic gains when compared to the non CRRM scheme. Reference [13] shows that implementing the CRRM in

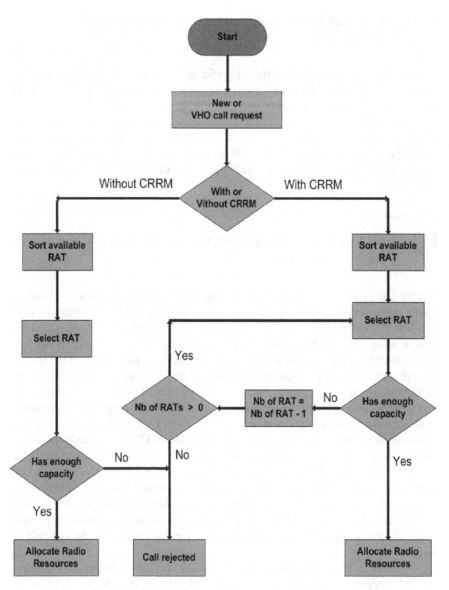

Fig. 13.2. RAT selection process with and without CRRM algorithm

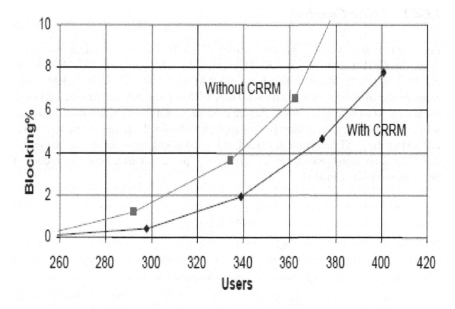

Fig. 13.3. Blocking probability with and without CRRM algorithm. Result presented from Ref. [23].

heterogeneous wireless networks can increase the throughput, performance, accessibility, survivability and modularity.

13.4 RRM and CRRM Interactions

CRRM was proposed to support the RRM decision related to the use of radio resources. Therefore, each RRM entity can take into account the availability of radio resources in other RRM entities of different RATs [2]. Therefore, CRRM entity will have information on the state of the RRM entities that are related to different RATs and it will be responsible for them.

When there is an admission for a new call or a necessary VHO call, CRRM and RRM entities will make a decision regarding admitting the call or rejecting it depending on the degree of interaction between them. Different possibilities of coupling between CRRM and the RRM entity of each RAT may coexist in a heterogeneous wireless network. These types have been proposed in [18, 2]. They can be categorized into:

13.4.1 Loose Coupling

Loose coupling approach is presented in Fig. 13.4. It is a minimum interaction between the CRRM and the local RRM entities. In this approach, CRRM entity is involved only in the decision for the initial RAT selection and VHO. CRRM entity receives the measurements from each local RRM entity that include the cell load for each RAT. Therefore, CRRM can take into account the availability of radio resources for each RAT. It selects the appropriate RAT for the requested new or VHO call. After the RAT has been allocated, the local RRM entity treats the requested call with admission control, packet scheduling, power control, congestion control and necessary horizontal HO.

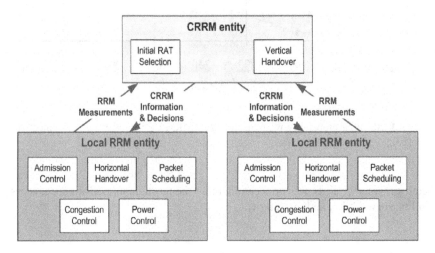

Fig. 13.4. Loose coupling

13.4.2 Tight Coupling

A higher interaction between the CRRM and the local RRM entities is shown in Fig. 13.5. In this approach, CRRM entity is involved in the decision for the initial RAT selection, admission control, horizontal and vertical HO and congestion control. In the tight coupling approach, CRRM entity did not stop by selecting the appropriate RAT for the requested new or VHO call but it becomes involved in deciding the specific cell for the selected RAT.

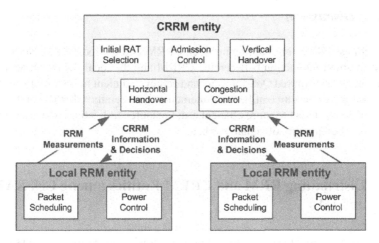

Fig. 13.5. Tight coupling

13.4.3   Very Tight Coupling

A very strong interaction between the CRRM and the local RRM entities is presented in Fig. 13.6. In the very tight coupling approach, CRRM entity is involved in the most decisions for local RRM entities. It selects the appropriate RAT and cell for requested call and it is also engaged in the packet scheduling policy for the call.

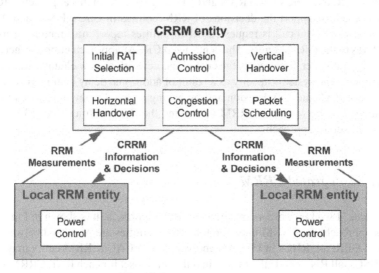

Fig. 13.6. Very tight coupling

13.4.4   Discussion

Different possibilities of coupling between CRRM and the local RRM entities can coexist. A small interaction between the CRRM and the local RRM entities is not able to guarantee required QoS for users and achieve efficient radio resource utilization. StrongRadio resource utilization interaction between the CRRM and the local RRM entities will lead to more efficient radio resource utilization; however, it may lead to a higher quantity of signal overhead.

13.5   Distributing RRM and CRRM Entities among CN, RAT and UTs

CRRM was proposed to support the interconnection between different RATs in heterogeneous wireless networks. The distribution of CRRM functions can be categorize into three approaches: centralized CRRM (CRRM server), integrated CRRM or distributed CRRM (terminal

13.5.1   Centralized CRRM

The centralized CRRM approach has been proposed in [18, 1]. In this approach, CRRM server is added to the core network, and the RRM entity is located in the BSCs in GERAN, RNCs in UTRAN and APCs in WLAN. In this approach, CRRM server is used to support the decisions of RRM entities in BSCs, RNCs and APCs when a new or VHO call is requested. RRM entities report information regarding their RATs to the CRRM entity. The centralized CRRM approach improves network scalability. However, it increases the cost of the network as new element was added and it reduces the system capacity as a result of additional delay coming by delayed selection of RAT and HO decision and by having a signal delay introduced by the communication between the CRRM server and the RRM entities. Fig. 13.7 shows the CRRM server approach.

13.5.2   Integrated CRRM

References [18, 1] propose an integrated CRRM approach, it is shown in Fig. 13.8. In this approach, the CRRM entity and the RRM entities are presented between the BSCs in GERAN, RNCs in UTRAN and APCs in WLAN. CRRM entity may exist in all BSCs, all RNCs and all APCs or in only one subset for each RAT. CRRM functions act immediately between RATs instead of working through the core network.

This approach is already implemented in some heterogeneous wireless networks such as Beyond 3G (B3G) network where IuR, IuR-G & IuR-W interfaces already contain the required components for supporting the CRRM entity. The main advantages of this approach are to limit the infrastructure changes of the network and use the existed functions that can achieve the required system performance. In addition, this approach will achieve the required functions for interconnection in the heterogeneous wireless networks without increasing the delay for initial and VHO RAT selection procedures. Furthermore, minimizing the delay in VHO will have a positive impact on power control and thus maximizing the system capacity. However, this approach may lead to scalability problems when the number of interconnections between RRM entities grows as a result of increasing the number of RRM entities.

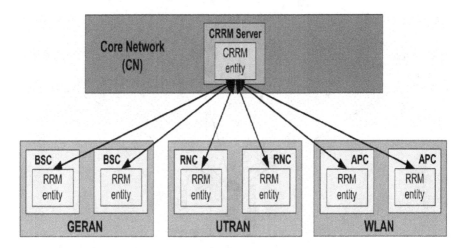

Fig. 13.7. CRRM server approach

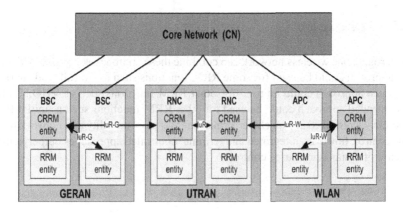

Fig. 13.8. Integrated CRRM approach

13.5.3 Distributed CRRM

Reference [14] proposes a terminal controlled CRRM approach. In this approach, CRRM functions are distributed in the UTs where the CRRM entity can make the decision regarding the suitable RAT that a new or VHO call can be allocated. The main advantages of this approach are providing users with the best possible connection and achieving higher QoS for UTs. However, distributed CRRM approach may lead to inefficient radio resource utilization. Fig. 13.9Radio resource utilization shows the terminal controlled approach.

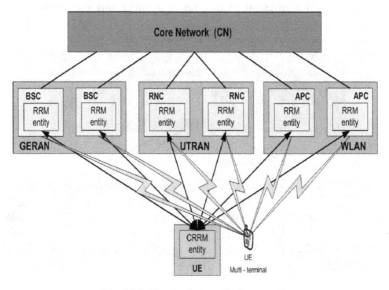

Fig. 13.9. Terminal controlled approach

13.5.4 Discussion

A heterogeneous wireless network can combine these approaches together. Centralized approach could be used for some RRM functions such as overall load sharing. Integrated approach could be used for other RRM functions such as VHO. Terminal controlled approach could be used for other RRM functions such as initial RAT selection. Centralized approach performs better for slow mobility and network functions control; however, integrated and terminal controlled approaches outperform in the case of dynamic resource control.

13.6   Distributed vs. Centralized CRRM Algorithms

With the increase in capacity, ability and power of mobile terminals, a distributed approach (terminal controlled) has been proposed in the literature to get benefits from mobile terminals ability where some of the management workload can be transferred from network to UTs.

A terminal controlled selection approach is presented in [3, 9]. The main idea of the terminal controlled approach is to minimize the management workload in the network equipment and distribute some of them to the UTs. UT keeps some information and measurement for its capacity, profile and preference. Therefore, UT can make decision on selection of the appropriate RAT by analysing the information and measurements. A fully terminal controlled RAT selection approach for heterogenous wireless networks is proposed in [16] without any change in the network infrastructure. This approach allows the UT to select the most efficient RAT that maximizes its satisfaction. By implementing distributed RAT selection algorithms, users will be able to choose the best RAT that achieves their preferences and satisfactions. Also, distributed algorithm reduces the signalling in the network. Therefore, the network capacity is increased. Moreover, there is no need for any change in the existed network infrastructure to implement this approach. However, distributed algorithm does not take into account the network benefits and policies. This may lead to inefficient radio resource utilization and it may cause an unbalanced network load.

On the other hand, a number of centralized RAT selection algorithms have been proposed in the literature for heterogeneous wireless networks. These algorithms improve the overall network stability. Centralized RAT selection algorithms have the benefit of considering more criteria in making decision. However, centralized algorithms do not guarantee the required QoS for the admitted calls. Moreover, they reduce the network capacity as a result of the introduced signal load or delay resulted by the communication between the network entities.

Distributed RAT selection algorithms do not consider networks preferences and policies. Centralized RAT selection algorithms do not guarantee the required QoS for the admitted calls and reduce the network capacity. A hybrid RAT selection algorithm (distributed with network assistance) seems to be a solution for this problem. In the hybrid approach, the network will provide the UTs some information that assists them in their decisions to select the most efficient RAT that maximizes their satisfaction and in the same time improve the efficient radio resource utilization. IEEE P1900 Standards Committee proposes an IEEE P1900.4 Protocol [10] that is able to provide the required support to the hybrid approach. This protocol is described in the next section.

A distributed RAT selection strategy at the UT by using IEEE P1900.4 protocol is proposed in reference [20]. It takes the NCCB RAT selection algorithm proposed in [19] for heterogeneous CDMA/TDMA scenario and implements it in a distributed approach using IEEE P1900.4 protocol. Results show that the distributed NCCB RAT selection algorithm using IEEE P1900.4 protocol performs better then the centralized load balancing RAT selection algorithm.

13.7 Conclusion and Future Works

CRRM is essential for heterogeneous wireless networks. An intelligent implementation strategy will improve the efficiently of radio resource utilization, increase users' satisfactionsRadio resource utilization and maximize network operator's revenue. In this paper, we discuss the need of CRRM in heterogeneous wireless networks. A comparison between implementing or not the CRRM in heterogeneous wireless networks is studied. Then, the interaction between RRM and CRRM entities is presented. After that, different approaches for the distribution of RRM and CRRM entities among the CN, RATs and UTs are studied. Finally, a comparison between centralized and distributed CRRM algorithms is presented. Centralized algorithms have the benefit of considering more criteria in making decision. However, they have a disadvantage in terms of reducing the network capacity. Distributed algorithms allow UTs to select the most efficient RAT that maximizes their satisfactions. They maximize the network capacity. However, distributed algorithms are inefficient because of the limitation of information at UT such as cell load, and network policies and preferences. A best solution could be by implementing a hybrid CRRM approach (terminal controlled with network assistance), where the network will assist the UTs in making RAT selection decision by providing information and policies related to the network. Future research works may include the following: provide an evaluation between different centralized and distributed CRRM algorithms in terms of blocking probability, dropping probability, throughput and users' satisfactions probability, propose an intelligent hybrid CRRM approach to support the RAT selection in heterogeneous wireless networks, simulate the proposed intelligent hybrid approach in different scenarios and do an analytical approach for the proposed intelligent hybrid approach using Markov Model.

Acknowledgements. This work is sponsored by the Centre for Real-Time Information Networks (CRIN) in the Faculty of Engineering & Information Technology at the University of Technology, Sydney (UTS). This paper is an extended version of the ACASE'12 conference paper [6].

References

1. 3GPP TR 25.881 v5.0.0: Improvement of RRM across RNS and RNS/BSS (Release 5) (December 2001)
2. 3GPP TR 25.891 v0.3.0: Improvement of RRM across RNS and RNS/BSS (Post Rel-5) (Release 6) (February 2003)
3. Adamopoulou, E., Demestichas, K., Koutsorodi, A., Theologou, M.: Intelligent Access Network Selection in Heterogeneous Networks. In: 2nd International Symposium on Wireless Communication Systems (ISWCS 2005), Siena, Italy, pp. 279–283 (September 2005)

4. Agusti, R., Sallent, O., Perez-Romero, J., Giupponi, L.: A Fuzzy-Neural Based Approach for Joint Radio Resource Management in a Beyond 3G Framework. In: First International Conference on Quality of Service in Heterogeneous Wired/Wireless Networks (QSHINE 2004), Dallas, Texas, USA, pp. 216–224 (October 2004)

5. AL Sabbagh, A., Braun, R., Abolhasan, M.: A Comprehensive Survey on RAT Selection Algorithms for Beyond 3G Networks. In: International Conference on Communications, Networking and Mobile Computing (ICCNMC 2011), Dubai, UAE, pp. 834–838 (January 2011)

6. AL Sabbagh, A., Braun, R., Abolhasan, M.: Interaction of Radio Resource Management in Heterogeneous wireless Networks. In: 1st Australian Conference on the Applications of Systems Engineering (ACASE 2012), Sydney, Australia, pp. 34–35 (February 2012)

7. AL Sabbagh, A., Braun, R., Abolhasan, M.: A Comprehensive Survey on RAT Selection Algorithms for Heterogeneous Networks. World Academy of Science, Engineering and Technology (WASET) (73), 141–145 (2011)

8. AL Sabbagh, A.: A Markov Chain Model for Load-Balancing Based and Service Based RAT Selection Algorithms in Heterogeneous Networks. World Academy of Science, Engineering and Technology (WASET) (73), 146–152 (2011)

9. Bari, F., Leung, V.C.M.: Automated Network Selection in a Heterogeneous Wireless Network Environment. IEEE Network 21(1), 34–40 (2007)

10. Buljore, S., Merat, V., Harada, H., Filin, S., Houze, P., Tsagkaris, K., Ivanov, V., Nolte, K., Farnham, T., Holland, O.: IEEE P1900.4 System Overview on Architecture and Enablers for Optimised Radio and Spectrum resource usage. In: 2008 IEEE Symposium on New Frontiers in Dynamic Spectrum Access Networks, Chicago, Illinois, USA (October 2008)

11. Garg, V.K.: Wireless Communications and Networking. Morgan Kaufmann Publishers, San Francisco (2007)

12. Karabudak, D., Hung, C., Bing, B.: A Call Admission Control Scheme Using Genetic Algorithms. In: The 19th Annual ACM Symposium on Applied Computing (SAC 2004), Nicosia, Cyprus (March 2004)

13. Lincke-Salecke, S.: The Benefits of Load Sharing when Dimensioning Networks. In: 37th Annual Simulation Symposium (ANSS-37 2004), Arlington, VA, USA, pp. 115–124 (April 2004)

14. Magnusson, P., Lunds, J., Sachs, J., Wallentin, P.: Radio Resource Management Distribution in a Beyond 3G Multi-Radio Access Architecture. In: IEEE Global Telecommunication Conference (GLOBECOM 2004), Dallas, TX, USA, pp. 3472–3477 (November - December 2004)

15. Murray, K., Mathur, R., Pesch, D.: Network Access and Handover Control in Heterogeneous Wireless Networks for Smart Space Environments. In: 1st International Workshop on Managing Ubiquitous Communications and Services (MUCS 2003), Waterford, Ireland (December 2003)

16. Nguyen-Vuong, Q.T., Agoulmine, N., Ghamri-Doudane, Y.: Terminal-Controlled Mobility Management in Heterogeneous Wireless Networks. IEEE Communications Magazine 45(4), 122–129 (2007)

17. Nicopolitidis, P., Obaidat, M.S., Papadimitriou, G.I., Pomportsis, A.S.: Wireless Networks. John Wiley & Sons Ltd., Chichester (2003)

18. Perez-Romero, J., Sallent, O., Agusti, R., Diaz-Guerra, M.A.: Radio Resource Management Strategies in UMTS. John Wiley & Sons Ltd., Chichester (2005)

19. Perez-Romero, J., Sallent, O., Agusti, R.: Network Controlled Cell Breathing in Multi-Service Heterogeneous CDMA/TDMA Scenarios. In: 2006 IEEE 64th Vehicular Technology Conference (VTC 2006), Montreal, Canada (September 2006)

20. Prez-Romero, J., Sallent, O., Agusti, R., Nasreddine, J., Muck, M.: Radio Access Technology Selection enabled by IEEE P1900.4. In: 16th IST Mobile and Wireless Communications Summit, Budapest, Hungary (July 2007)
21. Ramrez, S.L., Genovs, M.T., Navarro, M.F., Skehill, R., McGrath, S.: Performance Evaluation of Policy-Based Admission Control Algorithms for a Joint Radio Resource Management Environment. In: 2006 IEEE Mediterranean Electrotechnical Conference (MELECON 2006), Benalmdena, Mlaga, Spain, pp. 599–603 (May 2006)
22. Schiller, J.: Mobile Communications, 2nd edn. Addison Wesley, Harlow (2003)
23. Tolli, A., Hakalin, P., Holma, H.: Performance Evaluation of Common Radio Resource Management (CRRM). In: 2002 IEEE International Conference on Communications (ICC 2002), New York, USA, pp. 3429–3433 (April - May 2002)
24. Yu, F., Krishnamurthy, V.: Optimal Joint Session Admission Control in Integrated WLAN and CDMA Cellular Networks with Vertical Handoff. IEEE Transactions on Mobile Computing 6(1), 126–139 (2007)

Chapter 14
An Intelligent Model for Distributed Systems in Next Generation Networks

Pakawat Pupatwibul, Ameen Banjar, Abdallah AL Sabbagh, and Robin Braun

Abstract. Over the past twenty years, network technology has been improved rapidly in term of speed, performance, component, and functionalities. Therefore, a number of different types of network devices have been developed; this led to an increase in the complexity of network systems. Traditional network structures are inadequate to meet today requirements. It is centralized network which imposes on human operators to have a high experience on how to detect changes, configure new services, recover from failures and maximize Quality of Service (QoS). Therefore, network management involves heavy reliance on expert's operators. The adopted centralized network management is not suitable for new technologies emerging, which are complex and difficult to interact among heterogeneous networks that contain different types of services, products and applications from multiple vendors. As a result, the current network management lacks of efficiency and scalability; however, it has an acceptable performance generally. The centralized information model cannot stand and achieve the requirements from such complex, distributed electronic environments. This paper studies the need of distributed systems in next generation networks. Then, the paper presents three network structure paradigms: centralized, hybrid and distributed. After that, Software-Defined Networking (SDN) is described. Finally, the paper proposes a distributed approach for OpenFlow technology using a Distributed Active Information Model (DAIM) which supports an autonomic management of the distributed electronic environment.

14.1 Introduction

The increasing adoption of advanced technologies in communication networking, computing applications, and information modelling have played a significant role in

Pakawat Pupatwibul · Ameen Banjar · Abdallah AL Sabbagh · Robin Braun
University of Technology, Sydney (UTS), Sydney, Australia
e-mail: {Pakawat.Pupatwibul,ameen.r.banjar}@student.uts.edu.au,
　　　{abdallah.alsabbagh,robin.braun}@uts.edu.au

R. Klempous et al. (eds.), *Advanced Methods and Applications in Computational Intelligence*,　　315
Topics in Intelligent Engineering and Informatics 6,
DOI: 10.1007/978-3-319-01436-4_14, © Springer International Publishing Switzerland 2014

providing management services for large and complex systems. As the complexity of centralized system grows over time, an effective management requires monitoring, interpreting, and handling the behaviour of the managed resources in order to ensure required Quality of Service (QoS) and improve networks performance.

Currently, many network management systems pursue a platform-centred paradigm, where all of the computation is controlled at a central location. As an example, in traditional Simple Network Management Protocol (SNMP), a fully centralized management paradigm is used. Agents are accessed by applications via management protocol to monitor the system and collect the network information. Moreover, several researchers in the network management discipline believe that in most, if not all, network management problems can be addressed by using appropriate centralized systems and intelligence control [16]. However, in today's real networks, there are many network management complexity and limitations that cannot be adequately solved by a fully centralized approach such as lack of flexibility and information bottlenecks.

Centralized implementations are also inefficient to handle the huge number of high level decision makings. This paper compares the centralized paradigm with distributed system paradigm (decentralized), in which some or all of the intelligence and management control are locally distributed within the network entities. In wired networks, distributed system minimizes the complexity that occurs in layer 3 devices (e.g. routers) by distributing some roles into layer 2 devices (e.g. switches). OpenFlow is an example of a network that may be able to apply new distributed model on. In wireless networks, distributed system also minimizes the complexity that occurs in the core network and the Radio Network Controllers (RNC) such as in Beyond 3G (B3G) network by distributing some management functions (e.g. decision making for allocation of radio resources) into the Evolved Nodes B (eNodes B) such as in Long Term Evolution (LTE) network or into Base Stations (BSs) such as in Mobile Worldwide Interoperability for Microwave Access (WiMAX) [1],[20].

A new information model named: Distributed Active Information Model (DAIM) is presented to allow the local decision making process, that will essentially contribute to complex distributed network environments [5]. An implementation of DAIM model is expected to meet the requirements of the autonomic components of the distribution networks, such as self-management. An autonomic system in this context means that, each distributed device can draws its own strategies for adaptation driven by the goals of the system [7]. The distributed autonomic system adapts the network for needs of dynamic changing in business, and reduces operations and management complexities. Benefits that can be achieved through implementing the DAIM model include control of any network device, which OpenFlow enabled from any vendor, and rapidly configure, and update the hardware in the entire network. The approach accelerates business innovation by allowing network operators of Information Technology (IT) to program the network in real time to meet the business needs and specific requirements of the users. Accomplish this approaches is a challenging task. This is because the network control plane mechanisms take several years to fully design, and even longer to spread widely, a new control protocol. Also, it should consider that characteristics of incremental properties, complexity of new network operators, and missing some functions in elements of the network [22]. Another part of the approach is that, distributed system needs to have a well-defined Network Requirement Database (NRD), in order to maintain self-management, self-configuration, self-preservation and recovery [5]. In addition, current networks have many restrictions, including difficulties to meet the business and technical needs over the past few decades, while industry have developed protocols for the networks to provide a high performance, reliability and greater connectivity, and security more stringent. Moreover, there are difficulties to add or remove any device or configure these devices which must be touched by an IT person that need to configure many switches, routers and firewalls using the device-level management tools [19].

As a result of these limitations, networks today are relatively constant as it seeks to reduce the risk of interruption of service. To stay competitive, next generation network must provide a higher value than ever, and provide the best customer service. A promising solution for these requirements is the autonomic distribution system [22].

The approach has not been introduced before. This is because, in the classical routing or switching, fast forward packets (data plane), and high-level routing decisions (control plane) occur on same device. In addition, vendors' devices are closed and not accessible. Moreover, if the device is not described by Management Information Base (MIB), the device does not exist. Therefore, each device has a MIB in ASCII format that you need to access and edit to achieve the new requirements [10]. From vendors' side, they have lack of standard and open interfaces, and there are limitations in the ability of network operators to design the network to meet different individual requirements. This makes a gap between market requirements and network capabilities [19].

In response, the industry has created Software-Defined Networking (SDN) architecture and develop standards associated with it. The OpenFlow is an implementation of the SDN architecture that separates data and control plane functions. The data path is still residing on the switch, while the high-level routing decisions are separating in different device called controller, usually a standard server. The OpenFlow enabled switch and the controller to communicate over OpenFlow protocol, which defined messages, such as packet-received and send-packet-out [3]. As a result, companies get the programmability, automation, and the control of the network, which enable them to build highly scalable, flexible networks that can be easily adapted to different changing environment. Recent development techniques enable dynamic reprogramming of the devices through the data flow [3]. The DAIM model based-OpenFlow gives the distributed system environment a sustainable information model, which collects, maintains updates and synchronizes all the related information. Each device has decision-making ability on the basis of the information that is collected to autonomously adapt any changing environments [5]. The autonomic approach of distribution system leads to rapid innovation through the ability to provide network capabilities and new services without having to configure individual devices or wait for the launch from the seller. In recent times, there is a growing movement led by both industry and academia, aim to design mechanisms to reach a control model in which the separation of control plane from data plane and built it as a distributed system [22].

Thus, this paper describes the distributed system as a next generation network, which could meet the self-x paradigm. In addition, the paper covers a number of functionalities such as the DAIM model, support autonomic management for distributed system, and OpenFlow capabilities. The remainder of the paper is organized as follows. In Sect. 14.2 we demonstrate the needs of distributed systems. In Sect. 14.3 we present the different network structure paradigms. We present more details of SDN and how to implement the DAIM model to OpenFlow in Sect. 14.4, Sect. 14.5, and Sect. 14.6. Finally, conclusion and future works are shown in Sect. 14.7.

14.2 The Needs of Distributed Systems

Traditional static networks are struggling to meet the rapidly growing requirements of today's enterprises, carriers, and end-users. In addition, as the network scale in size and complexity, a distributed networking system is needed to ensure good quality of network services and performance. A distributed network system can refer to an application that executes a set of protocols to correspond the actions of multiple processes on a network [13]. Moreover, all components cooperate together to operate a single or small set of related tasks. The devices that are in a distributed system can be physically link together and connected by a local network. They can also be geographically distant and connected by a wide area network [14]. A distributed system can consist of any number of possible configurations, such as personal computers, minicomputers, mainframes, workstations, and so on.

A distributed system aims to make a network environment work as a single computer in order to cope with the extremely significant demand of users in both data storage and processing power. Examples of distributed systems may include Distributed File Systems (Hadoop), P2P Network, Cloud Network, Grid Computing, Web Server and Indexing Server, and Pervasive Computing.

There are several advantages such as the ability to connect remote users with remote resources in an open and scalable way. Regarding open, we mean that each component is continually open to interaction with other components. Whereas scalable, we mean that the system can easily be altered to accommodate changes in the number of users, computing entities, and resources [13].

Therefore, a distributed system can bring many benefits given the combined capabilities of the distributed components, than combinations of stand-alone systems [15]. However, it is not easy for a distributed system to be useful; it should provide system reliability as well. This is a very difficult goal to achieve due to the complexity of the interactions between simultaneously running components.

Ref. [4] indicates the concerned of reliability in distributed computing systems. The following table 14.1 summarises the characteristics of a reliable distributed system.

Table 14.1. Characteristics of a reliable distributed system

Characteristic	Description
Fault Tolerant	It can recover from network failures without performing incorrect actions.
Highly Available	It can restore operations, instructing it to resume network services even when some components have failed.
Recoverable	Failed components can reboot themselves and re-join the system, after the cause of failure has been recovered.
Consistent	The system can execute corresponding actions of multiple components often in the case of concurrency and failure. This underlies the ability of a distributed system to act as a non-distributed system.
Scalable	It can operate properly even some aspect of the system is scaled to a larger size. For example, if the number of users or servers increases, the overall load on the system should not have a significant effect.
Predictable Performance	The ability to provide desired responsiveness in a timely manner.
Secure	The system authenticates access to data and network services

Some of the key computing trends driving the need for a distributed system paradigm include [19]:

14.2.1   Change of Traffic Patterns

Regarding the enterprise data center, traffic patterns have changed dramatically. In contrast to client-server applications, the bulk of the traffic occurs between one client and one server, applications today access various databases and servers, and thus creating a flurry of "east-west" machine-to-machine traffic before returning data to the users' device in the classic "north-south" traffic pattern. Meanwhile, network traffic patterns are changed by users as they push for access to corporate content and applications connecting from any type of device anywhere and anytime. Moreover, many enterprise and carrier managers are contemplating a utility computing model, which might include a public cloud, private cloud, or a mix of both. This may result in additional traffic across the wide area network.

14.2.2   The Consumerization of IT

Users are increasingly employing mobile technology such as smart phones, tablets, and laptops to access the corporate network. This can cause pressure for IT to accommodate these personal devices in a fine-grained manner, while protecting intellectual property as well as corporate data and meeting compliance mandates.

14.2.3 The Rise of Cloud Services

The highly demand of enterprises for both public and private cloud services has result in unprecedented growth of these services. Today's enterprises want the agility to access applications, infrastructure, and resources on demand. Providing self-service provisioning in either a public or private cloud requires flexible scaling of storage, computing, and network resources, basically from a common viewpoint and with a common suite of tools.

14.2.4 Huge Data Demand More Bandwidth

Dealing with today's mega data-sets requires efficient parallel processing on thousands of servers, which all need direct connections to each other. This emerging trend of mega data-sets has led to a constant demand for additional network capacity in the data center. Administrators of large-scale data center networks face the daunting task of managing the network to previously unimaginable size due to ever increasing network complexity.

14.3 Network Structure Paradigms

This section briefly reviews different communication network paradigms and some concepts of each one, also how they have been built and how they work. There are varieties of network topologies that can be categorized into three groups: centralized (star), distributed (grid/mesh) and combination of these two called Hybrid which is group of stars paradigm connected together [2]. So, this section will compromise between these three network structures, to form reliable network, which is capable for implementing DAIM model to support autonomic network management.

14.3.1 Centralized Network Structure

The first paradigm is the centralized network (star) paradigm which has a single node as a core node, and multiple nodes connected to that core node where each node has an operating system and applications (see Figure14.1, respectively). Illustrate the centralized network. The core node is able to configure all connected nodes using an Operating System Communication Application (OSCA)[18]. The centralized paradigm has a major drawback such as increasing the number of connected nodes will affect the core node which has a limitation of process power [17]. In addition it is obviously weak, if destruction happened to that single central core; will lead to destroy the communication between end nodes [2]. However, the main advantage of the centralized architecture is that one core node can be responsible for managing all connected nodes, and thus managing the entire network from a single point [17].

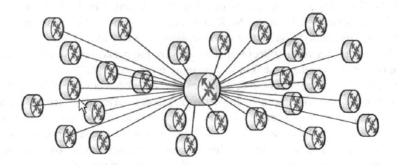

Fig. 14.1. Centralized network structure (star network)

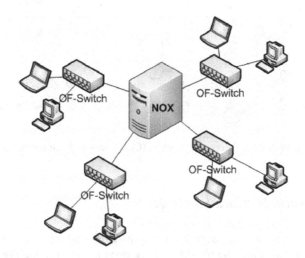

Fig. 14.2. OpenFlow network is logically centralized but can be physically distributed

The centralized structure of OpenFlow contains one NOX controller (control plane) and a number of OpenFlow switches (data forwarding plane) as shown in the Fig.14.2, NOX is network operating system, runs on a single server, which manages the forwarding decisions of multiple OpenFlow switches. It also has components developed as network requirements called network's applications to handle the network events, and able to access the network traffic, and generate traffic [12].

The combination of OpenFlow enabled switch and NOX controller, work as a router to deliver the packets from source to destination with high-level routing decisions made by the NOX controller. The data-forwarding plane is depending on the flow table, which contains flow entries to match with the first packet within a flow, happened on OpenFlow switch [23].

If the first packet of a new flow could not find a match in the flow entry, the switch will send directly to the NOX controller for high-level routing decisions. That packet passed to interested network applications to decide: whether to forward the flow on the network; collect statistics; modify the packets in the flow or view more packets within the same flow to gain more information [12].

A number of different flow-based applications have been developed in the NOX controller, they are able to control the network, reconstruct topology, track the computers whatever changing on topology, provide network access controls, and manage network history [12]. However, the centralized network structure is not feasible to implement the DAIM model because this network structure relies heavily on a single point and the approach is to distribute the self-management capabilities.

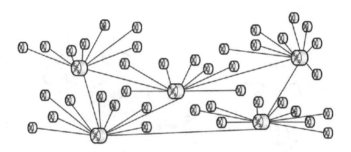

Fig. 14.3. Hybrid network structure (Combination of star and mesh)

14.3.2 Hybrid Network Structure

The second paradigm is the hybrid network paradigm which is a combination of star and grid paradigms. The hybrid network does not usually require a complete reliance on a single point [2]. Referring to figure 14.3, which shows a hierarchical structure of numbers of stars connected together as a massive star with an extra link to make a loop.

OpenFlow undertakes a logically centralized controller, which can be physically distributed as shown in Fig. 14.4. It contains three sites of OpenFlow network; each site has its own controller to serve the local requests. However, current implementation depends on a centralized single controller, which has drawbacks as mentioned in centralized network, including lack of scalability. This hybrid network structure can distribute the event-based control plane for OpenFlow environment. In addition, it provides the scalability and at the same time keeping the power of centralized network [23].

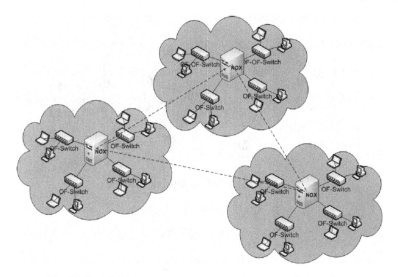

Fig. 14.4. Hybrid OpenFlow structure with number of controllers

People have introduced HyperFlow as a distributed control plane for OpenFlow, which is an application for NOX controller to communicate those distributed NOXs. In addition, each OpenFlow switch makes a local decision (relies on local flow table and local controller). HyperFlow use to passively synchronize state upon whole OpenFlow networks controllers which can enable local decision by individual controller to all packets. Thus, significantly reduces the response time of control plane for data plane requests [23].

Each controller in HyperFlow network has the ability to control the whole network because it has a coherent view of the entire network. If any controller fails, all affected switches have to reconfigure by themselves to join the other nearest controller [23]. Although this hybrid network structure has distributed control planes, however, it is not efficient to implement the DAIM model because it still relies on a central point which can control the network, even if that not usually relays on central nodes.

14.3.3   Distributed Structure Network

Lastly, it is expected that the current and next generation network should rely on distributed paradigm (mesh/grid) as presented in figure 14.5. The distributed paradigms

rely on number of distributed nodes connected together, and not rely on centralized point to communicate with all nodes. The distributed paradigm is continuous connection and it has ability to reconfigure itself within entire network [11].
For example, the flow packets can rout from node to node until they reach their
destination.

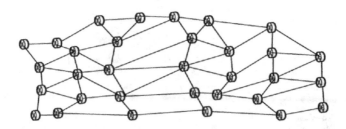

Fig. 14.5. Mesh or Grid network structure

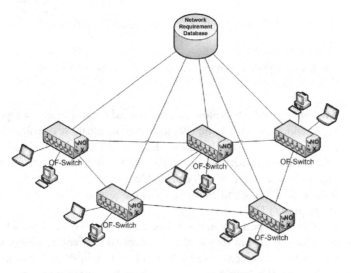

Fig. 14.6. Distributed switches which able to manage the whole network

This section is just a hypothesis to the paper approaches. Given that having a
distributed structure, it is able to implement DAIM model to support autonomic
network management (see Fig.14.6). This distributed structure contains some of
OpenFlow switches embedded with mini-NOX inside each switch. The switches
are connected together, while each switch is connected to a Network Requirement
Database (NRD).

Sending packets in this structure typically uses the OpenFlow environment. The main idea is to provide distributed autonomic management using mini-NOX, which is connected to each other. Each mini-NOX is connected to the NRD to obtain all network information needed to achieve autonomic functionalities as well as obtaining information of the entire network. Moreover, each mini NOX actively synchronises with the NRD, and hosts are connected to their local switch, which is responsible to fully serve all packets within its site, unless failure happened. If the failure happened then hosts should reconfigure to the nearest switch instead of that failed switch. The new switches can actively synchronise with NRD to know all the information and the requirements to serve connected hosts, which have been adapted themselves from other site.

The mini controllers can publish events to NRD and actively synchronizes with NRD, so that other controllers can reconstruct the whole information about the network. Individual switch can serve any coming packets locally or from other switches. As a benefit of NRD, it gives the ability to any local change within individual switch as a distributed switch to deploy autonomic functions such as self-configuration. Thus, the distributed system structure is feasible to deploy the DAIM model, which has the ability to synchronize controllers and the whole network events together, to achieve the self-management as well as enabling all autonomic functions.

14.4   Software-Defined Networking (SDN)

The future of networking domain will rely more and more on software. SDN, standardised by a non-profit industry called the Open Networking Foundation (ONF), is an emerging network architecture that seeks to transform traditional static networks into flexible programmable platforms by decoupling the network control and data planes. In addition, network intelligence and state are logically centralised, and the underlying infrastructure is abstracted from the applications, which treats the network as a logical or virtual entity. With SDN, carriers and enterprises can gain unprecedented automation, programmability, and network control that will enable them to create highly scalable and flexible networks in order to meet the changing business needs[19].

Fig. 14.7 shows a logical view of the SDN structure, where network intelligence is (logically) centralised in software-based SDN controllers, which maintain a global view of the network. Therefore, the network appears to the policy engines and applications as a single logical switch. SDN provides vendor independent control over the entire network from a single logical point. This can greatly simplify the network design and operation for enterprises and carriers. Moreover, SDN also simplifies the network devices themselves as they do not need to understand and process all of the standard protocols, but accepting instructions from the SDN controllers instead.

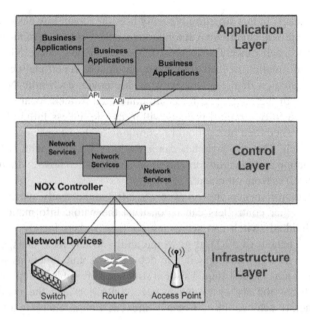

Fig. 14.7. Software-Defined Network Architecture

Most importantly, network administrators can programmatically configure this simplified network abstraction instead of having to hand-code thousands of lines of configuration among thousands of distributed nodes. This migration of control gives network operators the flexibility to manage, configure, secure, and optimise network resources via the SDN control software dynamically. As a result, they can efficiently alter the network behaviour in real-time as well as deploying new network services and applications.

OpenFlow-Based SDN architecture can present several substantial benefits including [19]:

• Centralised management and control of network devices from various vendors;

• Improve the management and automation by using common APIs to abstract the underlying network details from the provisioning systems and applications;

• Ability to provide new network capabilities and services without having to configure individual devices or wait for vendor's release;

• Programmability by administrators, enterprises, users, and software vendors using common programming environment;

• Increase network reliability and security as a result of centralised and automated management of networking devices, low configuration errors, and uniform policy statement;

• Advance network control as applications exploit central network state information to seamlessly adapt network behaviour to user needs

SDN requires some method in order for network control to communicate with the switch data path. One such mechanism is "OpenFlow" which is a standard interface for controlling computer networking switches.

14.5 Distributed Active Information Model (DAIM)

Autonomic communication relies heavily on a functional information model that provides source data to drive both decision-making process and information mining processes. The DAIM model consists of two major parts (1) O:MIB, (2) hybrid O:XML. In this section, a new information model is required to cope with the dynamics in distributed ACNs. In addition, an active object-oriented management information base (O:MIB) is proposed as a theoretical framework for the rest of the research with the hope to replace the current MIB. The corresponding programming language hybrid O:XML is explored as a practical technology to implement O:MIB, with platform-independent Java agents (e.g. Jade and JadeX). However, the details of O:XML will not be mentioned in this paper.

O:MIB Theory

DAIM model can be applied with distributed communication networks to enable autonomic functions. One of the most significant barriers when dealing with large-scale and complex distributed systems is insufficient centralised service management. Because the development of agent-based in the field of Distributed Artificial Intelligent (DAI) has grown rapidly, autonomous decentralised systems (ADSs) and multi-agent technology are by far the best solution for complex network environments. The DAIM model consists of adaptation algorithms for adapting the intelligent agents and information objects to be applied to large-scale distributed electronic systems. The main purpose of this model is to re-engineer the structure of network information model, so that the new structure can effectively cope with the next generation communication networks. It also aims to redesign the traditional MIB structure by adopting the object-oriented principles, which is required to fulfil management services such as configuration management, topology discovery, activating application process, and assigning resource process. Fig. 14.8 shows a characteristic comparison between the standard MIBs and proposed O:MIBs.

Furthermore, O:MIB can play as a part of the distributed information model to enable autonomic software agents that act as the network elements (other routers, switches, hosts, etc). These autonomic agents (AAs) inherit the surrounding agent's behaviour and also make local decisions based on the state of the network. The agents of distributed O:MIB technology will allow the richness of self-organized management. For example, dynamic software configurations, service activation, and service discovery. Moreover, DAIM model is developed specifically with embedded smart algorithms for distributed elements to improve the efficiency of local execution abilities.

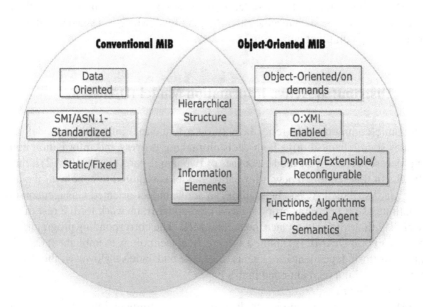

Fig. 14.8. Comparison between traditional SNMP MIB and O:MIB

These large amount of heterogeneous O:MIBs need to be well organised in a way to favour the distributed environment. This implies a distributed intelligent holonic system in order to efficiently manage and implement the O:MIBs in the hierarchical telecommunication system. The main characteristics of the holonic O:MIBs are that it's object-oriented MIBs with methods embedded, and exists on a holonic-level. For example, they are embedded into individual electronic devices such as mobile phones, printers, and even further in the sub-level of devices (chip-level). Braun and Chiang (2008) state that the holonic system is not only a component-based communication architecture, but also a universal way to construct distributed MEs in various levels. The proposed holonic agent-based O:MIB consists of three parts: (1) conventional MIB; (2) user-accessible provisioning; (3) methods/operation. Intelligent algorithms and functions are embedded in each holonic subsystem to fulfil any network tasks for agents to cooperate together and share synchronised information.

The necessary contents of an O:MIB class should cover four divisions of information: QoS parameters; device information; service information; application information and dependency information related to devices and services. Chiang et. al. (2007) designed this O:MIB class for each category of network components. When an object is required for activation during run time, any instance of the O:MIB_Class is directly created by java codes via the keyword "new":

OMIB_ClassOMIB_object = new OMIB_Class( );

The O:MIB model is expected to be used in peer-to-peer networks, mobile technology, and wireless ad-hoc sensor networks (WASNs) as well as to address other complex issues. O:MIB adopts the object-oriented principles to manage the MIB objects. It has multiple distributed agents that remain in every network component and node, which functions with its own O:MIB as a way to activate applications when required. These network components can also analyse the important data, learning the systems environment, calculate situations, and perform adapting capability. Therefore, a full understanding of autonomic communication will be obtained. Object or element is the basic information unit of the O:MIB. Each important element comprises [8]:

• Attributes: It specifies the information values that represent the characteristics of the managed object identifiers (OIDs).

• Method behaviours: An action helping to achieve autonomic communications. This can includes the self-awareness function in real time and intensive and spatial data.

• Algorithms: These are the algorithms that will support a specific network task to be embedded into O:MIB domains. It also represents a set of predefines uses of the total available method calls. For example, humidity of the network environment, temperature monitoring, and predicting the level of raising alarms and risks in autonomic communication networks.

• Messages: In a response of the on-demand requests, local messaging daemon action can invoke messages in order to obtain general information like network topology or mapping discovery.

Fig. 14.9 indicates the implementation process of O:MIB via O:XML. The software agents remain on each node having O:XML employed to populate the recorded data to the corresponding agents. In addition, JAVA agents are also involved because it is platform independent, and due to other agent development tools are mainly based on JAVA technology as well. Each agent is defined from instantiated agents according to their electronic environment. The O:MIB algorithms are invoked by the instantiated agents, whereas information values are re-configured by java-based agents. Ultimately, the agent's life-cycle are accomplished while the program is operating.

The overall stages of this approach can be described as the follows [5]:

1. Observing the current agents on distributed nodes and noticing the environments.

2. Generating new agents when a new environment is identified through the adaptation and learning strategies.

3. Functioning the local O:MIB by invoking the algorithms and methods instructed into the local O:MIB systems by default.

4. Adapting the node in regard to the awareness of the system-level objectives.

5. Wrap up the agent's lifecycle until the next round of process is ready.

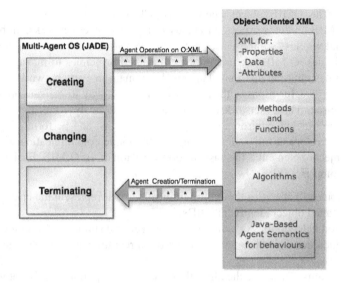

Fig. 14.9. Self-maintained process

This efficient O:MIB-based DAIM model approach is introduce to cope with managing autonomic communications in terms of self-configuring, self-adapting, self-optimizing, self-learning, self-awareness, and so on. This new information model scheme can also be applied into other Self-x properties in ACNs. The attributes of each information object in O:MIB-based DAIM model can be implemented in one O:XML file. This brings the possibility of embedding DAIM agents into portable communication devices as well as applying into real networks in the future such as wireless networks, including WASNs, MANET, Peer-to-Peer networks, and Mesh networks.

14.6   Implementing DAIM Model in OpenFlow: A Case Study

In traditional network configurations, if the circumstances should change or the requirements should change, then the network requires re-configuration again. For example, we are applying the classical compiler to executable paradigm as the following method. Firstly, we should get a software requirement. Then create the codes to meet the requirement. Finally, compile these codes into an executable program. In the case of OpenFlow-Based SDN, the process would be as the following: (1) get the system requirements from the business needs; then (2) create system configuration/code to meet business needs; finally (3) compile this into a SDN enabled configuration.

Thus, the challenge with network is that if the circumstances or requirements change, the network requires repeating all the above methodology. For example, new capacity change or the performance requirement change within the network, traffic load and traffic requirement change, therefore, this would be the problem. In order to overcome this issue, SDN decoupled the control plane and the data plan by developing components on top of the network operating system as network applications. However, these applications provide very low-level methods for interfacing with the network as the operators configure them. For example, application for discover the links in the networks by sending LLDP packets out of every switch interface, application for mapping network topology, and application for routing.

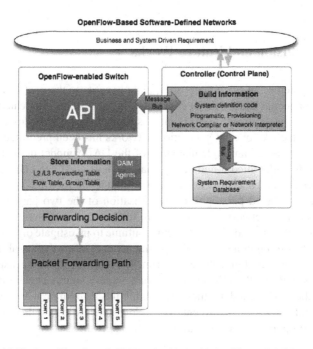

Fig. 14.10. OpenFlow-based SDN embedded with intelligent DAIM agents.

Hence, the alternatives could be implementing DAIM as a reactive interpreter network to enable autonomic behaviours – the paradigm is very similar to the SDN case. However, this new application model has some significant differences, which are based on the intelligent DAIM agents to collecting information, and the system requirement database driven from the business applications as shown in Fig. 14.10. The similarities of the proposed model and OpenFlow is that compiling the system configuration into some kind of intermediary binary code, which then hands over to a run-time environment (network operating system), which would be similar to how the NOX operates as an OpenFlow controller.

We are proposing that by creating a DAIM model on the networks could give effect to the reactive interpreter networks, so it would be a truly computational intelligent environment. Where the DAIM agents reside in the network elements, which would be OpenFlow switches in the case of OpenFlow. The actual variables in the OpenFlow table's entries, embedded within the OpenFlow switches, would be the properties of DAIM agents. These agents would then can modify or adapt their variables' values so as to implement the requirements of the network driven by the business needs. Therefore, the DAIM model would extend across all these network elements and could be thought of as reactive distributed interpreter, which is interpreting the system requirements to enable the infrastructure to provide for the business needs.

14.7 Conclusion and Future Works

Distributed systems can bring many significant benefits to the next generation networks. This paper has studied the needs of distributed systems. Three network structure paradigms: centralized, hybrid, and distributed, have been described. Then, the SDN architecture for next generation networks has been presented. Moreover, the paper has proposed an implementation of the DAIM model in next generation networks. It also identified a number of characteristics that can help to determine whether a given network structure paradigm should be realized in a centralized paradigm, a distributed paradigm, or hybridisation of the two paradigms to support the autonomic network management in the next generation networks. Future research works may include the following: continue to investigate on network structure approaches through a series of experiments and select the suitable choice of network structure paradigm, these experiments should extend the list of system requirements that has the efficiency to support autonomic network management; implement the DAIM model which can enable the autonomic management functionalities; and develop an intelligent hybrid Common Radio Resource Management approach to support the next generation wireless networks.

Acknowledgements. This work is sponsored by the Centre for Real-Time Information Networks (CRIN) in the Faculty of Engineering & Information Technology at the University of Technology, Sydney (UTS). This paper is an extended version of the ACASE'12 conference paper [21].

References

1. AL Sabbagh, A., Braun, R., Abolhasan, M.: A Comprehensive Survey on RAT Selection Algorithms for Heterogeneous Networks. Journal of World Academy of Science, Engineering and Technology (WASET) 73, 141–145 (2011)

2. Baran, P.: On Distributed Communications Networks. IEEE Transactions of Communications Systems 12(1), 1–9 (1964)
3. Bays, L.R., Marcon, D.S.: Flow Based Load Balancing: Optimizing Web Servers Resource Utilization. Journal of Applied Computing Research 1(2), 76–83 (2011)
4. Birman, K.P.: Reliable Distributed Systems: Technologies, Web Services, and Applications. Springer-Verlag, New York Inc. (2005)
5. Braun, R., Chiang, F.: A distributed Active Information Model Enabling distributed Autonomics in Complex Electronic Environments. In: Third International Conference Broadband Communications, Information Technology & Biomedical Application, pp. 473–479. IEEE (November 2008)
6. Chiang, F., Braun, R.: Self-adaptability and vulnerability assessment of secure autonomic communication networks. In: Proc. of Managing Next Generation Networks and Services Conf., pp. 112–122 (2007)
7. Chiang, F., Braun, R.: Towards a management paradigm with a constrained benchmark for autonomic communications. In: Proceedings of the Conference Computational Intelligence and Security, pp. 250–258 (2007)
8. Chiang, F., Mahadevan, V.: Towards the distributed autonomy in complex environments. In: International Conference on Information and Multimedia Technology, pp. 169–172 (2009)
9. Chiang, F., Fernandez, H., Braun, R., Agbinya, J.: Integrating Object-Oriented O: XML Semantics into Autonomic Decentralised Functionalities. In: 7th International Symposium on Communications and Information Technologies, pp. 768–773 (2007)
10. DenHartog, M.: How to Read and Understand the SNMP MIB. DPS Telecom (2008), http://www.dpstele.com/white-papers/snmp-mib/offer.php (viewed September 25, 2011)
11. Escribano, J.F.: Grid and Mesh Technologies, IT Entrepreneurship (2008), http://www.econectados.com/wp-content/uploads/grid_mesh.pdf (viewed April 30, 2012)
12. Github: NOX Introduction. noxrepo Tech. Rep. (2008), https://github.com/noxrepo/nox-classic/wiki/NOX-Introduction (viewed October 20, 2011)
13. Google Code University: Introduction to Distributed System Design, http://code.google.com/edu/parallel/dsd-tutorial.html
14. IBM: TXSeries for Multiplatforms: Concepts and Planning, 5th edn. International Business Machines Corporation (November 2005)
15. Manjula, K., Karthikeyan, P.: Distributed Computing Approaches for Scalability and high performance. In: Conference Proceedings of Distributed Computing, vol. 2, pp. 2328–2336 (2010)
16. Meyer, K., Erlinger, M., Betser, J., Sunshine, C., Goldszmidt, G., Yemini, Y.: Decentralizing Control and Intelligence in network Management. In: Proceedings of the 4th International Symposium on Integrated Network Management, Santa Barbara, CA (May 1995)
17. Morkved, T.: Peer-to-Peer Programming with Wireless Devices. Information and Communication Technology thesis, University of New South Wales, Sydney Australia & Agder University College, Grimstad Norway (2005)
18. NathRK, P.: Internet Technology with Client Server Architecture (2010), http://www.data-e-education.com/ E079_Centralized_Network_Architecture.html (viewed April 17, 2012)
19. Open Networking Foundation: Software-Defined Networking: The New Norm for Networks. ONF White Paper (April 2012)

20. Perez-Romero, J., Sallent, O., Agusti, R., Diaz-Guerra, M.A.: Radio Resource Management Strategies in UMTS. John Wiley & Sons Ltd., Chichester (2005)
21. Pupatwibul, P., AL Sabbagh, A., Banjar, A., Braun, R.: Distributed Systems in Next Generation Networks. In: Conference Proceedings of 1st Australian Conference on the Applications of Systems Engineering, ACASE 2012, p. 32 (February 2012)
22. Rodriguez, H.F.: Active MIB, An Object Oriented Solution For Network Management. Master thesis, Chalmers University of Technology, Sweden (May 2007)
23. Tootoonchian, A., Ganjali, Y.: HyperFlow: A Distributed Control Plane for OpenFlow. In: Proc. of INM/WREN, San Jose, CA (April 2010)

Chapter 15
The Study of the OFDM and MIMO-OFDM Networks Compatibility – Measurements and Simulations

Michał Kowal, Ryszard J. Zieliński, and Zenon Chaczko

Abstract. The article presents the results of measurements and simulations of intrasystem compatibility of the wireless networks operated in accordance with technical documentation IEEE 802.11g and IEEE 802.11n. Simulations were carried out using an advanced model of the MIMO-OFDM system. The simulations have been preceded by measurements in the anechoic chamber. The results of these measurements were used to define the input parameters of the simulator. The results of analyzes confirmed the usefulness of presented MIMO-OFDM system simulator to performance prediction of the wireless networks in the absence and presence of interference from other networks.

15.1 Introduction

The beginning of the twenty-first century is facing to the rapid development of wireless local area networks (WLAN). Now it can be said, that there is a time of information society. Everybody uses mobile phones, emails, instant messaging and other form of telecommunications. The people would like to be connected everywhere and all the time. The manufacturers of devices offer preconfigured devices, which are able to operate just after switching them on. The process of creating wireless network is very easy and it does not require knowledge about wireless network – just plug and play. The market is dominated by devices constructed with accordance to the specification IEEE 802.11n [3], but in the neighborhood there is a lot of wireless networks operating in accordance to the older documentation IEEE 802.11g [2]. As

Michał Kowal · Ryszard J. Zieliński
Wroclaw University of Technology, Wroclaw, Poland
e-mail: {michal.kowal,ryszard.zielinsk}@pwr.wroc.pl

Zenon Chaczko
University of Technology, Sydney, Australia
e-mail: zenon.chaczko@uts.edu.au

R. Klempous et al. (eds.), *Advanced Methods and Applications in Computational Intelligence*,   335
Topics in Intelligent Engineering and Informatics 6,
DOI: 10.1007/978-3-319-01436-4_15, © Springer International Publishing Switzerland 2014

we know, between systems, "g" and "n" there are number of similarities and differences. Both systems operate at 2.4GHz. The "n" standard can work well in the 5GHz band as "a" standard. There is a big difference in the maximum bit rate. For the "g" it is 54Mbps and for the "n" with four transmitting and receiving antennas up to 600Mbps. In the "g" standard occurs only one spatial stream and in the "n" standard there can be up to four streams. The bandwidth occupied by the radio channel in both systems is 22MHz, but the bandwidth in the "n" standard can be also doubled. As was mentioned, the older and newer networks work in the same ISM (Industrial Scientific Medical) band. The free operational channels not exist. This may cause the compatibility problems. The cooperation of these networks in the same area was subject of preliminary studies and very interesting results of this work have been already published [5]. The compatibility problems observed during the experiments are good starting point for the further research. For more detailed analysis of this phenomenon it is necessary to control all elements of the telecommunication system. Such possibilities gives the system simulator, which usefulness in predicting the performance of the wireless networks has been demonstrated in this paper.

15.2   The MIMO Technology

Systems with more than one antenna are known from many years. Systems with the transmitter with one antenna and the receiver with more than one antenna are called SIMO (Single Input Multiple Output). The signal from the transmitter is sent through the propagation channel to the receiver. On a path between ends of transmission there are a lot of obstacles, that can change an amplitude and a phase of the signal. The receiver uses two or more antennas to receive transmission. The antennas are spaced from each other to prevent correlation of signals. Signals from different antennas can be used for the detection and correction of the transmission errors. The systems with more than one transmitting antenna and one receiving antenna, called MISO (Multiple Input Single Output), can be rarely found. The equipment supporting the IEEE 802.11n specification offers significantly faster data transfer (theoretically up to 640Mbps) and the greater range. These systems, in addition to the multiplexing OFDM (Orthogonal Frequency Division Multiplexing) use MIMO (Multiple Input, Multiple Output) technology. The MIMO systems combine the advantages of SIMO and MISO systems. To understand how this technology works, one can consider a system with two transmitting antennas and two receiving antennas (See Fig.15.1).

It shall be assumed that a channel is constant during the transmission and their transmittance does not change. The channel can be described by:

$$\hat{y}_1 = \hat{h}_{11}x_1 + \hat{h}_{12}x_2 + w_1$$
$$\hat{y}_2 = \hat{h}_{21}x_1 + \hat{h}_{22}x_2 + w_2$$

(15.1)

where $\hat{x}$ denotes the transmitted signal, $\hat{y}$ – the received signal and $\hat{w}$ – the white Gaussian noise. In the matrix notation the equations (15.1) can be rewritten as:

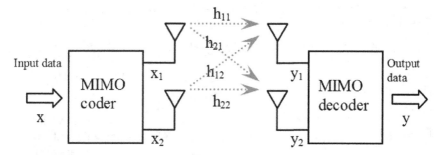

Fig. 15.1. MIMO system with two transmitting and two receiving antennas (2,2)

$$\hat{Y} = \hat{H} \cdot \hat{X} + W \tag{15.2}$$

The H matrix is defined as:

$$\hat{H} = \begin{bmatrix} \hat{h}_{11} & \hat{h}_{12} \\ \hat{h}_{21} & \hat{h}_{22} \end{bmatrix} \tag{15.3}$$

The equation (15.1) can be considered as the system of two equations with two unknowns. This system has unambiguous solution if there is no correlation between elements of the H matrix. The signal on the path between the transmitter and the the receiver can be refracted, diffracted and reflected many times. As a result of this - signals, that are received by different antennas, are different. If the received signals are not correlated, there are two independent radio channels between two end points of the transmission. Of course in the system might be more than two antennas. So, if the condition of orthogonality is met, then the path between the transmitter and the receiver will be independent.

15.3   The Measurement Procedure

A lot of performance measurements of the wireless network operated in accordance with the recommendation IEEE 802.11g and IEEE 802.11n has been done [9]. The measurements were carried out in an anechoic chamber to eliminate the influence of other random wireless networks. This kind of measurement set-up was chosen because it is very difficult to find the propagation environment free of other operating wireless networks. In the anechoic chamber two wireless networks were built: one victim (interfered) and one disturbing network (interfering). The measurements were carried out for the two scenarios. The set-ups for both scenarios were presented in Fig.15.2 and Fig.15.3.

Two access points were used to build each network. One of them worked in the default configuration as an regular access point, while the second was used as a wireless client. The location of the equipment used to build the victim network (AP_2,

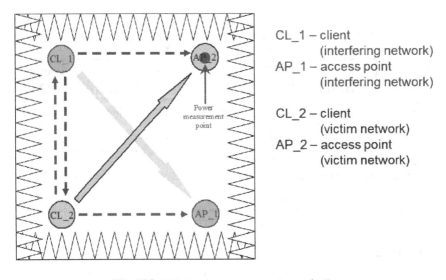

Fig. 15.2. Measurement test set-up (scenario 1)

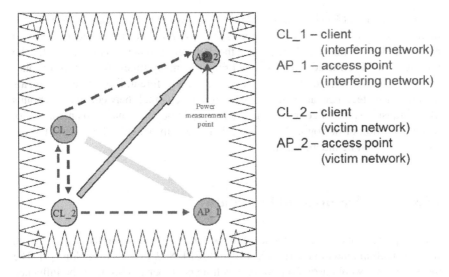

Fig. 15.3. Measurement test set-up (scenario 2)

Client_2) was fixed during the measurement period. The location of the access point (AP_1) of the interfering network was unchanged, while the location of the interfering client network (Client_1) was changed. The different environmental conditions were simulated by changing the distance between the victim and interfering stations.

In scenario 1 devices were placed in the anechoic chamber in such a way which allows to achieve the same signal field strength near the device AP_2. This refers to the situation where two independent wireless networks work in the same area and their signal field strength at the victim receiver is equal to each other. To ensure that this conditions have been met, the signal field strength measurement were done by the spectrum analyzer. The measured signal field strength was equal to around -67dBm (Fig.15.4).

In scenario 2 (Fig.15.3) the CL_1 device was moved to create situation when the signal of the interfering network is smaller than the signal of the victim network. As in scenario 1 the signal field strength was measured by spectrum analyzer (Fig.15.5), the signal field strength of victim network was around -67dBm, when the signal of interfering network was around -77dBm. This kind of interfering scenario is more frequently observed in real environment, where the signal field strength of interfering network is smaller than signal field strength of victim network.

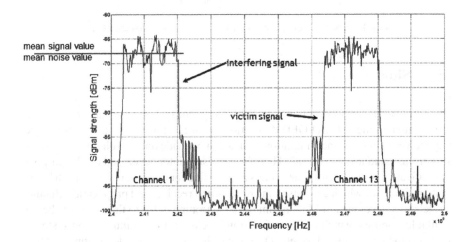

Fig. 15.4. Signal field strength during measurements (scenario 1)

For all transmission configurations in both networks the transmission took place in the direction from the client device (transmitter) to the AP (receiver). During the measurement the Iperf software was used and the transmission speed of the victim network was observed. In each configuration of the network the operation channel of the victim network was set to 13th, while the operation channel of the interfering network was changed from the 1st channel to the 12th channel. The duration of each measurement was set to 30 seconds, after which the results were averaged. The maximum output power of devices during tests was 1 mW for all configurations. For each configuration, the signal levels of both networks were recorded by the spectrum analyzer (Fig.15.4 and Fig.15.5). This results were used in MIMO-OFDM system simulator to meet the same conditions as in the anechoic chamber during measurements.

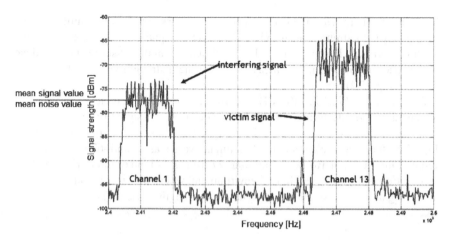

Fig. 15.5. Signal field strength during measurements (scenario 2)

15.4   Simulations

The simulator is used to simulate not only compliant systems, "g" and "n". We can also simulate other MIMO-OFDM solutions with it. However, it is worth to note that the verification of the simulation results can be made only in relation to existing systems as standard "g" and "n". Therefore, the simulator developed for the purpose of verification uses the same parameters as specified for the systems "g" and "n". The system simulator which consists of the transmitter, receiver and propagation channel models was developed in Matlab Simulink environment due to the large number of available libraries, which contains function blocks of telecommunication systems, such as modulators, encoders, etc. During developing of the model, it appeared that the set of these blocks is insufficient, therefore a lot of additional function blocks were created. Simulations can be performed in the presence or in the absence of the interference from other systems as well. The MIMO-OFDM system model consist of the autonomic transmitter, receiver and propagation channel model. This model is able to perform simulations with data rates from 6.5 to 540Mbps, depending on the chosen MCS (Modulation and Coding Scheme) and the structure of the propagation model of the radio channel. The receiver's model is able to perform synchronization and to make channel estimation. The model of the propagation channel contains among others the set of TGn models of the propagation sub-channels [1] which were used during work on the technical documentation of IEEE 802.11n. Each of the 6 models of the channel identified by the letters from A to F has different number of clusters and delays. For each type of the channel it is possible to set the distance between the antennas at the transmitter and receiver, which shall be used during designated correlation matrices. The path loss was added to the model of the channel to make possibility of determining the maximum range and system throughput in the simulations. It is worth noting that presented simulator is modelling transmission in the radio frequency domain, when the most of simulator described in the literature are modelling transmission in baseband [6, 7, 8]. This new approach allows to model the propagation phenomena occurring in the RF channel and the direct comparison of the simulation results with measurements of the real systems. In the Fig. 15.6 the signal spectrum measured in the real environment and the signal observed at the simulated RF output of the transmitter is shown - both signals are very similar.

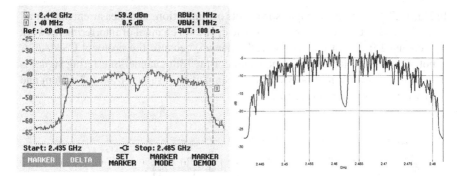

Fig. 15.6. The signal spectrum measured in the real environment (on the left) and the signal observed at the simulated RF output of the transmitter (on the right)

The presented simulator was used to determine the wireless networks performance for both scenarios which were described in details in the Chapter 3 and in [4]. The simulator has been enhanced with additional functionality, the ability to simulate the transmission in the presence of interferences. The simulation of the other networks presence was implemented in two ways. The interference may be generated as the Gaussian noise or as the signal of another wireless network. In the first case the unwanted signal was the Gaussian noise with average equal to zero and variance depended on the required noise power. The noise power was calculated for signal of 20MHz bandwidth. In the next step, depending on the level of overlapping operating channels of victim and interfering network, the relevant part of the interfering signal (Gaussian noise) was added to the victim signal in the channel model. In the second case, instead of the Gaussian noise, another previously recorded signal of the wireless network was used. The method of calculating the interfering signal power was analogous as in the first case. Comparison of simulation results for two kinds of interfering signals with the measurement results give the answer for question: which method gives results more correlated with the measurement results. Simulations were performed for the same input parameters as the measurements were carried out. In the simulation environment the power of wireless networks signals were set in accordance to signal levels values measured during the measurement using a spectrum analyzer and presented in Fig.15.4 and Fig.15.5. For each operational channel of the interfering network simulation was carried out five times and then the results were averaged. Simulations for IEEE 802.11n network were done for the 2x2 MIMO system and 20MHz channel bandwidth. Both, the simulator and real devices during measurements were operated in backward compatibility mode ("mixed mode").

15.5   The Results of the Simulations vs. Measurements

In Fig. 15.7 ÷ 15.10 the comparison results of simulations and measurements for all scenarios and networks are presented. The operation channel of interfering network was changed from 1st channel to 12th channel, because the protocol CSMA/CA is not yet fully implemented in the simulator (at 13th channel interfering signal is treated as an indication of busy channel). Simulation results for the "g" system and scenario 1 can be seen in Fig. 15.7.

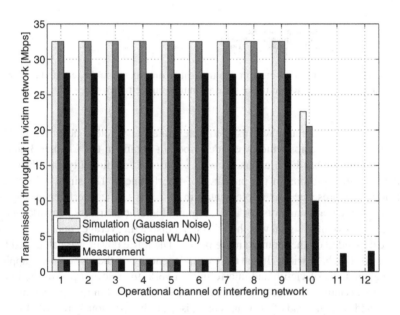

Fig. 15.7. Transmission throughput in victim network (IEEE 802.11g) – scenario 1

Color light grey was selected for the simulation results obtained when the interfering signal was the noise, dark grey color when the interference is the signal coming from another OFDM system, and the black color for the measurement results. It can be observed that the signal from the interfering channel of the number up to 9 does not affect the signal of the OFDM system. The transmission speed achieved in simulations is equal to the maximum theoretical IEEE 802.11g throughput in the absence of the interference. Only when the interfering signal is at 10, 11 or 12th channel, the throughput decreases. The results obtained for the scenario 2

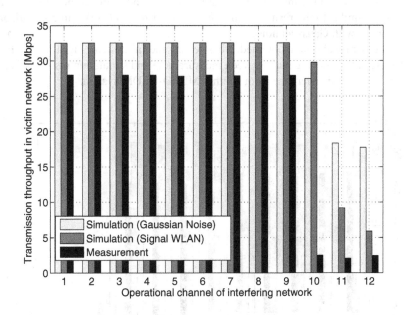

Fig. 15.8. Transmission throughput in victim network (IEEE 802.11g) – scenario 2

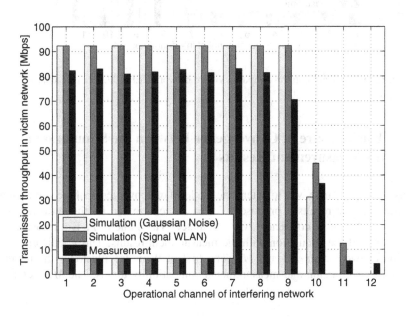

Fig. 15.9. Transmission throughput in victim network (IEEE 802.11n) – scenario 1

(Fig. 15.8) were too optimistic, especially in the case when as an interfering signal Gaussian noise was used. Generally for IEEE 802.11g networks simulation results for the variant when the interfering signal was the another wireless network signal were more similar to the results of measurements of real networks, in comparison to the variant with Gaussian noise.

The similar results were obtained, when testing the "n" system (Fig. 15.9 and Fig 15.10). Here, the convergence between measurement and simulation results is even greater for signal interference from another MIMO-OFDM system.

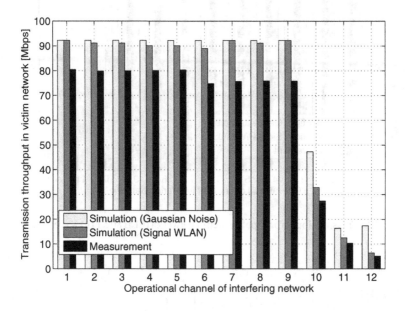

Fig. 15.10. Transmission throughput in victim network (IEEE 802.11n) – scenario 2

15.6 The Measure of Convergence between the Simulation and the Measurement Results

In this section a simple way to determine the level of convergence of the simulation and measurement results is proposed. For each scenario and for the networks "g" and "n" a value of M coefficient was calculated. The value of M is a measure of convergence of the simulation with the measurement results. This coefficient is a function of two factors $M1$ and $M2$. The first of them $M1$ illustrates the results obtained by simulation and measurements compatibility trends. The $M1$ factor is defined as cross correlation coefficient:

$$M1 = \frac{\sigma_{xy}}{\sigma_x \sigma_y} \qquad (15.4)$$

The standard deviation of the throughput achieved in simulator σ_x and in measurements σ_y were calculated in accordance with the equation (15.5) and (15.6):

$$\sigma_x = \sqrt{\frac{1}{n}\sum_{i=1}^{n}\left(R_{b\_simulated}(i) - \frac{1}{n}\sum_{k=1}^{n}R_{b\_simulated}(k)\right)^2} \qquad (15.5)$$

$$\sigma_y = \sqrt{\frac{1}{n}\sum_{i=1}^{n}\left(R_{b\_measured}(i) - \frac{1}{n}\sum_{k=1}^{n}R_{b\_measured}(k)\right)^2} \qquad (15.6)$$

where:

- $R_{b\_simulated}(i)$ is the transmission bit rate achieved in simulator for the measurement No. i,
- $R_{(b\_measured)}(i)$ is the measured transmission bit rate for the measurement No. i,
- n - number of measurements.

The cross-covariance between simulated and measured transmission bit rate was calculated as follows:

$$\sigma_{xy} = \sum_{i=1}^{n}\left[\left(R_{b\_simulated}(i) - \frac{1}{n}\sum_{k=1}^{n}R_{b\_simulated}(k)\right) \cdot \right. \\ \left. \cdot \left(R_{b\_measured}(i) - \frac{1}{n}\sum_{k=1}^{n}R_{b\_measured}(k)\right)\right] \qquad (15.7)$$

It takes values from 0 to 1, the larger value of cross-correlation the greater convergence of trends. The value of M2 factor was calculated as the average mean relative error (15.8).

$$M2 = 1 - \sum_{i=1}^{n}\frac{|R_{b\_simulated}(i) - R_{b\_measured}(i)|}{R_{b\_measured}(i)} \qquad (15.8)$$

The value of the coefficient M2 is normalized the same way as the cross-correlation coefficient and will accept value close to 1, when difference between simulated and measured throughput is small, and 0 if it is large.
Coefficient M is calculated as an average of two coefficients M1 and M2.

$$M = \frac{M1 + M2}{2} \qquad (15.9)$$

In Table 15.1 values of M coefficient for all considered scenarios and networks are presented.

It is easy to notice that the smallest values of M factor were obtained for scenario 2 for IEEE 802.11g network, which confirms the worst convergence of simulation and measurement results (significant differences for channels 10, 11, 12 - see Fig. 15.8). Originally simulator has been designed to match compatibility with

Table 15.1. Values of M coefficient for all considered scenarios and networks

	IEEE 802.11g		IEEE 802.11n	
	Scenario 1	Scenario 2	Scenario 1	Scenario 2
Gaussian noise	0.887	0.456	0.929	0.892
WLAN signal	0.897	0.574	0.920	0.956

IEEE 802.11n and therefore the results for these networks are very good. The IEEE 802.11g technical specification was added later. Obtained results of the simulation are too optimistic, therefore, works to implement this standard in the simulator will be continued.

15.7 Conclusion

The summary results of simulation and measurement confirms that it is possible to determine the impact of disturbances on the radio interface of the wireless networks using the developed simulator of MIMO-OFDM system. The greatest convergence between the results of measurement and simulation was obtained in case of IEEE 802.11n networks when the interfering signal was an another wireless network signal. Consistency of simulation results with measurements shows that the way of inserting noise into the channel is correct. The simulator can be used to predict coverage and performance of the wireless networks not only in the absence of interference from other networks, but also to take into account their impact. It is extremely important in the times of widespread availability of wireless networks.

References

1. Erceg, V., Schumacher, L., Kyritsi, P.: TGn Channel Models IEEE 802.11-03/940r4 (May 10, 2004)
2. IEEE Standard for Information technology - Telecommunications and information exchange between systems?Local and metropolitan area networks - Specific requirements Part 11: Wireless LAN Medium Access Control (MAC) and Physical Layer (PHY) Specifications Amendment 5: Enhancements for Higher Throughput (IEEE Std 802.11n?-2009)
3. IEEE Standard for Information technology. Telecommunications and information exchange between systems. Local and metropolitan area networks. Specific requirements Part 11: Wireless LAN Medium Access Control (MAC) and Physical Layer (PHY) specifications Amendment 4: Further Higher Data Rate Extension in the 2.4 GHz Band
4. Kowal, M.: The performance of the MIMO-OFDM radio interface in presence of interferences. - PhD. Thesis in polish, Wroclaw University of Technology (2011)
5. Kowal, M., Kubal, S., Zielinski, R.J.: EMC in 802.11n - selected aspects. Przeglad Telekomunikacyjny, Wiadomosci Telekomunikacyjne 81(4), 286–289 (2008) (in polish)

6. Panda, M., Patra, S.K.: Simulation Study of OFDM, COFDM and MIMO-OFDM System. Sensors & Transducers Journal 106(7), 123–133 (2009)

7. Xuehua, J., Peijiang, C.: Research and Simulation of MIMO-OFDM Wireless Communication System. nternational Forum on Information Technology and Applications (2009) 978-0-7695-3600-2/09 25.00 Cop. 2009 IEEE

8. Zhang, L., Zhao, Q.: Simulation and Analysis of MIMO-OFDM System Based On Simulink, 978-1-4244-8223-81101 Cop. 2010 IEEE

9. Zielinski, R.J., Kowal, M., Chaczko, Z.: Intrasystem compatibility of OFDM and MIMO-OFDM networks in measurements and simulations. In: 14th International Asia Pacific Conference on Computer Aided System Theory, Sydney, Australia, February 6-8 (2012)

Part III
Intelligent System Applications

Chapter 16
EMC between WIMAX 1.5GHz and WLAN 2.4GHz Systems Operating in the Same Area

Ryszard J. Zieliński, Michał Kowal, Sławomir Kubal, and Piotr Piotrowski

Abstract. The chapter presents the data rates measurements of WiMAX operating in 1.5GHz and WLAN 802.11g conducted in a reverberation chamber. The goal of the measurements was to perform a compatibility study between these systems working simultaneously. The motivation to perform the compatibility study was that these systems were chosen for operation in underground mining environment during research on telecommunication system for mine excavations. A short description of these measurments as well as the general concept of the above-mentioned systems are also shown. All tests were performed in the reverberation chamber which is widely used for electromagnetic compatibility studies. Beside the data rates measurement results, the chapter also presents a testbed, procedures for performance tests and a short decription of the reverberation chamber.

16.1 Introduction

In recent years one can observe a rapid development of technology and prevalence of electronic equipment, including transceivers (e.g. terminals and mobile base stations, wireless system terminals and access points). As a consequence, unwanted emissions of electromagnetic fields and strong interactions between devices and systems occur. This is observed especially in the radio systems, which are sources of both the desirable (in-band) as well as the unwanted (out-of-band) radio emissions. At the same time, they are prone to interferences from other systems, which limits their performance. These unwanted emissions may be a result of their shared operation in the same frequency band while working in the immediate vicinity of each other. Such adverse events can be easily observed in the 2.4GHz frequency

Ryszard J. Zieliński · Michał Kowal · Sławomir Kubal · Piotr Piotrowski
Wroclaw University of Technology, Wybrzeze Wyspianskiego 27, 50-370 Wroclaw
e-mail: {ryszard.zielinski,michal.kowal,slawomir.kubal,
    piotr.piotrowski}@pwr.wroc.pl

R. Klempous et al. (eds.), *Advanced Methods and Applications in Computational Intelligence*,   351
Topics in Intelligent Engineering and Informatics 6,
DOI: 10.1007/978-3-319-01436-4_16, © Springer International Publishing Switzerland 2014

band, which was designed for industrial, scientific and medical applications (ISM) and is commonly used by many wireless systems. Both in-band and out-of-band interference can significantly degrade transmission quality or, in some cases, make it impossible, especially when the effect of receiver blocking by a strong radio signal takes place. Immunity against the influence of external fields on electronic devices and reduction of electromagnetic emission are important elements of increasing the likelihood of smooth coexistence of all the devices and systems. This can be achieved through compliance with emission limits on radiated power levels (either in-band or out-of-band). The regulator (Administration) in each country sets the maximum levels for the equivalent radiated power in each spectrum band for radio transmission.

The problem of coexistence on the same area of systems based on IEEE 802.11 and IEEE802.16 standards is still up-to-date due to growing popularity of hybrid networks and wireless modules. This topic is widely discussed in literature, both as a possible source of interference and in the context of mutual coexistence methods. One can find many ways to avoid those problems, usually based on time scheme (Time Divison Operation based on Time Division Multiplexing), code scheme or power control. Also, dynamic frequency allocation using RSSI measurement or cognitive radio functionality are quite popular. Schemes mentioned above can be found among others in [1],[8],[9],[10],[19]. Nevertheless all of them are dedicated to IEEE802.11 systems working in 2.4GHz ISM band and IEEE802.16 in 2.3 and 2.5 GHz.

16.2   Wireless Communications in Mine Excavation

Why is the detailed investigation between WiMax 1.5GHz and WLAN 2.4GHz so important? To answer this question the results of the other project shall be presented. In years 2007 to 2010 the possibility of applying wireless systems to a build communications system for underground mine excavation was investigated. One of the objectives was to examine the propagation conditions in the underground corridors. Among the many system available on the market, four were selected for investigation: WiMax 1.5 GHz band (Airspan), WiMax 3.5 GHz band (ExcelMax from Axxcelera), WLAN 900 MHz band (SuperRange9 from Ubiquiti Networks) according to Spec. IEEE 802.11 b/g (bandwidth 5, 10 and 20 MHz) and WLAN 2.4 GHz (WRT350N with Notebook Adapter WPC300N from Linksys) according to Spec. IEEE 802.11 n. The typical WiMax throughputs are shown in Tab. 16.1. The WiMax system under test was operating with 3.5MHz bandwidth.

Poland is the fourth largest producer of copper in the world. All the propagation measurements were done in one of its mines at the depth of 600 m below the surface of the earth. The plan of the undergound mine coridors with two investigated areas (a long straight walkway and the corridors grid, which is a very difficult area in terms of propagation of the electromagnetic waves) is presented in Fig. 16.1. The investigations led to the conclusion, that the best results can be obtained with the use

Table 16.1. Typical throughputs of WiMax

Modulation / Coding	QPSK 1/2	QPSK 3/4	16QAM 1/2	16QAM 3/4	64QAM 2/3	64QAM 3/4
1.75MHz	1.45	2.18	2.91	4.36	5.82	6.55
3.5MHz	2.91	4.36	5.82	8.73	11.64	13.09
7MHz	5.82	8.73	11.64	17.45	23.27	26.18
14MHz	11.64	17.45	23.27	34.91	46.55	52.36
20MHz	16.26	24.40	32.53	48.79	65.05	73.19

of WiMax 1.5GHz and 2.4GHz WLAN. The topology of the underground wireless communication system is shown in Fig. 16.2. The system consists of a backbone based on the WiMax 1.5 GHz base stations (WBS), that are located approximately within 300 - 400m from each other. The WiMax terminals (WT) collaborate with these stations. Each terminal is directly connected to a WLAN access point (AP). These hotspots cooperate with the WLAN terminals (T).

Fig. 16.1 Plan of the investigated area underground with the two different marked areas

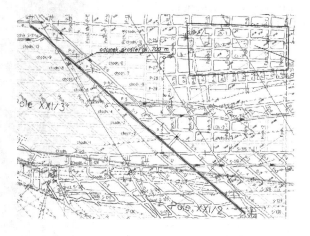

The proper operation of these systems requires locating the WiMax terminal close to the WLAN access point. Fig. 16.3. presents the WiMax terminal (at the top) with the WLAN access point attached (bottom). Both devices s hall operate in the same time and area, i.e. in adverse conditions, which was the reason for the EMC investigations.

It is very difficult to carry out EMC investigations underground. Therefore, another environment shall be used to simulate such adverse propagation conditions. Experience shows that similar conditions can be achieved in the reverberation chamber. Typically, it is used to test the sensitivity or emission of the equipment. But it can also be used to study properties of wireless systems in the extremely difficult propagation conditions and to test the compatibility between systems.

Fig. 16.2 Topology of wireless underground communications system (WiMax Base Station - WBS, WiMax Terminal - WT, WLAN Access Point - AP, WLAN Terminals - T)

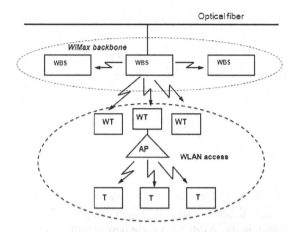

Fig. 16.3 The WiMax terminal (top) and WLAN AP (bottom) located close to each other on a single mast during the measurement in the mine excavation

16.3 Reverberation Chamber

The electromagnetic compatibility study in real conditions is not unambiguous because of the possible additional impact of other devices and systems. In addition, the selection and configuration of the test environment can significantly affect the obtained results. A reverberation chamber is a laboratory environment, where the statistical measurements of te radiated power of a radio equipment in an environment with multipath propagation can be performed. These chambers are now commonly used in EMC device testing (both immunity to electromagnetic fields tests and measurements of unwanted emissions). The chamber can also be used to test the electromagnetic compatibility of radio systems, wherein the propagation conditions in the interior of the chamber are extremely unfavorable due to the very large number of reflections of radio waves and the occurrence of resonances. The usefulness of the chamber for this type of testing is described i.e. in [2],[5],[17].

The reverberation chamber is a space limited by walls made of materials with very high conductivity. Radio waves radiated in this closed space, due to multiple reflections from walls and equipment, create an environment of three-dimensional standing wave. Typically, the chamber has a shape of a cuboid. For achieving the large number of modes with different resonant frequencies it is recommended, that chambers are constructed in this way that each of its linear dimensions was not a multiple of any of the other dimensions and all of three dimensions are of the same order.

During research a statistically homogeneous distribution of the electromagnetic field should be provided inside the chamber - in its test set. So, it is necessary to apply techniques to change its parameters, shape, location of walls or pieces of equipment to change directions of waves reflections and thereby change the field distribution. For this purpose, one or more stirrers are used, what is the most commonly technique applied in the chambers. The continuous stirrer rotation during the test causes the redistribution of field minima and maxima inside the chamber.

The stirrer (its shape, size, position in the chamber) has a big impact on getting adequate homogeneity of the field inside the chamber. It should be made, as well as walls, from material with high conductivity (e.g. aluminum). The stirrer should also be large relative to the size of the chamber and positioned in asymmetric way relative to the chamber walls. Linear dimensions of the chamber should be large enough to provide adequately low the Lowest Useable Frequency (LUF) and also to ensure free-hold in the interior of the test equipment, antennas, field probes and metal stirrers.

The typical capacity of the reverberation chambers ranges from 70 to 100m^3, therefore, the lowest useable frequency is about 200MHz. Measurements below 200MHz require chambers with larger than the typical linear dimensions. The resonant frequency depends on the dimensions of the chamber and for the rectangular-shaped can be determined from the following relationship:

$$f_{ijk} = \frac{c_0}{2}\sqrt{\left(\frac{i}{l}\right)^2 + \left(\frac{j}{w}\right)^2 + \left(\frac{k}{h}\right)^2}$$ (16.1)

where

- l,w,h - chamber dimensions [m] (for chamber used: l - length 7.76m, w - width 4.3m, h - hight 3.05m);
- i,j,k - integer constans;
- c_0 - wave velocity in chamber.

The lowest useable frequency of the chamber is determined as approximately triple value of the first resonance frequency (f_{001}). Its exact value also depends on the stirrer operation and the quality factor Q of the chamber.

The field strength present in the reverberation chamber is greater than in other test beds, which is related to the effect of temporary energy accumulation, which depends on the Q factor. The quality of the chamber is also dependent on the frequency and the test equipment. Therefore the loaded chamber - with additional test equipment and the device under test - has greater homogeneity of the field, but lower levels of the induced field due to the lower value of the Q factor. The quality factor Q determines the ability of the chamber to accumulate an energy. It is determined by the losses introduced directly by the chamber. These losses depend on the material properties of the walls, the floor and the ceiling and on the quality of the connection of individual elements of the chamber. Additional equipment such as antennas, measuring equipment, peripherals and the device under test can also affect the value of the Q factor [7],[15],[18]. For sufficiently high frequencies, this factor could be calculated as:

$$Q = \frac{3}{2} \frac{V}{S\sigma} \tag{16.2}$$

where:

- V - the chamber capacity [m^3];
- S - the total area of internal walls [m^2];
- σ - the penetration depth of the wave [m] - it depends on electrical parameters of the walls.

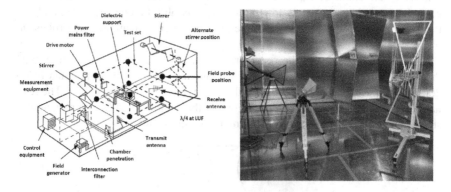

Fig. 16.4. Reverberation chamber

Due to the statistical nature of the electromagnetic environment in the interior, the reverberation chambers are increasingly used in measurements of the transmission efficacy of the radio systems and the immunity to radio signals emitted by other radio communication systems [3],[13],[16]. However, it is necessary to reduce the quality factor of the chamber by placing inside it elements absorbing the electromagnetic energy. These elements reduce the effect of accumulation of the energy and change the Power Delay Profile (PDP) [6],[14]. So, during the tests of the wireless systems, it reduces the possibility of the receiver blocking by too strong signals and excessive inter-symbols errors. The scheme of the reverberation chamber and antennas placed inside it were shown in Fig. 16.4.

16.4   Systems under Tests

Based on the preliminary research conducted in the hallways of buildings of the Wroclaw Technical University and then during the test in real conditions of the excavation mine [12] the two systems has been chosen for testing in the reverberation chamber. One of them was Airspan's WiMAX system, operating in the 1.5GHz band, and the second one was WLAN operating in 2.4GHz band. Based on the architecture shown in [11] the devices of these two systems are placed next to each other. Therefore, to achive proper operation of both systems they should not interact with each other. Therefore the transmission properties of both systems were tested in the same conditions, by sending data streams between each pairs of computers connected via tested system and measuring the transmission rates. The automatic settings option were chosen in both systems during the tests, allowing to dynamically adjust the transmission parameters of both systems to the conditions of propagation.

Table 16.2. Parameters of the tested systems [21]

	Airspan BS	Airspan CPE	802.11g
Freq. Band	1426.5-1524MHz	1426.5-1524MHz	2400MHz
Air Interface	Adaptive TDMA	Adaptive TDMA	CSMA/CA
Architecture	Point to Multipoint	Point to Multipoint	Point to Multipoint
Duplex	TDD	TDD	N/A
RF Channel Sizes	5MHz, 3.5MHz, 1.75MHz	5MHz, 3.5MHz, 1.75MHz	20MHz
Modulation	64QAM, 16QAM, QPSK, BPSK	64QAM, 16QAM, QPSK, BPSK	64QAM, 16QAM, QPSK, BPSK
Coding Rates	1/2, 2/3, 3/4	1/2, 2/3, 3/4	1/2, 2/3, 3/4
Transmit Power	+27dBm	+24dBm	+17dBm
Receive Sensitivity	-104dBm @ 1.75MHz, -100dBm	-104dBm @ 1.75MHz, -100dBm	-70dBm for 54Mb/s
Antenna Beam Width	60°	Azimuth - 60°, Elevation - 30°	N/A
Antenna Gain	10.5dBi	10.5dBi	9dBi

16.5   Testbed

The results of the measurements are highly dependent on the chambers load (additional equipment). So, measurements of properties of each system in the reverberation chamber were performed for the case of the simultaneous installation of both systems in the chamber. The tests performed for each system with other devices turned off were a reference to compatibility tests of both systems operating. The study was performed for several configurations, changing the orientation and antenna settings and the number and location of the absorbers.

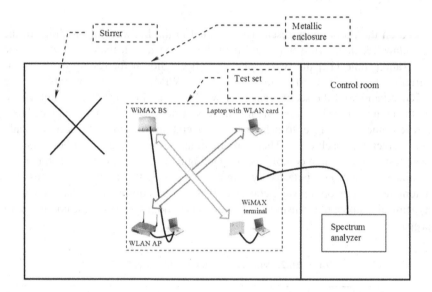

Fig. 16.5. Testbed in reverberation chamber

The observation of the interaction between WLAN and WiMAX systems was limited to record the transmission rate changes in both systems during their simultaneous transmission. Both systems operate in different frequency bands (Airspan WiMAX in the 1.5GHz band and the WLAN system in the 2.4GHz band). So, their simultaneous work should not degrade the quality of transmission, unless the out of band emissions of transmitters of both systems will be sufficiently low - acceptable by each of the systems. Fig. 16.5 shows the measurement site in reverberation chamber. Three laptops were used for testing, two of them were sources of data transmitted by the terminals of each of the systems (Iperf clients). The third one with Iperf server was the recipient of two streams of data. The measurements were provided for four different set-up configurations:

1. WiMAX terminal and WLAN AP next to each other;
2. WiMAX terminal and WLAN AP separated by absorber;
3. WiMAX terminal on the floor with one absorber;
4. WiMAX terminal on the floor with two absorbers.

The WiMAX base station in all configurations was covered with two absorbers. The WLAN AP with the antenna were placed on the small dielectric table during all measurements. Only the position and the radiation direction of the WiMAX terminal were changed. All configurations of the test set-up were shown in Fig. 16.6. The change of the devices placement (and adding additional absorbers) was only supposed to change the electromagnetic field distribution in the chamber.

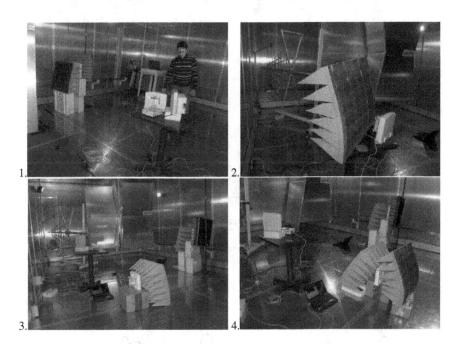

Fig. 16.6. Configurations of the testbed

16.6   Measurement Results

The study was conducted for four scenarios associated with the different placement of both devices of the systems in the chamber and with different number and placement of absorbers. The data transfer rate for each system was monitored every second at a constant slow stirrer circulation from 0 to 360 degrees [20].

16.6.1　Reference Measurements

In order to get an overview of systems performance some reference measurement were conducted. Both the placement of devices as well as the testing procedure were exactly the same as in the first scenario. However during WiMAX performance test the WLAN AP was turned off and accordingly, test of the WLAN was performed while WiMAX devices were not operating. As shown in Fig. 16.7 and Fig. 16.8 both systems were able to transmit data for almost entire stirrer rotation. The average transmission rate for WiMAX was about 0.9Mb/s whilst for WLAN it was about 9.3Mb/s. One have to bear in mind, that reverberation chamber is extremly strict environment. That is why abolute values of trasmission rates are less meaningful.

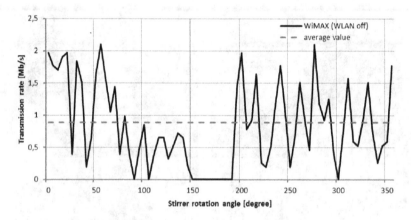

Fig. 16.7. The results of the reference measurements for WIMAX

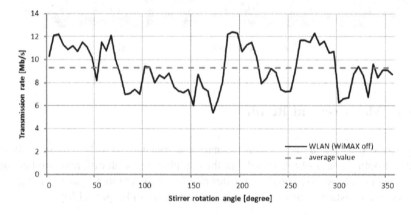

Fig. 16.8. The results of the reference measurements for WLAN

16.6.2   Measurement Results for WiMAX System

The results of average data rate for each scenario for WiMAX measurements are shown in Tab. 16.3 . The data rate for each scenario as a function of the stirrer position is shown in Fig. 16.9 and Fig. 16.10. Taking into account the reference results one can say that operation of the WiMax system was not disturbed by WLAN. The achieved average transmission rates for all configurations were very similar to each other as well as to reference value. The high variance of transmission rates was not caused by the interfering system but by the changes of EM field distribution in the chamber.

Table 16.3. Averaged measurement results for WiMAX

Average transmission rate [Mb/s]			
Configuration 1	Configuration 2	Configuration 3	Configuration 4
0.966	0.880	0.815	0.869

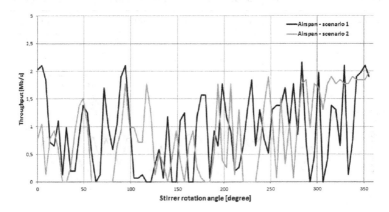

Fig. 16.9. The measurement results of WiMax for the first two scenarios

16.6.3   Measurement Results for WLAN System

The results of average data rate measurements for each scenario for WLAN are shown in Tab. 16.4. The data rate for each scenario as a function of the stirrer position is shown in Fig. 16.11 and Fig. 16.12. As in the earlier case the operation of the system was not disturbed by WiMAX and high variance of transmission rates was not caused by interfering system but by changes of EM field distribution in the chamber. The achieved average transmission rates for all configurations were similar

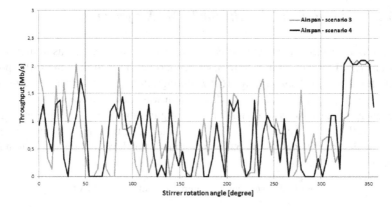

Fig. 16.10. The measurement results of WiMax for the scenarios 3 and 4

Table 16.4. Averaged measurement results for WLAN

Average transmission rate [Mb/s]			
Configuration 1	Configuration 2	Configuration 3	Configuration 4
8.45	10.60	10.71	13.11

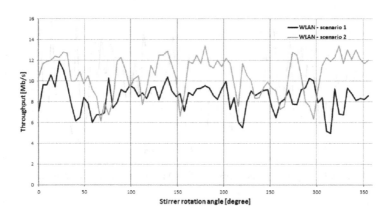

Fig. 16.11. The measurement results of WLAN for the first two scenarios

to each other as well as to reference value. A better performance for Configuration 4 could be caused by degradation of chamber's Q factor. It was a result of increasing the chamber's load by adding two additional absorbers. The detailed explanation of above mentioned phenomena can be found in [15].

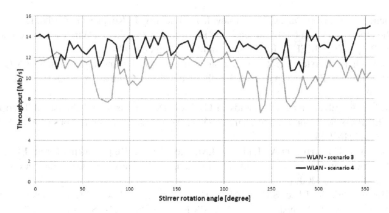

Fig. 16.12. The measurement results of WLAN for the scenarios 3 and 4

16.7   Summary

The propagation environment in the reverberation chamber is very demanding for wireless data transmission systems due to the very large number of wave reflections from the walls of the chamber. Such an extremely difficult environment makes possible to check the efficiency of transmission techniques and technical solutions of radio equipment. Only systems with above-average properties can operate in such conditions. All the tests conducted in the chamber confirmed the excellent transmission properties of the WiMAX system in the hard propagation conditions, which emerged during the test in the mine environment. Although the achieved transmission rate differ significantly from the maximum (about 11Mb/s) it is essential that the system enables transmission of the data for each configuration and each measurement scenario. The operating WLAN system had no apparent effect on the work of the WiMAX system and vice versa. In any configuration the WiMAX system achieved similar results, confirming the good electromagnetic compatibility between these systems.

References

1. Andreev, S., Dubkov, K., Turlikov, A.: IEEE 802.11 and 802.16 Cooperation Within Multi-Radio Stations. Wireless Personal Communications 58, 525–543 (2011) ISSN: 0929-6212, doi:10.1007/s11277-010-0134-1
2. Backstrom, M., Lunden, O., Kildal, P.S.: Reverberation Chambers for EMC Susceptibility and Emission Analyses. Review of Radio Science 1999-2002, 429–452 (2002)
3. Belhouji, A., Decroze, C., Carsenat, D., Mouhamadou, M., Reynaud, S., Monediere, T.: A MIMO WiMAX-OFDM based system measurements in real environments. In: Antennas and Propagation, EuCAP 2009, pp. 1106–1109 (2009) E-ISBN: 978-3-00-024573-2

4. Genender, E., Holloway, C.L., Remley, K.A., Ladbury, J., Koepke, G., Garbe, H.: Use of Reverberation Chamber to Simulate the Power Delay Profile of a Wireless Environment. In: Proceedings IEEE EMC Europe, Hamburg, September 8-12 (2008)
5. Heinrich, R., Karsten, U.: Reverberation chambers - Design and application for EMC. In: Proc. 17th Int. Zurich Symp. Tech. Exhib. Electromagn. Compat., pp. 190–193 (February 2006)
6. Hill, D.A.: Electromagnetic Fields in Cavities: Deterministic and Statistical Theories. IEEE Press Series on Electromagnetic Wave Theory. John Wiley & Sons (2009) ISBN: 978-0470465905
7. Holloway, C.L., Hill, D.A., Ladbury, J.M., Koepke, G.: Requirements for an Effective Reverberation Chamber: Unloaded or Loaded. IEEE Trans. on Electromagnetic Compatibility 48(1) (2006)
8. Kim, J., Park, S., Rhee, S.H., Choi, Y.-H., Chung, Y.-U., Hwang, H.Y.: Coexistence of WiFi and WiMAX Systems Based on PS-Request Protocols. Sensors 11, 9701–9716 (2011)
9. Kim, J., Park, S., Rhee, S.H., Choi, Y.-H., Hwang, H.: Energy efficient coexistence of wiFi and wiMAX systems sharing frequency band. In: Kim, T.-h., Lee, Y.-h., Kang, B.-H., Ślęzak, D. (eds.) FGIT 2010. LNCS, vol. 6485, pp. 164–170. Springer, Heidelberg (2010)
10. Kondo, T., Fujita, H., Yoshida, M., Saito, T.: Technology for WiFi/Bluetooth and WiMAX Coexistence. Fujitsu Sci. Tech. J. 46(1), 72–78 (2010)
11. Kowal, M., Kubal, S., Piotrowski, P., Zielinski, R.: Real time people location and VOIP services efficiency in underground mining environment. In: Grzech, A., et al. (eds.) Information Systems Architecture and Technology: Service Oriented Distributed Systems: Concepts and Infrastructure, pp. 337–346 (2009) ISBN 978-83-7493-477-0
12. Kowal, M., Kubal, S., Piotrowski, P., Zielinski, R.: Operational characterictic of wireless WiMax and IEEE 802.11x systems in under-ground mine environments. Electronics and Telecommunications Quarterly (International Journal of Electronics and Telecommunications), 81–86 (2010) ISSN 0867-6747
13. Olano, N., Orlenius, C., Ishimiya, K., Ying, Z.: WLAN MIMO throughput test in reverberation chamber. In: Antennas and Propagation Society International Symposium, AP-S 2008, pp. 1–4 (2008) E-ISBN: 978-1-4244-2042-1
14. Orlenius, C., Franzen, M., Kildal, P.S., Carlberg, U.: Investigation of Heavily Loaded Reverberation Chamber for Testing of Wideband Wireless Units. In: IEEE International Symposium on Antenna Propagation, pp. 3567–3572 (2006)
15. Pomianek, A., Staniec, K., Joskiewicz, Z.: Practical remarks on measurement and simulation methods to emulate the wireless channel in the reverberation chamber. Progress in Electromagnetic Research 105, 49–69 (2010)
16. Primiani, V.M., Moglie, F., Recanatini, R.: On the use of a reverberation chamber to test the performance and the immunity of a WLAN system. In: Electromagnetic Compatibility (EMC 2010), pp. 668–673 (2010) ISSN: 2158-110X
17. Rajamani, V., Bunting, C., Freyer, G.: Why consider EMC testing in a reverberation chamber. In: Electromagnetic Interference & Compatibility, INCEMIC 2008, pp. 303–308 (2008) ISBN: 978-81-903575-1-7
18. Staniec, K., Pomianek, A.: On Simulating the Radio Signal Propagation in the Reverberation Chamber with the Ray Launching Method. Progress in Electromagnetics Research B 27, 83–99 (2011)

19. Thomas, N.J., Willis, M.J., Craig, K.H.: Analysis of Co-existence between IEEE 802.11 and IEEE 802.16 Systems. In: 2006 3rd Annual IEEE Communications Society on Sensor and Ad Hoc Communications and Networks, SECON 2006, September 28, vol. 2, pp. 615–620 (2006)
20. Zielinski, R.J., Kowal, M., Kubal, S., Piotrowski, P.: Compatibility study between WiMax 1.5GHz and WLAN systems - selected aspects. In: 14th International Asia Pacific Conference on Computer Aided System Theory, Sydney, Australia, February 6-8 (2012)
21. www.airspan.com

Chapter 17
An Anticipatory SANET Environment for Training and Simulation of Laparoscopic Surgical Procedures

Christopher Chiu and Zenon Chaczko

Abstract. Surgical simulation environments are required to train laparoscopic surgeons to become familiar with the technology, before they operate on patients. The chapter examines how a Sensor-Actor Network (SANET) incorporated into the laparoscopic training facility can be augmented with the Belief-Desire-Intention (BDI) agents for autonomous operation of the sensor facility along with the training environment. In addition to neural network processes, path trajectory estimation can be superimposed as a visual aid in a variety of training scenarios. The algorithms used in this experiment provides a balanced approach to monitor the training surgeon's level of progress and feedback as they perform the simulated training procedures.

Keywords: Biomimetic Methodologies, Extended Kohonen Maps, Anticipatory Systems, Laparoscopic Surgical Simulation, Wireless Sensor Networks (WSN), Sensor Actor Networks (SANETS).

17.1 Introduction

Current technology advances in haptic feedback, along with realistic graphic presentation systems, enable doctors to diagnose patient health in detail [2]. This level of interactivity and immersion, while providing an accurate picture for surgeons to perform their skill with the greatest of care, has resulted in a greater level of complexity in knowledge and skill for a surgeon to be professionally qualified and accredited [5].

In addition, surgical training technologies such as Simbionix LAP Mentor and SimSurgery provide commercial medical-grade hardware for public hospitals and private medical clinics [4], and it is these types of systems that would benefit from

Christopher Chiu · Zenon Chaczko
Faculty of Engineering & IT, University of Technology, Sydney, Australia
e-mail: {Christopher.Chiu,Zenon.Chaczko}@uts.edu.au

R. Klempous et al. (eds.), *Advanced Methods and Applications in Computational Intelligence,*   367
Topics in Intelligent Engineering and Informatics 6,
DOI: 10.1007/978-3-319-01436-4_17, © Springer International Publishing Switzerland 2014

Sensor-Actor Network (SANET) technology [13]. SANET systems can be easily customized to enhance the processing of sensory data from a variety of sources and provide fine-grain actuator control in the following ways [2]:

- **Redundancy:** SANETS are wireless-based through low-powered RF communications to multiple devices, thus allowing the capturing of analogue data in key regions of space [10]. The reduction of SANET development costs means that a robust network of sensors can be easily aggregated for multiple medical readings, and ensures that the failure of some sensors will not impact on the overall health of the system;
- **Real-time feedback:** SANET technologies incorporate microcontroller technologies that are capable of processing data at efficient process cycle-to-watt ratios, allowing for medical readings to be read collaboratively within the network for processing at the gateway node. The nodes incorporate embedded microprocessor hardware to consolidate the sensor data for near real-time presentation of patient health information so surgeons are able to act upon critical situations [5]; and
- **Anticipatory analysis and heuristic processing:** The sensory data is analyzed via machine-learning neural-network heuristics [7, 11], combined with data from historical patient information and previous surgical operation data to provide forecasting of potential impediments to the surgical operation. The preemption of complications in a medical operation in a quantifiable manner reduces the risk of complications and improves post-operative care.

This chapter investigates how SANET technologies, combined with target-tracking heuristic functions, can improve the methodology of the training and planning of laparoscopic surgical procedures [2]. The development of a middleware infrastructure is essential to ensure sensory systems adhere to medical-grade specifications and standards, while enabling the development of hybrid biomimetic intelligence for path planning and obstacle avoidance [5]. The data presented to the surgeon must not be obtrusive or complicated when they perform their task, while also being intuitive enough to provide real-time feedback while an operation takes place. These competing quality attributes must be balanced to ensure the surgical training resource will augment existing practice [3], while providing future surgeons specializing in laparoscopy a safe way to practice and harness their skills - prior to undertaking operations with physical patients in a supervised manner [2].

17.2 Modeling of Laparoscopic Surgery Using an Agent-Based Process

The simulation model has been implemented using the Belief-Desire-Intention (BDI) software agent paradigm by Rao and Georgeff [13]. The visualization parameters and coordinate maps used to model laparoscopic surgery consists of MRI data-sets obtained from surgical training data [5], which is then imported into the middleware heuristic environment. The agent-based architecture of the SANET simulation system consists of three core agent entities [2]:

- **Organic tissue:** The organic tissue is represented as a matrix of discrete coordinates imported into the simulation environment. The agent will have the properties of cell type, volumetric capacity and tissue density, and can be classified as a critical or non-critical zone;
- **Critical zone:** The obstacle in free space is represented as a single coordinate point. The agent is classified as a critical zone and can reside within the organic tissue coordinate points; and
- **Laparoscopic end-effector:** The surgical operation is performed with the apparatus. The agent consists of normal usage specifications; consisting of manufacturer data sheets and physical properties of each device. It is monitored and controlled using a active-tracking function via self-organizing map algorithms [7].

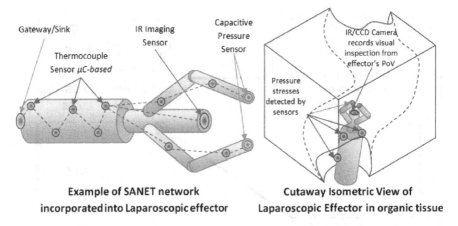

Example of SANET network
incorporated into Laparoscopic effector

Cutaway Isometric View of
Laparoscopic Effector in organic tissue

Fig. 17.1. A Laparoscopic End-effector integrated with SANET network devices

As shown in Figure 18.1, these agent entities will combine the interactive elements of the agents, and synthesize the final result for the surgeon's perspective [2]. To provide an example of the dynamic interactivity between the agents, consider the case for a probe that touches the surface of an organ. For instance tissue, with its unique properties and characteristics, will respond differently to the actions of the end-effector as contact is reached with one another; while from the effector's perspective, the attachment type of effector will determine the nature of the operation taking place [2].

The Virtual Assistant Surgical Training (VAST) System by Feng and Rozenblit et. al. [5] sets a benchmark from traditional surgical training systems that are traditionally direct feedback-input loop based. The VAST system is a knowledge-based construct, where a supervisory control system assists with the user input of the system, to provide guidance and control of the surgeon operator's input as they perform a training procedure [3]. The purpose of this supervisory layer is to combine sensory input with a knowledge-based inference system to enhance the trainer experience as the surgical simulation takes place. The purpose of this research is to enhance the VAST supervisory control system with the BDI agent-based prototype framework [2].

17.2.1 Applying BDI Principles a Knowledge-Based System

The BDI Agent Paradigm compliments the need for developing a proactive laparoscopic surgical simulator, as the inference analysis processes designed for the VAST knowledge-based simulator can be applied to an agent context. The BDI agency concept by Rao and Georgeff [13] is designed with a learning feedback loop modeled using the BDI agent methodology depicted conceptually in Figure 18.2.

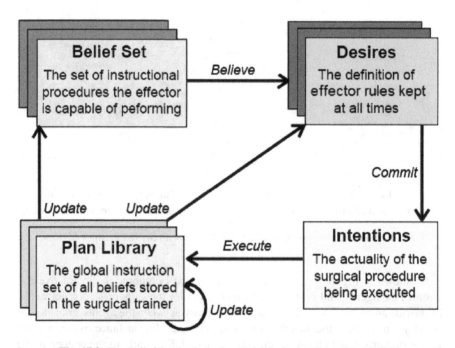

Fig. 17.2. The Belief-Desire-Intention (BDI) Software Agent Model [13]

The Belief-Desire-Intention agent components are elaborated from the context of a laparoscopic surgical simulation as follows:

- **Plan library**
  The *Plan Library* contains the global instruction set of all beliefs available to the laparoscopic surgical simulator. As a simulation prototype, only a subset of the

library is implemented for obstacle navigation. However, this can be easily expanded to contain new procedures or theories; an example is when a new surgical discovery is shared for the mutual benefit of the entire medical community.

- **Belief set**

  The *Belief Set* contains the set of instructional procedures that the laparoscopic effector is capable of performing. These rules bound the rational software agent within the patient simulation space, and the specifications of the surgical tools being used to undertake the operation. The belief set can be interchanged with different beliefs in the library set, as long as compatibility is maintained with the surgical procedure type. An example would be the common medical procedure such as suturing: a principle for applying Jenkin's Rule for thread lengths [8, 9] (i.e. that the required length of the surgical thread is four times the wound length) is only applicable for surgical operations firstly requiring a surgical incision to the patient.

- **Desires**

  The *Desires* define the precautionary safety rules that are observed while a belief is realized. The desires inherent during the laparoscopic procedure are done in conjunction with the target procedure in mind; the desire to ensure a minimum distance is maintained between the end-effector and critical organs should not conflict with the navigation of the effector itself [5]. Desires are linked with the plan library so prioritization ensures desires with higher-priority (i.e. ensuring minimum distance is maintained between an organ and the end-effector) overrides lower-priority desires (i.e. moving the end-effector towards a different region in a cavity).

- **Intentions**

  The *Intention* is the actuality of the surgical procedure taking place. In a knowledge-based system context, this is where the BDI interfaces with the simulation environment between the surgical operator and the medical simulation space. The intention is executed as the surgeon executes a move or activity, for which the Plan Library is updated to the next state of Belief Set. An example is upon the completion of a suture, the simulator will update the surgeon as to what surgical knots can be accomplished as stored in the library set, and a range of thread material that can be used for the suture.

17.3 Application of Heuristics in a Laparoscopic Surgical Domain

The VAST knowledge-based system also contains a feedback response, which is encapsulated within the hybrid view generator. The visual and physical haptic feedback provided by the generator is accomplished through a CCD camera and magnetic kinematic sensors. The human-computer interface displays the CCD vision of the operation, overlaid with the hybridized view of the movements of the end-effector in a given historical timeframe, and the predicted path to execute for future guidance. This predicted path overlay will be the focus of applying obstacle detection heuristics in a SANET BDI-Agent simulation environment.

17.3.1 Extended Kohonen Map (EKM) Techniques

To apply obstacle navigation in a surgical simulation context, the research of Low et. al. [10, 11] has been utilized when employing active function heuristics for autonomous actuator control. However, key differences between robotic actuation and surgical simulation need resolution; this is primarily concerned with the datasets being processed by the heuristic engine, and the total training samples used for multivariate regression.

As shown in Figure 18.3, the general EKM heuristic methodology consists of an inner loop to perform the weights adaption of the neuron weights and initialize the control matrix [14]. The outer loop closes the EKM process, where the EKM dictates the actuator's motion to update the position. Finally, the new inputs of the sensors reflect the change in the actuator's position.

The EKM heuristic procedure [7] is adapted for a navigational process in a surgical simulator, as elaborated below:

- **Feature mapping**
  Stage 1: Define the initial state described by the input vector $u_{(T)}$ in input space U;
  Adaptation: The map self-organizes to partition the laparoscopic end-effector's sensory input data. This includes the furthest geographic position of the effector's end-point and the position of the effector's joints relative to the base. For the SANET simulation prototype, the geographic position considered in the experiment is the end-point location only. The feature map is an essential part of the unsupervised learning process, when the end effector is configured to reach home in a region of space when frequently encountered stimuli are detected. This achieves the objective of navigating around vital organs and blood vessels along the actuator's path trajectory.
- **Multivariate regression**
  Stage 2 to 4: Adapt a new clustering or routing sequence of control vectors $c(t), t = 0, \ldots, T - 1$ in the sensory space C, and calculate the integration matrix weights $w^c_{(win)}$
  Adaptation: The uninterrupted mapping from U to C is done by training a multilayer perceptron (MLP) as required. Customization in the input matrix of the inner loop is required, due to the different nature of a medical simulation exercise. The neuron map must be an appropriate dimension size due to the resource limitations of current microcontrollers contained within the wireless SANET network. However, the map size can be easily expanded to accommodate new advances in medical science and technology.

u_T, u_{S1}, u_{Sn}	(1.) Get input from sensors: End-effector kinematics, Proximity Sensor.
For neural weight matrix: $w^T_{1,1} \ldots w^T_{i,j}$	(2.) Create neuron weights and apply distance.
$a_{i,j} = G\left(w^T_{win}, w^T_{i,j}\right)$	(3.) Calculate Activation Energy.
$D\left(u_T, w^T_{win}\right) = \min_{i,j-1\ldots n}\left(u_t, w^T_{i,j}\right)$	
$e_{i,j} = a_{i,j} - \left[b^{S1}_{i,j} + \ldots + b^{Sn}_{i,j}\right]$	(4.) Calculate Integration matrix, along with maximum value of matrix and initialize control matrix.
$e_{max} = \max_{i,j-1\ldots n}\left(e_{i,j}\right)$	
For $w^C_{1,1}\ldots w^C_{i,j}$: $D\left(e_{max}, w^c_{win}\right) = \max_{i,j-1\ldots n}\left(e_{max}, w^C_{i,j}\right)$	
$c = \begin{cases} w^C_{win} \cdot u^C_T & \text{If Abs Val} < \text{Param} \\ w^C_{win} \cdot w^C_{win} & \text{Otherwise} \end{cases}$	(5.) Respond to surgical simulation, and act upon output and get new input, producing a new control vector. Repeat until target is reached.
$v = u'_T - u_T$	
Control weights adaption:	
$\triangle w^C_{i,j} = \eta G(k,i)\left(c - w^C_{i,j}v\right)v^{Transpose}$	
Neuron weights adaption:	
$\triangle w^{Sensor}_{i,j} = \eta G(k,i)\left(v - w^{Sensor}_{i,j}\right)$	

Fig. 17.3. Extended Kohonen Map Heuristic Algorithm

- **Unsupervised learning execution**
  Stage 5: The resultant goal state elaborated by $u(T) \in U$ adapts the network structure for a desired target state.
  Adaptation: The initialization of the control matrix provides a guiding path overlaid on the head-user display. Once the surgeon decides to execute the next phase of action, they can either follow through with the guided path, or make a detour as he or she sees fit. The outer loop is thus closed once the sensor map is refreshed to reflect the changed coordinate space, and the feature map is re-executed. This loop is terminated once the target destination is reached, or if manually halted by the end-user.

17.3.2 BDI *Agent Integration Process*

The basis of the system architecture is implemented using the open-source JADEX BDI Agent Engine by the University of Hamburg, Germany [1, 12]. JADEX provides a standardized BDI platform to customize the prototype medical-simulation framework, as the agent-based project provides full visibility of the agent run-time structure in the Java programming language. A screen-shot of the JADEX front-end is shown in Figure 18.4; with the left pane showing the available agents active in the environment and sensor statistics, and the right pane showing a virtual perspective of the of the laparoscopic end-effector from the perspective of the JADEX run-time engine.

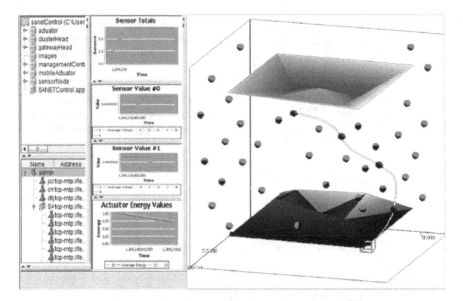

Fig. 17.4. The SANET Middleware Environment incorporating the JADEX BDI Agent Engine

The following BDI-agent components [1] have been adapted to suit the behavioral dynamics of a SANET information network:

- **JADEX rules**
  Consisting of the rule engine and the execution instructions, the beliefs available to the laparoscopic surgical simulator are limited to obstacle and avoidance thresholds for the current prototype. As Java is an object-oriented language, this suits the domain of surgical medicine as the inheritance and polymorphism can be exploited to identify common relationships between rules. As an example, no-fly zone parameters and maximum proximity thresholds would belong under the family of obstacle avoidance strategies [8, 9].

- **JADEX XML**
  The data binding and representation component of JADEX contains the information set in which the BDI engine accesses to govern its actions, with a code sample included in Figure 18.5. This means that the Extended Markup Language format of JADEX XML encapsulates more than just generic agent. It contains specific values specific to the laparoscopic surgical environment, for which this

```xml
<applications>
<application name="Default Configuration">
<spaces>
<env:envspace name="my3dspace" type="3dspace" width="1.0" height="1.0" depth="1.0">
  <env:objects>
  </env:objects>
  <env:processes>
    <env:process type="runPlanMobileActuator">
    </env:process>
    <env:process type="runManagementPlan">
    </env:process>
  </env:processes>
  <env:dataproviders>
    <env:dataprovider name="Sensor_Count">
      <env:source name="$sensorNode" objecttype="sensorNode"></env:source>
      <env:data name="time">$time</env:data>
      <env:data name="sensor_count">$sensorNode.totalCount</env:data>
    </env:dataprovider>
    <env:dataprovider name="Sensor_AttribVal0">
      <env:source name="$sensorNode" objecttype="sensorNode"></env:source>
      <env:source name="$sensorNode_attribVals0" objecttype="sensorNode"></env:source>
      <env:data name="time">$time</env:data>
      <env:data name="id">$sensorNode.getId()</env:data>
      <env:data name="sensorNode_attribVal0">$sensorNode.propertyValue0</env:data>
      <env:data name="sensorNode_meanAttribVal0">$Function.mean($sensorNode_attribVals0)</env:data>
    </env:dataprovider>
    <env:dataprovider name="Actuator_Energy">
      <env:source name="$actuator" objecttype="actuator"></env:source>
      <env:source name="$actuator_energies" objecttype="actuator" aggregate="true"></env:source>
      <env:data name="time">$time</env:data>
      <env:data name="id">$actuator.getId()</env:data>
      <env:data name="actuator_energy">$actuator.chargeState</env:data>
      <env:data name="actuator_meanEnergy">$Function.mean($actuator_energies)</env:data>
    </env:dataprovider>
  </env:dataproviders>
  <env:dataconsumers>
    <env:dataconsumer name="actuator_energy_chart" class="XYChartDataConsume">
      <env:property name="dataprovider">"Actuator_Energy"</env:property>
      <env:property name="title">"Actuator Energy Values"</env:property>
      <env:property name="labelx">"Time"</env:property>
      <env:property name="labely">"Energy"</env:property>
      <env:property name="maxitemcount">500</env:property>
      <env:property name="legend">true</env:property>
      <!-- Defines a multi series (as many series as seriesid's) -->
      <env:property name="seriesid_0">"id"</env:property>
      <env:property name="valuex_0">"time"</env:property>
      <env:property name="valuey_0">"actuator_energy"</env:property>
      <!-- Define a normal series. -->
      <env:property name="seriesname_1">"Average Energy"</env:property>
      <env:property name="valuex_1">"time"</env:property>
      <env:property name="valuey_1">"actuator_meanEnergy"</env:property>
    </env:dataconsumer>
  </env:dataconsumers>
  <env:observers>
    <env:observer name="SANET Control Space" view="view_all" perspective="icons">
      <env:plugin name="evaluation" class="EvaluationPlugin">
        <env:property name="comp_0">(AbstractChartDataConsumer).getChartPanel()</env:property>
        <env:property name="comp_0">(AbstractChartDataConsumer).getChartPanel()</env:property>
      </env:plugin>
    </env:observer>
  </env:observers>
</env:envspace>
</spaces>
<components>
<component type="ManagementCentre" number="10"/>
<component type="Actuator" number="10"/>
<component type="MobileActuator" number="10"/>
<component type="SensorNode" number="10"/>
<component type="ClusterHead" number="10"/>
<component type="GatewayHead" number="10"/>
</components>
</application>
</applications>
```

Annotations:
- Processes: Encapsulate Belief Set
- Data Providers: The available data for the belief
- Data Consumers: The eventual data used for the belief
- Observers: The actualization of the environment
- Components: Population of the total number of BDI agents

Fig. 17.5. Code Sample of BDI-agent Data Binding in XML-Script

prototype has been limited to end-effector diameter, precision of the Cartesian positioning sensors and the type of component connected to the end-effector itself. The XML script can be modified in future to extend new rules as the medical trainer deems necessary.

- **JADEX processes**
  The JADEX processes execute the workflow plans in the ruleset library. A screenshot of the JADEX process in execution is shown in Figure 18.4, where the the agent behaviors are dictated according to medical surgery contexts in a 3D space. Each agent is dedicated a synchronized thread, whereby agents requiring a higher level of processing priority such as the actuator agent are given precedence over sensor agents further away from the actuator's current location. The prototype environment is limited to a single development machine, with two-cores available for utilization by the JADEX agent engine.

- **JADEX Android library**
  The JADEX Android library is being investigated for its suitability in a surgical training scenario. Theoretically, the existing middleware code can be ported to run on the Android Davlik virtual machine with minor modifications to the system infrastructure. In practice, the third-party libraries written for JADEX also need to be designed for Android, complicating the migration process and is out of this project scope. Therefore, the current prototype does not run on mobile devices, although a mobile-based middleware infrastructure is envisaged in future version releases.

17.3.3 Distributed Processing by Integrating JADEX with EKM Heuristics

The integration of a heuristic engine with an agent-based framework such as JADEX is a non-trivial exercise, due to each framework having its own intrinsic, prerequisite requirements. Furthermore, the SANET simulation prototype must also contain laparoscopic end-effector specifications to govern the simulation parameters [9, 15]. An example is the coordinate normalization rules set by VAST [5]: sensor data by the magnetic kinematic sensor must be calibrated and synchronized to the image data. For the simulator prototype, these surgical simulator rules are captured linguistically using the XML data binding feature in JADEX.

In addition, realism is required to enhance the depth of immersion for the surgical operator. This means that while the JADEX agent framework visualizes the interaction between the body tissue and the end-effector as discrete agents, this does not take the surgeon into consideration as a trainee of the simulator. Therefore, a SANET visualization service was implemented to complete the end-user experience from a surgical training point-of-view. This visualization service screen-shot is shown in Figure 18.6, with the heuristic and agent configuration settings in the left panel and an enhanced-3D graphics service displayed on the right panel.

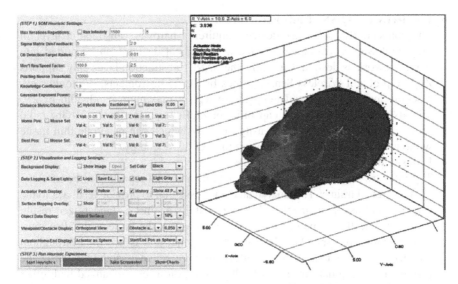

Fig. 17.6. The operation of xEKM Heuristic Service within Middleware Environment

The key aspects in the integration process to build the SANET Middleware Environment are detailed below:

- **Context drivers for an actor-based framework**
  The key context driver for developing the middleware environment for laparoscopic surgical training is to enhance the existing VAST knowledge-based system. For this reason, the context to design a supervisory system places an emphasis on the control and guidance to the trainee surgeon. Agent-based systems including JADEX focus on the interaction of agents in the information space, while the SANET infrastructure needs to treat the responsibilities of the simulator from an actor-based perspective. The obstacle navigation actor for the prototype design encapsulates the heuristic framework with the BDI Agency construct; thus resulting in the connection of the input and response factors in knowledge-based system.

- **Visual haptics and immersion**
  The training environment must consider the end-user in its operation, which for this case is the surgical trainee. The visualization service must consider the operation taking place from the surgeon's point of view, rather than the interaction of discrete agents in virtual space. An example is the difference in the display panel of Figure 18.4 and that of Figure 18.6. Obstacles can be considered as discrete points in space, although physical surfaces need be represented as the sum of all obstacles. This is because a surgeon perceives his or her environment in an analogous space of surfaces such as blood vessels, tissues and muscle; not as a digitized representation as how an agent would view the simulated world.

- **Connectivity to SANET infrastructure**
  The connectivity to SANET devices requires a hardware abstraction layer in the middleware environment, so that new medical devices can be ported into the framework as desired. Currently, the SANET abstraction layer only supports ZigBee-protocol Wireless Sensor Motes by Texas Instruments, to monitor gyroscopic measurements and temperature with a maximum bandwidth of 250 kilobits/second. The wireless capability of ZigBee operates at the 2.4Ghz radio spectrum with a low-power intensity of 1mW, which can penetrate thin tissue without the harmful doses of exposure to microwave energy.
- **Dataset integration and interchange**
  The infrastructure needs to exchange data for external reporting to be assessed by the surgical trainer. This analytical data is essential to provide feedback to the surgeon, and assess their manual dexterity skills against a standardized result. While outside the scope of this current iteration of the project, the data must be exportable to a specified data reporting format that is accepted throughout the medical community. A unified information interchange model will allow the results of training data from one health institution to be exchanged with other institutions when new medical breakthroughs arise, regardless of what medical training infrastructure they utilize.

17.4   Evaluation of the SANET Middleware Environment

An experiment of the heuristic analysis on the SANET middleware environment has been conducted, to evaluate the trajectory estimation function for surgical training contexts. In particular, it is important to determine if the training infrastructure produces results consistently in line with the observed outcomes produced for robotic contexts. Furthermore, a qualitative review of the development process will determine the merits of the middleware environment as a supervisory surgical training tool.

17.4.1   Heuristic Experiment

The experimental specifications are described as follows:

1. The SANET network composition consists of:

   a. The physical motes consist of 20% of the total population, with the TI Zig-Bee wireless motes providing Cartesian coordinates to the experimental framework;
   b. The simulated motes consist of the remaining 80% of the agent population, with density and spread determined in a pseudo-random fashion via the Mersenne-twister method.

2. The SANET middleware infrastructure:

   a. The software platform is a Windows 7 Enterprise operating system running the Java Runtime Environment Standard Edition 7.0;

   b. The hardware specification is a desktop computer with an Intel Pentium Dual-core Processor (2.10Ghz) and with 4GB of memory.

The experimental procedure executing the EKM heuristic platform is elaborated below:

1. The estimation methods used for the experimental case study:

   a. The first is conducted using the random walk function, where the actuator will pseudo-randomly redirect to an alternate course when an obstacle is reached;

   b. The second is conducted using Extended Kohonen Maps as elaborated in the previous sub-chapter.

2. The estimation of a target trajectory reaching its destination, with a simulated actuator passing through pseudo-randomly generated obstacles in a SANET environment - in a virtual area of 20cm x 20cm.

3. The same obstacle configuration is used for 20, 40, 60, 80, 100, 120, 140 and 160 agents, uniquely generated for every total number of agents, is the same for each experimental method.

4. 25 experimental iterations is conducted for each method, the average success rate of all iterations is calculated. A successful trajectory target determined if no obstacle is impacted to within a distance of 0.5cm, and if the target is reached at a predefined home position of an area by 1cm x 1cm.

5. The number of agents composes the wireless sensor network population in the SANET environment; the experiments are conducted with a population of 20, 40, 60, 80, 100, 120, 140 and 160 agents in total.

As observed in Figure 18.7 for both estimation methods, the general trend is that the trajectory estimation decreases as the number of obstacles increases. In particular, the Extended Kohonen Map Function (Figure 18.7 *Below*) is approximately a 10% improvement in estimating a navigation path compared to the Random Walk Function (Figure 18.7 *Above*). The importance of these results lies in that a SANET agent-based middleware environment can faithfully reproduce the estimation results without significant overhead. This prototype demonstration shows that a knowledge-based system can incorporate a unsupervised learning heuristic successfully, without notable degradation in the obstacle detection efficiency. Future tasks to consider will be to improve the detection efficiency for realistic contexts; as the increased resolution of sensor sensitivity due to SANET technology improvements, will result increase the total number of obstacles monitored by the middleware-based simulator.

Trajectory Estimation using Random Walk Function *(Above)*

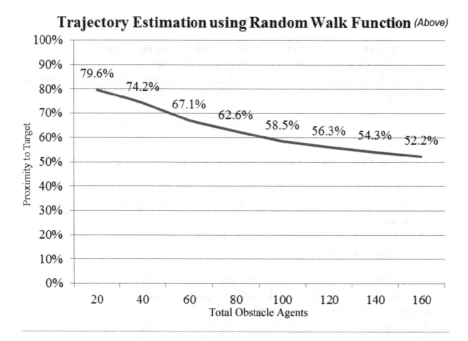

Trajectory Estimation using Extended Kohonen Maps *(Below)*

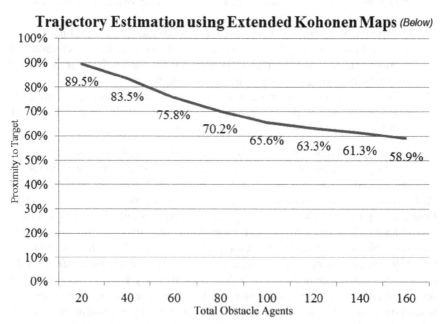

Fig. 17.7. *(Above)* Estimation using Random Walk Function; *(Below)* Estimation using Extended Kohonen Maps

17.4.2   Analysis and Further Work

A summary of the experimental analysis is contained in Table 18.1, comparing Standard Training frameworks with a SANET Knowledge-based framework. These frameworks are being analyzed according to their component framework adaptability, and their learning and cognitive adaptability.

- **Component framework adaptability**
  The ability to provision new surgical training services, and provide for new facilities to analyze medical training data. This data can be used for to assess laparoscopic surgical proficiency, or exchange with other medical research institutes for further analysis.
- **Learning and cognitive adaptability**
  The capability to integrate new heuristic methods into the system by means of library configuration. While the current prototype incorporates only an obstacle navigation heuristic, different heuristic methodologies may be applied in future for training purposes.

The performance and resource considerations need to be factored in determining the prototype's capability to be adapted for laparoscopic surgical training environments. It is universally accepted that knowledge-based information systems will be enhanced as engineering research improves over time [4, 6]. So while a SANET agent-based middleware will be an essential design construct for future iterations of the system framework, one must have a broad mind as to its implementation strategy. The core points of the future experimental work are examined below:

- **Assessing alternative actor design philosophies**

  - Artificial Intelligence, the branch of study in which the BDI agent model belongs, is undergoing a renaissance since its foundations in the 1950's. The BDI agent model used for the laparoscopic simulator does have notable performance and resource requirements, as the agents in the BDI framework execute in a multi-threaded manner on a single platform. A true actor design will not just emulate the behavior of its model representation, but also its physical characteristics.
  - The laparoscopic simulation should be distributed by its design [16]; in that the vision analysis, the knowledge-based engine and the kinematic control of the end-effector should be independently operating in its own environment [15]. This allows the sharing of processing resources among different devices, and has the added feature of redundancy to compensate for system failure.
  - An example would be if the knowledge agent device had a sudden, unexpected storage medium failure. To compensate for this catastrophic failure, the kinematic agent device would be able to take-up the additional role of knowledge-assistance until a hardware replacement is scheduled.

Table 17.1. Evaluation Summary of Medical Training Frameworks

	Standard Training Framework	**SANET Middleware Knowledge-based Framework**
Component Framework Adaptability	• Standard Training Frameworks used for laparoscopic surgical simulation have a direct loop between the feedback and response processes between the trainer and the surgical environment. The direction actions of the surgeon will result in a desired or non-desired reaction by the simulation framework. • Components are designed with a one-to-one mapping, such that the trainer is only able to execute a process iteratively every loop, or multiple processes executed in serial after a higher priority process is run. Training modules are coupled directly to the simulator system.	• SANET Middleware Frameworks for surgical simulation have an supervisory control system between the input stage of the feedback and response loop. This provides an interactive layer between the physical environment and the surgical operator, and improve sensory awareness of the environment through contextual feedback. • Components are thus able to be designed with a one-to-many mapping, so that the supervisory control system dedicated to one task module can be remapped with a different one as the user sees fit. An example would be if the surgeon has mastered gross dexterity of manipulating an endoscope, a more advanced module on fine-grain control can be loaded.
Learning and Cognitive Adaptability	• As the learning framework is directly coupled to the surgical simulator engine, superimposition of guidance and control of the operation is difficult. This is because any guidance information must be provided to the same response as the feedback information as only the single feedback exists between the trainee and simulator. • The merging of this data increases information overhead and complicates the logical responsibilities of training of the surgeon and to recognize the faults and identify errors in their procedure.	• The decoupling of the learning framework results in a separate data feed for feedback, and another for guidance control. This ensures a dedicated information stream provides feedback of the actual operation taking place, while providing guidance within a separate stream in a non-invasive manner. • Separation of the learning data simplifies the information space, while simplifying the logical responsibilities. The knowledge supervisor is to provide guidance and train the surgeon, while the simulator framework ensures immersion and depth to simulate a laparoscopic operation in a reliable manner as possible.

- **SANET integration and data exchange harmonization**
  - Surgical training simulation systems each have their own proprietary formatting standards for data analysis [6]. This data is important for analytical recording of the surgeon's training process, to analyze their technique and methodology for assessment and make improvements to their method. If this information is to be shared within the same health institution, it is assumed that the same simulation framework is to be used to access the data. However, the problem arises when data is exchanged with other institutions to learn different operation techniques from surgical specialists in other cities or countries.
  - Building a global knowledge database containing the body of surgical training works requires the development of a common data interchange format. This format must not only establish universally accepted standards on data format types, but the kind of data that is to be included in this format, for instance the coordinate geometry type used to measure an end-effector in space.
  - Hence, as long as the data is compliant with the standardized surgical training format, any institution can access this data regardless of the surgical training platform utilized in their premises. This will ultimately aid the medical science and research community to share ideas and strategies that will benefit patient healthcare standards.

17.4.3   Enabling Multi-dimensional Heuristic Contexts for Laparoscopic Surgical Simulations

The EKM Code has been applied to a laparoscopic surgical context, to illustrate how an unsupervised learning heuristic for robotic navigation systems can be used for laparoscopic end-effector navigation through obstacles. By applying a neural-network heuristic to an agent-based architecture, the modeling of agents as interactive components of a laparoscopic simulator seeks to emulate the physical environment as a virtualized representation in the SANET middleware infrastructure. As this representation is from a multi-dimensional perspective, the Figure 18.6 shows how the physical representation (X, Y and Z coordinate space) can be converged with non-physical constructs (such as risk against safety zones). An example of a non-physical construct is the obstacle collision risk determined by the deviation of the surgical trainee from the path generated by the EKM heuristic. The results from determining such risks can be quantified into examination results of the trainee's level of skill.

It must be emphasized that there is no single heuristic that will completely solve the problems posed in a laparoscopic surgical simulator environment. Rather, the agent-based framework is designed to be decoupled from the heuristic process. So as new research innovations into unsupervised learning leads to an optimized obstacle detection algorithm, these innovations can be easily experimented and benchmarked against existing heuristics within the SANET middleware framework. Therefore, medical researchers and scientists can focus on the surgical training context of heuristic algorithms, rather than considering the agency implementation and technical design limitations.

17.5   Conclusion

The combining of SANET middleware framework paradigms to a surgical knowledge-based construct has been presented in this chapter. The architectural hybridization of the training framework has enabled the adaptation of an unsupervised learning heuristic as a simulated laparoscopic training methodology. The primary benefit of the architecture is that this integration strategy has resulted in a seamless transition of a heuristic framework to be applied to surgical navigation training. Furthermore, the flexibility of the actor-based framework allows for integrating new heuristic methods to aid surgical training, without compromising the existing laparoscopic surgical simulator. Further work to be done is to incorporate this prototype by integrating the SANET framework as the supervisor of the VAST knowledge-based framework. This will require further analysis into the integration strategy and information space, ensuring VAST remains consistent with its primary responsibility to train surgeons on how to use laparoscopic tools effectively.

References

1. Braubach, L., Pokahr, A., Lamersdorf, W.: JADEX: ABDI-Agent System Combining Middleware and Reasoning, Software Agent-Based Applications. Platforms and Development Kits, Whitestein Series in Software Agent Technologies and Autonomic Computing, pp. 143–168 (2005)
2. Chiu, C., Chaczko, Z.: Collaborative Target Tracking for Laparoscopic Surgery Simulation utilizing Anticipatory. In: Proceedings from the Australian Conference on the Applications of Systems Engineering, pp. 83–84 (2012)
3. Feng, C., Haniffa, H., et al.: Surgical Training and Performance Assessment Using Motion Tracking, Engineering of Computer-Based Systems (2006)
4. Feng, C., Rozenblit, J.W., Hamilton, A.J.: A Computerized Assessment to Compare the Impact of Standard, Stereoscopic and High-definition Laparoscopic Monitor Displays on Surgical Technique. Surgical Endoscopy 24(11), 2743–2748 (2010), doi:10.1007/s00464-010-1038-6
5. Feng, C., Rozenblit, J., Hamilton, A.J.: A Hybrid View in Laparoscopic Surgery Training System, Engineering of Computer-Based Systems, Tucson, Arizona (2007)
6. Hamilton, E.C., Scott, D.J., et al.: Comparison of Video Trainer and Virtual Reality Training Systems on Acquisition of Laparoscopic Skills. Surgical Endoscopy Journal 16, 406–411 (2002)
7. Kohonen, T.: Self-Organizing Maps, 3rd edn. Springer, New York (2000)
8. Korndorffer, J., Dunne, J., et al.: Simulator Training for Laparoscopic Suturing Using Performance Goals Translates to the Operating Room. Journal of the American College of Surgeons 201(1), 23–29 (2005)

9. Leonard, J.: Sensor Fusion for Surgical Applications. In: 15th AESS IEEE Dayton Section Symposium, pp. 37–44 (1998)
10. Lien-Pharn, C., Rozenblit, J.W.: A Multi-modality Framework for Energy Efficient Tracking in Large Scale Wireless Sensor Networks. In: 5th Annual Conference on Artificial Intelligence, Simulation, and Planning in High Autonomy Systems, Gainesville, Florida (1994)
11. Low, K.H., et al.: An Ensemble of Cooperative Extended Kohonen Maps for Complex Robot Motion Tasks. Neural Computation 17(6), 1411–1445 (2005)
12. Pokahr, A., Braubach, L., Lamersdorf, W.: JADEX Implementing a Belief-Desire-Intention Infrastructure for JADE-based Agents. Exp JADE Journal, 76–85 (2003)
13. Rao, M., Georgeff., P.: BDI-agents: From Theory to Practice. In: Proceedings of the 1st International Conference on Multi-agent Systems - ICMAS (1995)
14. Ritter, H., Schulten, K., Denker, J.S.: Topology Conserving Mappings for Learning Motor Tasks: Neural Networks for Computing. In: American Institute of Physics: 151st Conference Publication Proceedings, Snowbird, Utah, pp. 376–380 (1986)
15. Shrivastava, S., Sudarshan, R., et al.: Surgical Training and Performance Evaluation using Virtual Reality based Simulator. In: Virtual Concept Conference, France (2003)
16. Wei, G., Arbter, K., Hirzinger, G.: Real-time Visual Servoing for Laparoscopic Surgery, Controlling Robot Motion with Color Image Segmentation. IEEE Engineering in Medicine and Biology Magazine 16(1), 40–45 (1997)

Chapter 18
Towards Ubiquitous and Pervasive Healthcare

Jan Szymański, Zenon Chaczko, and Ben Rodański

Abstract. In recent years, Wireless Sensor Networks (WSNs) have attracted a considerable attention from both academia and industry. There is a number of possible applications and great expectations, but at the same time there are some serious issues and challenges related to Wireless Sensor Networks. One of the areas of WSN applications is healthcare and there are a number of health related issues that WSN aims to address. Amongst the main such issues are: the aging population that causes a growing pressure on economy and the healthcare system supported by a declining number of working-age people as well as an increase in chronic diseases, which includes obesity frequently attributed to a lack of fitness and weight management due to busy lifestyles. Staying physically and mentally healthy is of the greatest importance to every individual and to the society. There is a growing interest in a new approaches to the support of the overstressed healthcare system. The fundamental concept is to shift some of the responsibilities from the clinicians, health centres and hospitals of the traditional system to the patients and their home environment. People themselves are able and should play a greater role in monitoring and maintaining their own health, provided that they are supported by an adequate technology and have a proper knowledge of how to use it. There are already a number of applications of computing and communication technologies related to healthcare including (but not limited to) the pervasive health monitoring, mobile telemedicine applications, the intelligent emergency management services, health aware mobile devices, the medical inventory management, the pervasive access to health information and the lifestyle management . In this chapter, we discuss a several of new generation of services and applications in the area of pervasive and ubiquitous healthcare that are enabled by sensor networks. We shall provide an overview of the new trends and introduce innovative ideas around the self-adapting ambient intelligence combined with specialised system requirements.

Jan Szymański · Zenon Chaczko · Ben Rodański
Faculty of Engineering & IT, University of Technology, Sydney, Australia
e-mail: {jan.szymanski,zenon.chaczko,ben.rodanski}@uts.edu.au

R. Klempous et al. (eds.), *Advanced Methods and Applications in Computational Intelligence*,   387
Topics in Intelligent Engineering and Informatics 6,
DOI: 10.1007/978-3-319-01436-4_18, © Springer International Publishing Switzerland 2014

18.1 Introduction

Pervasive Healthcare is a very broad and multidisciplinary subject. In this chapter and our research we focus mainly on wireless networking technologies enabling pervasive personal healthcare systems.

This paper consists of 4 parts.

- Part 1 we give an introduction and the definition of terms.
- Part 2 is the detailed description of Body Sensor Networks.
- Part 3 is the discussion of existing challenges facing Pervasive Healthcare Systems.
- Part 4 is the final conclusion and future works related to the subject.

18.1.1 Definitions of Terms

Following there is a number of terms explained for better understanding of the subject.

UBIQUITOUS COMPUTING

The term "ubiquitous computing" was coined by Mark Weiser of Xerox PARC in 1991. Weiser observed that humans consider walking around trees and staying close to a natural environment most relaxing while, at the same time, people find working with computers rather frustrating, and hence his remarks, that "The most profound technologies are those that disappear. They weave themselves into the fabric of everyday life until they are indistinguishable from it" [1]. This vision has become the foundations for the future work related to ubiquitous and pervasive computing. The term ubiquitous is something that is available anywhere anytime, while pervasive is something that is permeated in our environment.

PERVASIVE HEALTHCARE

(*Also known by the European Union term: ambient assisted living*) is a vision for the future of healthcare. The Body Sensor Networks based technology is expected to play a very important role in delivering pervasive healthcare to the masses [2]. Despite the fact, there is a significant demand and a number of prototypes already available, we seem to be still far away from providing an effective pervasive healthcare. There are a number of reasons for such a situation, including an inadequate and in-advanced technology, an immaturity of ethical and legal concepts as well as related sociological and psychological issues Pervasive Healthcare (or the European Union term: ambient assisted living) can be considered as an answer to health related challenges like increasing elderly population, increasing number of chronic and lifestyle diseases and increasing lack of medical professionals [4]. Pervasive Healthcare "has the potential to integrate health consideration and health promoting activities for patients and non-patients in their everyday conduct and provide added value to life quality for individuals" [3]. Body Sensor Networks are expected to play an important role in delivering pervasive healthcare to the masses.

WIRELESS SENSOR NETWORK (WSN)

Is a network made of number of small intelligent objects called motes or smart objects. Smart object is an item equipped at least with a form of sensor and/or actuator, a microprocessor, a communication device (radio transceiver) and a power source [5]. Additionally smart object might contain other modules for example data storage. Figure 1 shows the general architecture of WSN.

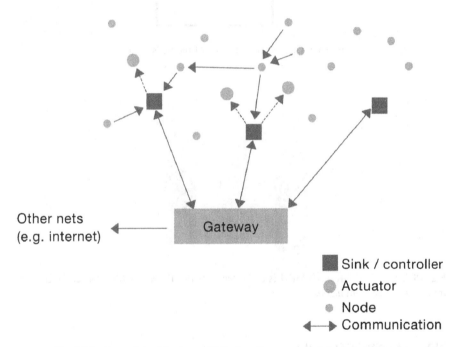

Fig. 18.1. General Architecture of Wireless Sensors and Actuators Networks

WSN NODES (MOTES, SMART OBJECTS)

The nodes that form the WSN, often called motes, are typically made of the following parts: sensors and actuators, a microcontroller(s), a radio transceiver and power supply sources. More advanced motes may include additional components such as:

a serial flash, aMicroSD card and an LCD glass module. Figure 2 shows a general architecture of a typical WSN node and Figure 3 depicts a commercially available WSN development kit.

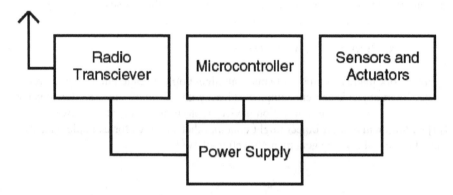

Fig. 18.2. General Architecture of the Sensor Node

Fig. 18.3. CC2530ZDK-ZNP-MINI is a ZigBee Network Processor Mini Development Kit from TI (Source from *ti.com*)

SENSORS AND ACTUATORS

A *transducer* is a device that converts energy from one form into another and a sensor is a transducer that converts energy in a physical world into electrical energy. An *actuator* is a device performing a reverse type of energy conversion (from electrical to another form). The symbolic representation of sensing and actuating is shown in Figure 4.

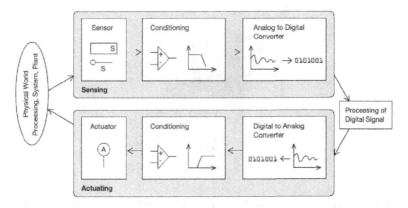

Fig. 18.4. Symbolic representation of data acquisition and actuation showing sensors and actuators

MICROCONTROLLERS FOR WSN

Microcontrollers selected for WSN have lower than other current consumption and ability to switch programmatically their power modes. Another desirable feature is the ability to run from different and scalable clock sources. The examples of popular microcontrollers used in WSN are:

- ATMega32 – 8 bit, Atmel, 16 MHz clock, 32kB flash, 2kB RAM, active current 3.5 mA @ 1MHz, power down < 1uA.
- MSP430F2274 – 16 bit, TI, 16MHz clock, 32kB flash, 1kB RAM, active current 270uA @ 1MHz, standby <0.7uA.
- EFM32G840F64 – 32 bit ARM-Cortex M3, Energy Micro, 32MHz clock, 64kB flash, 16kB RAM, active current 180uA/MHz, stop mode 0.6 uA.

RADIO TRANSCEIVERS FOR WSN

Typically an IC radio transceivers or a modules with integrated antenna are used. Most radio transceivers are programmable giving an option for different frequencies, modulations and power. An example module is shown in Figure 5.

POWER SOURCES FOR WSN

The requirements for power sources in WSN to last for a long time (usually years) and have a very small size at the same time are very difficult to satisfy. In most cases the power sources are batteries. With a very little progress in batteries technologies

Fig. 18.5. An example radio transceiver module A1101R09A from Anaren incorporates TI CC1101 transceiver chip, size 9x16x2.5 mm. (Source *anaren.com*)

(compare to other technologies) there is a growing interest in alternative sources in a form of super capacitors and energy scavenging from the environment.

SMART DUST

Smart dust was originally a concept for miniature wireless sensor networks and a project undertaken at University of California Berkeley. The Defence Advanced Research Projects Agency (DARPA) funded the project, setting as a goal the demonstration "that a complete sensor/communication system can be integrated into a cubic millimetre package" [6]. Tiny, ubiquitous, low-cost, smart dust motes have not yet been realized, however, some reasonably small motes are commercially available [6].

BODY SENSOR NETWORKS

When Wireless Sensor Network is deployed on the human body it is called Body Sensor Network (BSN). Body Sensor Networks are considered to be more challenging than other type of WSNs and extensive research is focusing on biocompatibility, signal propagation, power management [7] and other properties. The primary motivation for Body Sensor Networks (BSN) is to provide long term continuous sensing without activity restriction and behaviour modifications.

CONTEXT AWARENESS

Context information is anything that can characterize the situation of an object, such as a person, a device or a network [8].

SMART SPACES

Smart space is a space surrounded by technology that can sense and act, communicate, reason, and interact with people [2]. Examples of smart spaces are smart homes/houses, smart cars. A Smart Space refers to small intelligent devices embedded in a physical world and connected to the Internet [9]. The things in a physical world will be Internet-enabled via embedded technology and able to interact with one another to provide a smart space that adds intelligence to the environment [9]. One of the examples of Smart Spaces is Smart House with a number of implementations with a Gator Tech Smart House from University of Florida being a pioneering project [16]. Some of the features of it are smart blinds to control ambient light, smart bed to monitor sleep patterns, smart bathroom with sensors for measurement of weight, height and temperature and others. Cooperation can play a crucial role in such spaces as discussed in [9].

THE INTERNET OF THINGS AND SMART OBJECT

The Internet of Things is the next stage of the Internet, where a growing number of Smart Object (Things) is connected to the global internet for monitoring and control. The number of people using the Internet becomes significantly lower than a number of Smart Object and the amount of data transferred by Smart Objects becomes much larger then that transferred by people.

IP FOR SMART OBJECTS

6LowPAN (IPv6 over Low power Wireless Personal Area Networks) is a protocol defining IPv6 packets over IEEE 802.15.4 wireless networks. It uses MAC and PHY layer of IEEE 802.15.4 and IPv6 protocols for the layers above the MAC. 6Low-PAN is gaining more interest due to its promise of interoperability with existing IP networks, direct access from the internet ability and improved mobility over other WSN protocols.

STIGMERGY

It is known from the study of biologists, that some social systems in nature can present an intelligent collective behaviour although they are composed by simple individuals with limited capabilities and that is achieved through direct and indirect interactions of these individuals. The word stigmergy is used to describe the indirect interaction occurring between two individuals when one of them modifies the environment and the other responds to the new environment at a later time [10]. Stigmergy is an organizing principle in which individual parts of the system communicate with one another indirectly by modifying and sensing their local environment

[11]. The trace left in the environment by an action stimulates the performance of a next action, so subsequent action are built on each other, leading to systematic activity. The term "stigmergy" was introduced by French biologist Pierre Paul Grasse in 1959. He defined it as: "Stimulation of workers by the performance they have achieved" It is derived from the Greek words stigma "mark, sign" and ergon "work, action".

PERSONAL HEALTH DEVICES

There is no formal definition for the concept of a Personal Health Device (PHD), however in popular terms it is a device that is used for monitoring health, wellness and life activities in home/work environments as well as for various mobile applications. Figure 6 shows an example of a portable medical instrument, called a Neuromonics processor for treatment of tinnitus. The device has a wireless connectivity that uses IrDA mode of communication, however it is not networked.

Fig. 18.6. Example of portable medical instrument (Source: *neuromonics.com*)

18.2   Background Context

18.2.1   *Body Sensor Networks as Special* WSNs

Body Sensor Networks consists of a number of implantable, wearable, portable and ambient devices, each of them being capable of monitoring/sensing and/or displaying/actuating information [12]. BSNs are a special case of Wireless Sensor Networks and as they are deployed on human body they should feature the following attributes

[12]: Reliability, biocompatibility, portability, privacy and security, lightweight protocols, irretrievability, energy aware communication, prioritized traffic, RF radiation safety. Figure 7 shows general concept of Body Sensor Networks. Some examples of implantable devices are artificial pacemakers, cochlear implants, retinal implants. Examples for wearable sensors are ECG, EEG, accelerometers, inclinometers, blood pressure monitors, and blood oxygen saturation monitors (SpO2). Commercial-off-the-Shelf Personal Health Devices, smart phones, PDAs are the examples of portable devices, while the ambient sensors can be sensors for atmospheric pressure, ambient temperature, humidity and others.

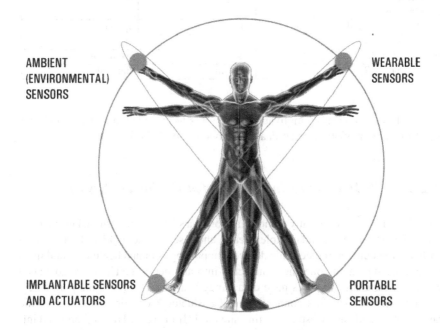

Fig. 18.7. The concept of Body Sensor Networks showing human being in the centre with a number of implantable, wearable, portable and ambient sensors/actuators.

The requirements for BSN are more restrictive than for other WSN including reliability, user friendliness, long lifetime, wearability creating extra challenges.

18.2.2   A Brief History of Body Sensor Networks

Body Sensor Networks consists of Body Sensors. First prototypes of Body Sensor Networks were reported about 10 years ago named WBAN [Jovanovic] and PAN [Zimmerman]. A group from Philips among first to use BAN instead of PAN listing the differences for example: Transmission range for WBAN is ~2m, while for

Table 18.1. Comparison of the Features of WLAN, WSN and BSN

	Traditional Networks	Typical WSN	BSN
Instance	WLAN	Smart Dust	Smart Ward
Coverage	50m	10m	1m
Density	Sparse	Dense	Dense
Data-centric	Address-centric	Data-centric	Data-centric
Large scale	No	Yes	Yes
Workloads	Unpredictable	Unpredictable	Partially Predictable
Error rates	Medium	High	**Must be very low**
Energy constraints	No	Yes	**Yes**
Hops	Single	Multihop	Optional
Infrastructure	Yes	No	No
Node Failure	No	Yes	**Prohibited**
Deployment	Random	Random	**Planned**

WPAN ~10m. The primary motivation for BSN is to provide long term continuous sensing without activity restriction and behaviour modifications

18.2.3   BSN Integration into Connected Healthcare System

Body Sensor Networks usually do not function as a stand-alone system, but rather as a part of comprehensive and complex health and rescue system [13]. There are currently a large number of electronic devices supporting personal healthcare and most of them are standalone ones not connected into any network. The more advanced (and more expensive) devices have data storage capability for logging parameters and some kind of wireless connectivity (like Bluetooth or Wireless USB) for the transfer of logged data. Despite existing standard (IEEE 1073, HL7, others) and initiatives (Continua Health Alliance, Medical Devices Plug and Play, others) there is still lack of interoperability between different devices.

18.2.4   Sensors and Actuators for BSNs

The common definition for a sensor is that it is a device that measures a physical quantity and converts it into a signal, which can be read by an observer or by an instrument [43]. Table 2 lists vital signs and their parameters and Table 3 lists a number of sensors used in WSN and their signals (sources [12], [14]).

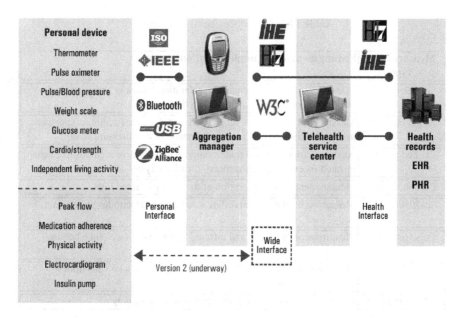

Fig. 18.8. The integration of technologies for Connected Healthcare (adopted from Continua presentation)

Table 18.2. Vital signs and their parameters

Vital Sign (nominal values)	Sampling rate	Quantization	Total bit rate
Electrocardiogram (60-80/min)	240 samples/sec	12-36 bits/sample	2.9-8.7 Kbps
Blood Pressure Sys<120, Dia<80	1 sample/minute	64 bits/sample	1 bps
Oxygen Saturation 95-99%	1 sample/sec	16 bits/sample	16 bps
Body core Temperature 36-37C	1 sample/minute	16 bits/sample	0.3 bps
Breathing rate 12-18/min	1 sample/sec	4 bits/sample	4 bps

18.2.5   *Wireless Technologies for* BSNs

The existing wireless technologies suitable for BSN are briefly summarized in Table 4 and Figure 9 shows examples of wireless modules of different technologies used for experimentation.

Table 18.3. Sensory devices used in BSN

Measurement Type	Sensing method and implementation example
Acceleration	MEMS sensor sensing dynamic acceleration (shock or vibration) and static acceleration (inclination or gravity)
Blood pressure	Measures the systolic (peak) pressure and diastolic (minimum) pressure using a stethoscope and a sphygmomanometer or pressure sensor
Blood sugar	Analyses drops of blood traditionally or non-invasive methods like near infrared spectroscopy, ultrasound, breath analysis
Carbon dioxide	Measurement of the absorption of the gas by infrared light
ECG/EEG/EMG	Measurement of potential differences between electrodes placed on a body
Pulse oximetry	Measurement of the red and infrared lights absorbance when passing through the body
Respiration	Measurement of oxygen dissolved in a liquid using two electrodes covered by a thin membrane
Temperature	Integrated Circuit uses infrared or resistance changing

Table 18.4. Technologies used in BSN

Wireless Technology	Coverage (Max.)	Bitrate (Max.)	Cost	Comments
6LowPAN	10m	250Kbps in 2.5GHz 25Kbps in sub-GHz	Low	Conforms to IEEE 802.15
Zigbee	10m	250Kbps in 2.5GHz 25Kbps in sub-GHz	Low	Conforms to IEEE 802.15
RFID	1m		Low	
Bluetooth	10m, 100m	900 Kbps	Low	
Bluetooth Low Energy	10m		Low	
Wireless LAN (IEEE 802.11)	100m	54Mbps	Low	
Cellular/3G	Wide area (nationwide)	Several Mbps	High, Carrier usage charge	

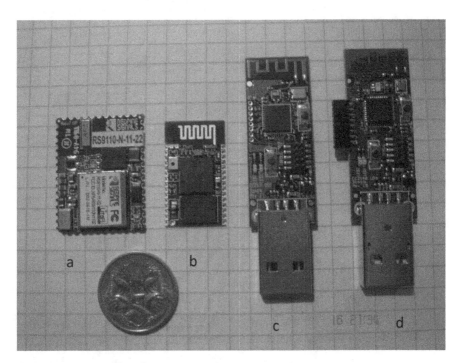

Fig. 18.9. Wireless modules of different technologies used for experimentation
a) RS9110-N-11-22 Wi-Fi module from Redpine
b) BTM182 Bluetooth module from Rayson
c) CC2531 ZigBee module from Chipcon (Texas Instruments)
d) Bluetooth Low Energy module from Texas Instruments

18.2.6   Connectivity Models for BSNs

There is a number of portable medical device currently on the market like pulse oximeters, blood test monitors, glucose meters functioning as the stand alone devices. They are able to measure usually one vital sign or other health related parameters and in some cases have some wired or wireless interface to PC for the purpose of data storage, visualization, trend monitoring. The autonomous Body Sensor Network system in the next step (Figure 10) as it combines a number of sensors. It is however not connected to global Internet, but sometimes an external access is possible. An example of such a system is an emergency monitoring system for people with cardiac problems with cellular connection to the service provide or personal cater. The model shown in Figure 11 is the extended internet model of connectivity with body sensor networks in moving between a numbers of smart spaces. The sensors of BSN can be access either directly or through Personal Health Device or through the smart object's gateways/routers anywhere from the Internet.

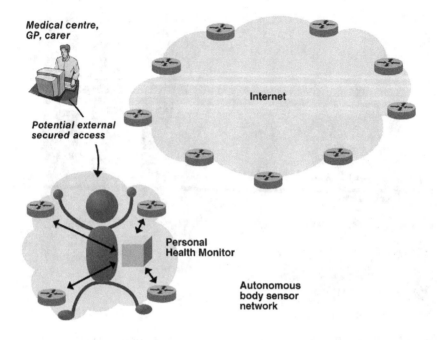

Fig. 18.10. Autonomous Body Sensor Network

18.3   Challenges for Ubiquitous and Pervasive Healthcare

Many authors [2], [15] list the following challenges:

Node-level challenges:

- *Sensor technology.* Despite a great progress in sensors technology mainly to the use of MEMS there is still an unsatisfied need for a number of medical sensors like blood pressure or ECG sensors. The existing sensor are too big and inconvenient to use and the prototypes of smaller devices either do not have satisfying parameters like the MEMS-based ECG sensors or are too expensive like the implantable blood pressure sensors, which need to be placed inside veins by chirurgical operation.
- *Small size (miniaturization):* For implantable, wearable and portable technologies physical size is important
- *Power consumption:* The technology is struggling with providing power sources with capacity enough for the long time reliable operation of BSN. There are different approaches to solve it including lower power electronics, energy harvesting, better batteries and optimizing software for lower power.

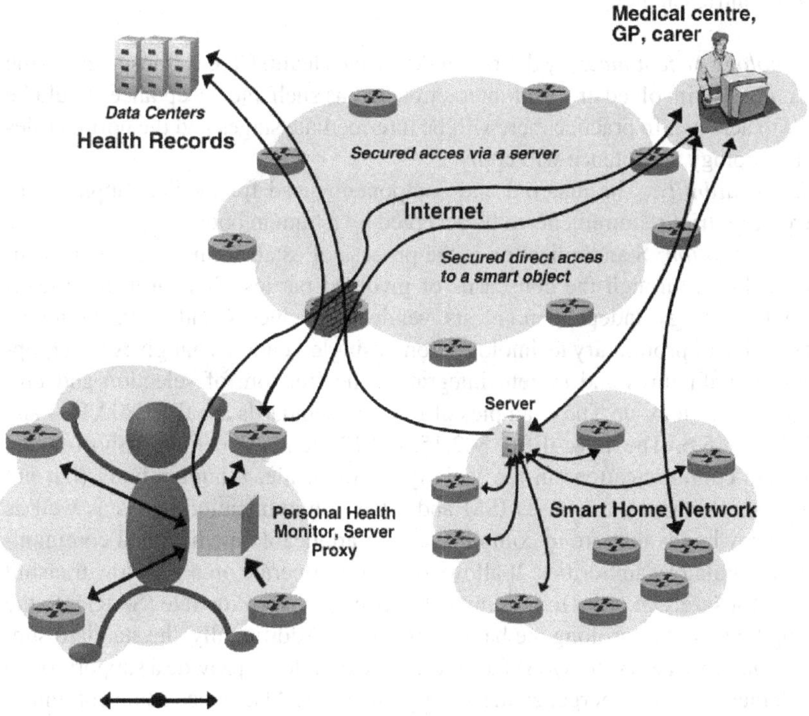

Fig. 18.11. Body Sensor Network as part of The Internet of Things

- *Cost.* It is usually needed to optimize the cost for a network of a very large number of nodes. In the case of BSN that requirement is less critical, however.

Network-level challenges:

- *Unreliable ("lossy") communication media:* BSN like other sensor networks work in the radio environment is often unreliable as being use by other nodes (collision) and other devices (interference) plus a number of other problems like fading, multipath can be taken into account. The network protocols have to include retransmission, channel switching, etc.
- *Network management:* Unlike with traditional WSN there is no option for a node failure with BSN and as such the requirements for network management are more strict.
- *Security:* Medical data has to be protected from unwanted access. The network security imposes an additional overhead on data packets as the encryption has to be implemented. The processing power of network nodes has to be increased to be able to deal with security.

Human–Centric Challenges:

- *Technology Acceptance:* By definition Pervasive Health Care is transparent to the user, doesn't involved any user interaction and as such and acceptance should be easy to achieve. In practice there will be intermediate stages and the normal rules of technology acceptance will apply.
- *Biocompatibility:* The material and components used for medical applications have very strict requirements to be accepted by a human body.
- *Standardization:* Standardization is the process of establishing a technical standard achieve through the consensus of involved parties. Standardization leads to a technology independent of its vendors, producers and users being an alternative to proprietary technology from a single vendor. That gives the equipment manufactures and system integrators the freedom of selection and prevents vendor lock-up.The examples of relevant standards are IEEE802.15.4 and IEEE802.15.6. The new IEEE 802.15.6-2012 standard defines "Short-range, wireless communications in the vicinity of, or inside, a human body (but not limited to humans). It utilises ISM and other communication bands as well as frequency bands that are in compliance with applicable medical and communication regulatory authorities. It allows devices to operate on a very low transmit power for safety in order to minimize the specific absorption rate (SAR) into the living body and to prolong the battery life" [17]. Additionally, the standard supports Quality of Service (QoS) facilities. For example, to provide a support for an implementation of emergency messaging functions. Also, since some communication systems need to transmit sensitive data, the IEEE 802.15.6-2012 provides mechanisms for strong security [17]. There are other standards used in BAN. These are: 802.15.4-based ZigBee and 6LoWPAN and Medical Implant Interface Standard (MICS). For WSNs or BANs, it is uncommon that to create a complete solution that is based solely on a single standard. In this chapter, we provide a brief comparison of different low power wireless standards and their suitability for BANs.
- *Interoperability:* Interoperability is defined as the ability of systems and equipment from different vendors to operate together. To achieve interoperability an agreement on technical architecture and the setup of test and conformance suites is necessary.

18.4   Conclusion and Future Work

Pervasive Healthcare (or Ambient Assisted Living) is a vision for the future of Healthcare. Despite of a great need and a large number of prototype systems that have already been built, we are still away from providing a truly effective solution for pervasive healthcare. In this chapter, we aimed to explain the need for Pervasive Healthcare, provide a review of the underlying technology and discuss the existing challenges towards practical implementations. Body Sensor Networks are expected to play an important role in delivering pervasive healthcare to the society. The expected impact of BSN is to provide monitoring services and assist the traditional healthcare with prevention, diagnosis and rehabilitation. The fundamental requirements of being available "anytime, anywhere and for anyone" and the fact that BSNs are deployed on human body make the implementation of BSNs is much more challenging than most of the solutions base on traditional WSNs. This aspect leads to two major directions of the investigation into improving the performance of BSNs. Firstly, in order for the systems to become smarter, far more research is required into the hardware and software for BSNs to gain more powerful communication (mainly network protocols) and more intelligent computing mechanisms. The second direction of research is to explore how BSNs can be assisted through cooperation with Smart Space concepts and how to provide cognitive-awareness for BSNs, humans and the environment. The authors believe that their present modelling and practical experimentation work supports the above directions well.

References

1. Abowd, G.D., Dey, A.K., Brown, P.J., Davies, N., Smith, M., Steggles, P.: Towards a Better Understanding of Context and Context-Awareness. In: HUC 1999, pp. 304–307 (1999)
2. Abraham, A., Grosan, C., Ramos, V.: Stigmergic Optimization. SCI. 1st edn. Softcover of orig. ed. 2006 edn. Springer (December 20, 2010)
3. Chaczko, Z., Klempous, R., Nikodem, J., Szymanski, J.: Applications of Cooperative WSN in Homecare Systems. In: International Conference of Broadband Communication, Information Technology and Biomedical Applications, Pretoria, Gauteng, South Africa (November 2008)
4. Dishongh, J.T., McGrath, M.: Wireless Sensor Networks for Healthcare Apliications. Artech House (2010)
5. Dargie, W., Poellabauer, C.: Fundamentals of Wireless Sensor Networks: Theory and Practice. Wiley (2010)
6. Yang, G.-Z.: Body Sensor Networks. Springer (2006)
7. Helal, S., Mann, W.C., El-zabadani, H., King, J., Kaddoura, Y., Jansen, E.: The Gator Tech Smart House: A Programmable Pervasive Space. IEEE Computer - Computer 38(3), 50–60 (2005)
8. Ichalkaranje, N., Ichalkaranje, A., Jain, C.L.: Intelligent Paradigms for Assistive and Preventive Healthcare. SCI. Springer (2006)
9. Lai, D.T.H., Palaniswami, M., Begg, R.: Healthcare Sensor Networks: Challenges Toward Practical Implementation. Taylor & Francis Inc. (2011)
10. Leitner, M.: Stigmergic Systems in Pervasive Computing, Doctoral Seminar, Linz (2006)
11. Mihailidis, A., Bardram, J.E.: Pervasive Computing in Healthcare, 1st edn. CRC Press (2006)
12. Navarro Felix, K.M.: A heterogeneous network management approach to wireless sensor networks in personal healthcare environments, UTS Theses (2008)
13. Varshney, U.: Pervasive Healthcare Computing: EMR/EHR, Wireless and Health Monitoring, 1st edn. Springer (2009)
14. Vasseur, J.P., Dunkels, A.: Interconnecting Smart Objects with IP: The Next Internet, 1st edn. Morgan Kaufmann (2010)

15. Warneke, B., Last, M., Liebowitz, B., Pister, K.S.J.: Smart Dust: communicating with a cubic-millimeter computer. California Univ., Berkeley, CA, ieeexplore.ieee.org/xpls/abs_all.jsp?arnumber=895117
16. Weiser, M.: The computer for the 21st century. Scientific American 265(3), 94–104 (1991)
17. Szymanski, J., Chaczko, Z., Rodanski, B.: Towards Ubiquitous and Pervasive Healthcare. In: 14th International Asia Pacific Conference on Computer Aided System Theory, Sydney (February 2012)
18. IEEE Standard for Local and metropolitan area networks – Part 15.6: Wireless Body Area Networks

Index

Printed in the United States
By Bookmasters